DYNAMIC SPATIAL MODELS

NATO ADVANCED STUDY INSTITUTES SERIES

Proceedings of the Advanced Study Institute Programme, which aims at the dissemination of advanced knowledge and the formation of contacts among scientists from different countries.

The series is published by an international board of publishers in conjunction with NATO Scientific Affairs Division

A	Life Sciences	Plenum Publishing Corporation
B	Physics	London and New York
C	Mathematical and Physical Sciences	D. Reidel Publishing Company Dordrecht and Boston
D	Behavioural and Social Sciences	Sijthoff & Noordhoff International Publishers B.V.
E	Applied Sciences	Alphen aan den Rijn, The Netherlands and Rockville, Md., U.S.A.

Series D: Behavioural and Social Sciences — No. 7

DYNAMIC SPATIAL MODELS

edited by

DANIEL A. GRIFFITH, B.S., M.A., Ph.D.
Assistant Professor of Geography,
State University of New York at Buffalo

ROSS D. MACKINNON, B.A., M.S., Ph.D.
Full Professor of Geography and Chairman
State University of New York at Buffalo

SIJTHOFF & NOORDHOFF 1981
Alphen aan den Rijn, The Netherlands
Rockville, Maryland, U.S.A.

Proceedings of the NATO Advanced Study Institute on
Dynamic Spatial Models
July 6-19, 1980
Bonas, France

ISBN 90 286 2721 9

Printed in The Netherlands.

PREFACE

The development of mathematical models and theories for space-time processes would seem to be a natural and timely next step in the evolution of formal spatial analysis. Events such as the energy crisis, international inflation cycles, and a perpetuation of acute interregional and international economic disparities have increased social awareness of locational interrelatedness, coupled with the possibility of selecting alternative feasible time paths of socio-economic development. An explanation of past time paths, together with the capacity of selecting, in a rational way, future time paths are important endeavors. To date, such explanations and means of selection have been piecemeal, couched in theories that often overlook prominent geographical and/or temporal relationships. In order to provide realistic explanations and useful selection techniques, statisticians, geologists, econometricians, quantitative geographers, and members of other professions have been attempting to come to grips with analytical and conceptual complications that immediately arise when the simultaneous consideration of phenomena in time and space is undertaken. Because this multidisciplinary work is highly specialized in nature, and is being carried out primarily in Europe and North America, it seemed appropriate to provide a forum for the intensive interaction among members of this community. Such interaction should help to circumvent a stringent polarization of ideas, perspectives, and methodologies, while encouraging cooperative efforts toward overcoming impasses encountered in space-time analyses. More specifically, the literature indicates that a need exists for (1) acquiring a more comprehensive understanding of space-time phenomena, (2) the identification of salient space-time problems, (3) an exchange of viewpoints held by scientists from different disciplines, cultures and ideologies, and (4) the ability to predict with a high degree of accuracy outcomes of spatial dynamics. Accordingly, the major scientific objectives of the Institute included (1) the exchange of ideas concerning dynamics from the perspective of mathematical spatial models and geographical systems analysis, (2) the exploration of the subject of space-time processes, (3) the enhancement of the development of mathematical models describing space-time processes, (4) the articulation of theories of space-time processes, (5) a survey of methods for treating analytically space-time data sets, and (6) the identification of areas for future study.

These four needs and six objectives furnished motives for the Institute. Geographers, economists, regional scientists and ecologists were assembled to study six prominent themes, namely frameworks for space-time analysis, spatial demographic change, ecological perspectives on dynamic spatial models, structural change, the statistical analysis of space-time series, and applications of optimal control theory to space-time problems. The format used to study these topics consisted of lectures about a given theme followed by a panel discussion of the lecture materials.

We are indebted to the Scientific Affairs Division of the North Atlantic Treaty Organization for funding the Institute, and to Drs. Robert Bennett and Bernard Marchand for serving in the capacity of advisors for the Institute. Special appreciation goes to M. and Mme. Simon and their staff at the Château de Bonas for providing an excellent physical environment, to the secretarial staff of the Geography Department, State University of New York at Buffalo, for helping in preparing materials for the institute, to Mrs. Debbie Kinda for the typing of the final manuscript, and to our wives for being understanding about our two-week sojourn at the Chateau de Bonas.

Daniel A. Griffith

and

Ross D. MacKinnon

State University of New York at Buffalo
November 1, 1980

TABLE OF CONTENTS

INTRODUCTION†

Until recently a majority of spatial models in geography have been static in nature, thus constituting an abstraction that refers to a single point in time, and is embedded in a timeless domain. Given a set of inputs, it produces a set of outputs. If changes are introduced to this input, be they changes in data or model structure, its output is simultaneously and instantaneously altered. One popular type of static model deals with equilibrium positions. For example, the linear programming transportation problem model takes a set of supply and demand constraints together with inter-point transport costs as inputs, and produces as output an optimal flow pattern. If, say, transport costs are changed then the flow pattern is changed automatically. This model ignores the transformation from an original flow pattern to the new flow pattern.

In an attempt to uncover information about this transformation, numerous studies have analyzed the relationship between changes in inputs and the resulting alterations in outputs. This approach is what has become known as comparative statics modelling. Moreover, the equilibrium distribution is first determined under certain conditions, and then new equilibria are determined after specific changes are implemented, with the differences between initial and final equilibria being attributed to those changes introduced. Unfortunately this approach renders little information about the temporal transformation process. Response times between a changed input and the materialization of its accompanying output are overlooked, meaning the response to input changes is still appearing at the same instant as the changed input. Further, if the input remains constant so does the output, resulting in a temporally invariant time series. The process of change itself is not addressed by this kind of model.

Change involves several prominent components. First, descriptions of change for some variable x provide the relationships between x_{t+1}, x_t and x_{t-1}. This is usually achieved in terms of a difference or differential equation, depending upon whether or not change is cast as being discrete or continuous. Second, time lags exist between input changes and output responses. Third, a movement takes place from the initial output to the resulting output, and this movement is characterized by equilibrating negative feedback or deviating positive feedback. A dynamic model must include all three of these components. Consequently, a dynamic spatial model may be defined as a model that describes the transformation from a spatial distribution at time t_0 to a spatial distribution at time t_k, $k \geq 1$, such that

$x_{i,t} = F(x_{j,t-m};\ m \geq 1,\ j = 1,2,\ldots,n)$ and the specification of the function F captures prevailing feedback mechanisms.

This book is concerned with dynamic spatial models. Its papers seek to improve the understanding of the evolution of phenomena through time under various degrees of spatial constraint. Its central idea is to pursue alternatives to "equilibrium" models in which the trajectories towards equilibrium, or the transformations of initial spatial distributions, are of little interest. Judging from the papers included in this volume, dynamic spatial modelling involves the following:

(1) the development of a clear set of axioms for metric space-time;

(2) the ability to describe the evolving space-time structures (for example by its autocorrelation properties);

(3) the understanding of constrained dynamic behavior, where the constraints arise from

  (a) artificial spatial lattices, as induced by discretization,

  (b) the existing spatial lattice (as with hierarchical diffusion),

  (c) an historically developing response surface (as in the change in urban fabric), or

  (d) a direct response to control policies involving spatial preferences; and,

(4) the development of unconstrained spatial structures through process-response models. (In this case the questions are: Given a set of processes and an undifferentiated surface, how will the process models lead to a real differentiation and along what trajectories will the ensuing development occur? How will the developing structures themselves feedback to influence the processes? This inevitably leads to non-linear models, of which density-dependent population growth leading to a logistic equation in biology is a familiar example. It also leads to questions concerning the stability of such systems under unsteady behavior.)

Recurrent questions concerned the nature of space and time and, in particular, the disciplines that seek to study spatial processes. The major theoretical contributions to this debate were made by Bernard Marchand, Hubert Beguin and Jacques Thisse. Beguin and Thisse's paper sought to axiomatize the role of space and time as metrics upon which phenomena are measured. In contrast, Marchand emphasized the manner in which spatial structure deriving from the past introduces both constraints and imperatives for current phenomena. Hence, if Beguin and Thisse were able to clarify the dangers of the space-centered view so characteristic of geography in the 1960's, Marchand demonstrated the important links of spatial and structural constraints which underpin a dialectic of social and spatial evolution. In turn this led Marchand to consider the specificity rather than the generality of historic place, spatial location and the phenomena and attributes under study.

Methodological distinctions lie at the center of several recurrent themes; in particular the differing interpretations of the role of space. In some papers, especially those dealing with migration (Huff) and housing search (Clark), space is partitioned into regions or neighborhoods and the processes specified in terms of variable characteristics of the areas. In other papers, the organization of space (its structural properties as they mould flows and influence interactions) is of central importance. Papers by Paelinck and Ancot, and Haining consider space as an homogenous nearest neighbor surface, Beckmann considers space as a potential surface, while the paper by Puu considers the development of an ordered surface of a central place lattice. Space appears in both discrete and continuous form, the former either as an approximation to a continuous surface or as a representation of space in its own right. Whatever the role played by space, a major theme here is the need to justify the representation in terms of the underlying processes and to develop mathematical structures that will enable a broader range of spatial structures to be encompassed. For example, developmental and stability analysis is applied to spatial structures by Puu and Marchand; Sonis and Griffith treat space-time structure aspects, the former in terms of the ergodic hypothesis, and the latter in terms of space-time autocorrelation functions. Similar problems of spatial correlogram analysis, process interpretation and estimation are also discussed by Sokal and Wartenberg, Haining and Marchand.

Several papers address specific problems in regional economics. For example, Paelinck and Ancot, and Beckmann consider the local diffusion of income changes, though both recognize in addition the importance of such influences as government policy. This influence, exogenously modelled in these two papers, is explicitly considered by Bennett in a study of the United Kingdom rate support grant, the subsequent discussion of which emphasizes the need to combine both elements. Claude Marchand, using spectral analysis, studies spatial and temporal fluctuations in economic activity in an urban system using unemployment data (an empirically useful surrogate for income, but also important in its own right). Barber and Ralston consider the problem of optimal network development, in relation to regional economics, centered around explicit micro-level models. Spatial analysis has always been dogged by the problem of relating the evolution of macro-spatial pattern to micro-level decision making and behavior. A series of papers strive to attack this problem ranging from numerical analysis, by Clark on housing search, MacKinnon and Rogerson on vacancy chains, Huff on population change and Sokal and Wartenberg on ecological diffusion, to abstract master equations of states derived from individual utility functions by Smith. Clearly such analyses are most useful when macro-level patterns can be derived from indisputable micro-level behavior.

Bifurcation and stability analysis prove to be a further emerging focus of many papers. The treatment of systems over the full global range of state and parameter values has become increasingly important following the seminal mathematical results of Thom, Zeeman, Poston and Smale. The papers by Curry, Smith, Casetti, Puu and Papageorgiou each encompasses analysis of such behavior. This range of papers includes discussion of vacancy chains, urban growth, and demographic movements. Clearly the range of phenomena to which this approach can be applied is very wide. However, the power of the technique draws largely from its very general and simple structure. A state space equation (or master equation) defines the dynamics, and response characteristics are examined over a range of control variables, parameters, or initial conditions. The resulting global analysis allows the identification of points of singularity or bifurcation, and this in turn allows the extension of dynamic analysis from rather restricted stable systems to a broad class of unstable, evolutionary, nonlinear and switching phenomena. Analysis of spatial systems using such models is now being widely adopted and offers a great potential for the future development of spatial analysis.

Optimization is also a theme of this volume. The use of objective functions to define utility, welfare and other criteria now form a more explicit part of much micro-level analysis (as in the papers by MacKinnon and Rogerson, Huff, and Smith). However, some research is now more specifically concerned with situations in which variables and components of the systems of interest are amenable to explicit control and manipulation. In the papers by Bennett, Fujita, and MacKinnon and Rogerson, explicit control variables such as the allocation of intergovernmental grants, land use controls and the creation of job vacancies enter explicitly, and the problem is to define the appropriate values of these variables as instruments of policy. This thrusts policy analysis directly into dynamic spatial models and also raises more explicit criteria of welfare and distribution.

Any collection of papers such as this will clearly present many divergent as well as convergent themes and perhaps future research will fruitfully develop as much from the differences between approaches as the relationships emphasized above. The success of this volume will be evaluated by the new research it will stimulate.

---

† We would like to express our appreciation to Robert Bennett, Robert Haining and John Thornes for their significant contribution to the summary appearing as part of this introduction.

REFERENCES

Cordey-Hayes, M., 1972. Dynamic Frameworks for Spatial Models, Socio-Economic Planning Sciences 6: 365-385.

Machlup, F., 1959. Static and Dynamics: Kaleidoscopic Words, The Southern Economic Journal, 26: 91-110.

Martin, R., Thrift, N. and Bennett, R. (eds.), 1978. Towards the Dynamic Analysis of Spatial Systems, Pion, London.

FUNDAMENTAL FRAMEWORKS FOR SPACE-TIME ANALYSIS

Since a dynamic spatial model must include descriptions of change, equations of motion are required. These equations describe the path of movement from an initial to a final spatial distribution. Clearly such equations must have reference to some mathematical framework. The three papers in this section provide an overview of some important properties for this framework.

Beckmann demonstrates that a treatment of continuous rather than discrete space leads to the notion of flow fields. These flow fields, in turn, necessitate the use of vector analysis. The focus then becomes the distribution in two-dimensional space of potential functions and their associated flows. Beckmann operates on the linear expenditure function, inspecting the cases of no autonomous expenditures, no neighborhood expenditures, constant autonomous expenditures, and finally neighborhood expenditures. Implications regarding flows through space and the interaction between price and demand are sketched. He cautions, however, that relevant relationships must be carefully posited, and not derived from analogies with mathematical physics, since the corresponding relations in physics are considerably different, and take on different kinds of substantive meanings.

Beguin and Thisse provide a formal definition of space and time in geography based upon topology and measure theory. In a general and rigorous fashion, properties of each concept are proposed and examined, the concepts are compared and an integration is proposed. One important conclusion they reach is that while space and time are treated separately in existing space-time studies, these two concepts must be joined together in order for meaningful results to be forthcoming. A similar finding is put forth by Griffith in a subsequent paper in this volume. The joint concept defined in this paper is incomplete, however. A bothersome property it had to retain, for instance, is that time is reversible. Thus the authors conceed that considerable additional work remains to be done, and suggest that their framework augments rather than precludes the use of others such as differential or difference equations to describe change.

Marchand's paper represents a time-geography perspective. Once again the introduction of time merely as a subscript or variable is questioned. Further, his arguments corroborate Beckmann's contention that time in a social context is more subtle, involved and ambiguous than its oversimplified counterpart that is employed in sciences such as physics. Marchand outlines various philosophical stances about time, contrasting time in history with time in forecasting.

These three papers emphasize that, among other properties, the time dimension utilized in dynamic spatial models must go beyond something as simple as subscripting some variable x with a t. Furthermore, direct analogies with mathematical physics often will produce inappropriate, misleading or erroneous results.

CONTINUOUS SPATIAL MODELS OF INCOME DIFFUSION AND COMMODITY TRADE

Martin J. Beckmann
Department of Economics
Technical University of Munich
and
Brown University

A natural starting point for demonstrating the nature of the continuous flow approach to spatial dynamic problems is the linear expenditure model discussed by Paelinck, Bennett and others in this volume:

$$y(t+1) = \underline{A}y(t) \tag{1}$$

where

$$\sum_k a_{ik} = 1 \quad (i = 1,2,\ldots,n) \quad . \tag{2}$$

Equation (2) states that the propensities to spend must add up to unity so that total expenditure equals income (one of Parkinson's laws says that "expenditure always rises to meet income"). There is no particular spatial structure assumed in (1), (2). Distance has no visible effect on expenditure.

Suppose however that we break down expenditure into two components: local and neighborhood expenditure on the one hand and expenditure over larger distances on the other hand. Thus in a central place system some expenditure will take place in the vicinity of the given location and some at central places of higher order assumed here to be a long distance away. The expenditure at higher order central places and conversely the expenditure by higher order central places in a given location will be treated here as autonomous.

We consider a grid or Cartesian coordinate system (i,k). The statement that income in the next time period equals expenditure in this period may then by written as follows:

$$y_{i,k}(t+1) = c_o y_{i,k}(t) + c_1[y_{i+1,k}(t) + y_{i,k+1}(t) + y_{i-1,k}(t) + y_{i,k-1}(t)] + z_{ik}(t+1) \quad . \qquad (3)$$

Here the first term on the right hand side represents local expenditure, the second term expenditure by locations in the immediate neighborhood and the last term autonomous expenditure. Notice that

c_o = the propensity to spend locally;

c_1 = the propensity to spend at any one of the nearest neighbor locations; and,

z_{ik} = autonomous expenditure by agents located a long distance from location (i,k).

A straightforward transformation yields

$$\begin{aligned} y_{ik}(t+1) - y_{ik}(t) = c_1[&y_{i+1,k}(t) - 2y_{ik}(t) + y_{i-1,k}(t) \\ &+ y_{i,k+1}(t) - 2y_{ik}(t) + y_{i,k-1}(t)] \; , \qquad (4) \\ &+ [c_o + 4c_1 - 1]y_{i,k}(t) + z_{ik}(t+1) \; . \end{aligned}$$

Change the time period from 1 to Δt, let $\Delta x_1 = \Delta x_2 = \Delta x$ and adjust the distance units to Δx_1, Δx_2;

$$[*]y_{ik}\Delta t + 2_{ik}\Delta t$$

similarly changes the distance unit to Δx and we write the last equation

$$\frac{\Delta y}{\Delta t} = \frac{c_1(\Delta x)^2}{\Delta t} \cdot [\frac{\Delta^2 y}{\Delta x_1^2} + \frac{\Delta^2 y}{\Delta x_2^2}] + [c_o + 4c_1 - 1]\, y + z \quad . \qquad (5)$$

Going to the limit so that

$$\lim \frac{(\Delta x^2)}{\Delta t} = 1 \quad ,$$

the expenditure equation becomes

$$\frac{\partial y}{\partial t} = c_1 \Delta y + [c_o + 4c_1 - 1] y + z \; . \tag{6}$$

Here

$$\Delta = \frac{\partial^2}{\partial x_1^2} + \frac{\partial^2}{\partial x_2^2} \tag{7}$$

denotes the Laplace operator.

At the boundary of a self contained region we obtain in a similar fashion the boundary condition

$$\frac{\partial y}{\partial n} = 0 \tag{8}$$

where n is the direction normal to the boundary.

Equation (6) describes the process by which income is diffused spatially. It is similar in form to the diffusion equation of physics that applies to heat transfer and to material transport in gases and liquids. We are not interested here in finding general mathematical solutions to this equation. Rather we shall discuss some properties and solve some spatial cases of interest.

CAST I: NO AUTONOMOUS EXPENDITURE

Consider first the cast that autonomous expenditure is absent: $z \equiv 0$. Suppose first that not all income is spent locally or in the neighborhood, such that

$$c_o + 4c_1 < 1 \quad . \tag{9}$$

In that case the expenditure gap will cause incomes to fall persistently. To see this, integrate (6) over the region

$$\iint \frac{\partial y}{\partial t} \, dx_1 dx_2 = c_1 \iint \Delta y \, dx_1 dx_2 + [c_o + 4c_1 - 1] \iint y \, dx_1 dx_2 \; . \tag{10}$$

Write

$$Y = \iint y \, dx_1 dx_2 \quad (11)$$

and observe the Gauss integral theorem [Courant and John (1965)]

$$\iint \Delta y \, dx_1 dx_2 = \int \frac{\partial y}{\partial n} \, ds \quad . \quad (12)$$

Equation (6) becomes

$$Y \frac{\partial Y}{\partial t} = c_1 \int \frac{\partial y}{\partial n} \, ds + [c_o + 4c_1 - 1]Y \quad , \quad (13)$$

or in view of the boundary condition (8)

$$\frac{\partial Y}{\partial t} = -[1 - c_o - 4c_1]y \quad , \quad (14)$$

which has the solution

$$Y(t) = Y(o) \exp\{-[1 - c_o - 4c_1]t\} \quad . \quad (15)$$

We conclude that aggregate income in the region falls to zero when the propensity to spend totals less than one. It remains constant when the total propensity to spend equals one. In that case the expenditure equation (6) has the simpler form

$$\frac{\partial y}{\partial t} = c_1 \Delta y \quad (16)$$

which is the standard diffusion equation of mathematical physics. Thus the local and neighborhood effects of income generation may be understood as operating in a way mathematically similar to physical diffusion. This is true even though our derivation differs fundamentally from that in the physical context.

The solution to the homogeneous equation (16) with its boundary condition (8) approaches a constant everywhere, the average income for the region. This will now be shown.

Let $\bar{y}$ be the average income, and let $\eta = y(x_1,x_2,t) - \bar{y}$ be the deviation from this average income. Clearly η satisfies (16) such that

$$\frac{\partial \eta}{\partial t} = c_1 \Delta \eta \quad .$$

Multiplying by η and integrating over the region yields

$$\iint \eta \frac{\partial \eta}{\partial t} dx_1 dx_2 = c_1 \iint \eta \Delta \eta \, dx_1 dx_2 \quad . \tag{17}$$

Now

$$\mathrm{div}(\eta \text{ grad } \eta) = \mathrm{grad}(\eta' \text{ grad } \eta + \eta\Delta\eta)$$

so that

$$\iint \eta\Delta\eta \, dx_1 dx_2 = \iint \mathrm{div}(\eta \text{ grad } \eta) dx_1 dx_2 - \iint |\text{grad } \eta|^2 \, dx_1 dx_2 \; . \tag{18}$$

The first integral of the right hand side vanishes by the Gauss integral theorem and the boundary condition, giving

$$\mathrm{div}(\eta \text{ grad } \eta) dx_1 dx_2 = \int (\eta \text{ grad } \eta)_n \, ds = 0 \quad . \tag{19}$$

Substituting (19) and (18) into (17) produces

$$\iint \eta \frac{\partial}{\partial t} \eta \, dx_1 dx_2 = - c_1 \iint | \text{ grad } \eta|^2 \, dx_1 dx_2 \quad . \tag{20}$$

The left hand side may be written

$$\frac{1}{2} \frac{\partial}{\partial t} \iint \eta^2 \, dx_1 dx_2$$

and the right hand side is negative unless $\eta \equiv$ a constant. This shows that the squared deviations from the average income decrease steadily. Income is diffused until it is uniform.

CASE II: NO NEIGHBORHOOD SPENDING

As a second special case suppose that no income is diffused to neighboring points because all spending is either local or at long distances. In that case equation (6) assumes the form

$$\frac{\partial y}{\partial t} = (c_o - 1)y + z \quad . \tag{21}$$

Equation (21) is no longer a partial differential equation but a family of linear ordinary differential equations applying to each location separately:

$$\frac{dy}{dt} + \lambda y = z \tag{22}$$

where $\lambda = 1 - c_o$. To solve equation (22) multiply by $e^{\lambda t}$:

$$\frac{d}{dt}(ye^{\lambda t}) = ze^{\lambda t}$$

$$ye^{\lambda t} - y(o) = \int_0^t ze^{\lambda s}\, ds \tag{23}$$

$$y(t) = y(o)e^{-\lambda t} + \int_0^t z(s)e^{-\lambda(t-s)} ds \quad .$$

Equation (23) shows how initial income and past autonomous expenditure determine current income.

In the special case that autonomous expenditure is constant over time

$$y(t) = \frac{z}{\lambda} + (y_o - \frac{z}{\lambda})e^{-\lambda t} \quad . \tag{24}$$

Here income approaches the equilibrium value

$$\bar{y} = \frac{z}{\lambda} = \frac{z}{1 - c_o} \tag{25}$$

given by the well-known multiplier formula. No equilibrium exists when the propensity to spend equals unity, or

$$c_o = 1$$

In that case equation (21) takes the form

$$\frac{dy}{dt} = z \tag{26}$$

and for positive expenditure income increases steadily.

This multiplier effect is present also in the general case of (6), but it applies only in the aggregate

$$\iint y\, dx_1 dx_2 = \frac{1}{1 - c_o - 4c_1} \iint z\, dx_1 dx_2 \quad . \tag{27}$$

To see this, let

$$z(x_1,x_2,t) = z(x_1,x_2)$$

be constant in time and integrate (6) over the region so that

$$\iint\frac{\partial y}{\partial t}\,dx_1dx_2 = c_1\iint\Delta dx_1dx_2 + [c_o + 4c_1 - 1]\iint y\,dx_1dx_2 + \iint z\,dx_1dx_2 \quad . \tag{28}$$

As shown before the first term on the right hand side vanishes in view of the Gauss integral theorem and the boundary condition. Thus

$$\frac{dY}{dt} = [c_o + 4c_1 - 1]\;y + Z \tag{29}$$

where

$$Z = \iint z(x_1nx_2)dx_1dx_2 = \text{a constant} \quad .$$

The solution of (29) is then

$$Y(t) = \frac{Z}{1-c_o-4c_1} + [Y_o - \frac{Z}{1-c_o-4c_1})\exp\{-(1-c_o-4c_1)t\}\ , \tag{30}$$

and the last term of the right hand side approaches zero, proving (27). It is assumed here that the propensities to spend locally and in the neighborhood are less than unity. Otherwise no equilibrium exists and income rises forever as seen before in (26).

It may be shown, also by means of the Gauss integral theorem, that in the case of constant autonomous expenditure $z = z(x_1,x_2)$, implying that income stabilizes not only in the aggregate but also locally. Thus the diffusion process (6) is a stable one.

CASE III: EXPANSION OF THE NEIGHBORHOOD REGION

Suppose that we extend the neighborhood to include points that are two distance units away. Then on the right hand side of (3) an additional term appears, namely

$$+ c_2[y_{i+2,k} + y_{i+1} + y_{i,k+2} + y_{i-1,k+1} + y_{i-2,k} + y_{i-1,k-1} + y_{i,k-2} + y_{i+1,k-1}] \quad .$$

The sum in brackets may be rewritten

$$\Delta\Delta y + 6\Delta y + 8y \quad ,$$

where Δ as before denotes the Laplace operator--first in terms of finite differences and, upon going to the limit, in terms of derivatives. The equation of income diffusion assumes the form

$$\frac{\partial y}{\partial t} = z + [-1 + c_o + 4c_1 + 8c_2]y + (c_1 + 6c_2)\Delta y + c_2 \ \Delta\Delta y \quad . \tag{31}$$

Our previous conclusions concerning steady state equilibria and stability remain valid provided the following boundary condition is added:

$$(\Delta \text{ grad } y)_n = 0 \quad .$$

In principle one would incorporate neighborhood effects over increasing distances, but only at the expense of increasing the order of the difference equation.

FLOWS IN SPACE

The diffusion equation (6) must be understood in terms of flows. What is the nature of the flow of expenditure? We shall argue that it is a money flow. Consider the flow vector

$$\phi = - \text{grad } y \quad . \tag{32}$$

Notice that

$$\Delta y = \text{div grad } y = - \text{div } \phi \quad .$$

Now the operator div denotes the net outflow from a location so that -div is the net inflow. Equation (6) assumes the form

$$\frac{\partial y}{\partial t} = [-1 + c_o + 4c_1]y - c_1 \text{ div grad } \phi + z \quad . \tag{33}$$

This may be interpreted as follows:

the change in income at a location is the result of:

1) the deficit of local and neighborhood expenditure as against local income,

2) the net inflow of money from the neighborhood, and

3) autonomous expenditure in the location by non-residents.

Equation (32) states an important principle: money flows in the direction of steepest income decrease. To put it loosely: money flows from the rich to the poor. This is to be interpreted not in terms of personal income but of regional income. Thus money flows from locations of high income to adjacent locations of lower income.

We turn now to the goods and services that are purchased with these money flows. Clearly each money flow must generate a physical flow in the opposite direction of goods purchased. In the case of services the flow involved is one of persons who must move themselves to the point where such services are rendered, or else of the service agents. In this paper we shall consider only the flow of goods. For a corresponding analysis of services the reader is referred to Beckmann (1980). The flow of goods in a spatial market can be understood only in connection with the spatial pattern of prices. Consider a single commodity and let $\Psi(x_1,x_2)$ be its flow assumed to be stationary. In the simplest case local demand and supply of the good are given for each location independent of price. Let $q(x_1,x_2)$ denote local excess demand. Now since div Ψ is net outflow which in turn equals excess supply, we have a first equilibrium condition

$$\text{div } \Psi + q = 0 \qquad \text{in } A. \tag{34}$$

Let the region A be self-contained and let flow across the regional boundary vanish, so that

$$\Psi_n = 0 \qquad \text{on } \partial A. \tag{35}$$

By the Gauss integral theorem [Courant and John (1965)]

$$\iint_A \operatorname{div} \Psi \, dx_1 dx_2 = \int_{\partial A} \Psi_n \, ds \quad . \tag{36}$$

In order that a spatial equilibrium should exist in the market of this commodity, it is therefore necessary in view of (34), (35) that

$$\iint q \, dx_1 dx_2 = v \quad . \tag{37}$$

Condition (37) states that aggregate demand must equal aggregate supply.

A second condition is needed to determine equilibrium, and this is based not on physical ideas but on economic ones. Consider the gain from trade that is realized by moving one unit of the commodity in the direction of Ψ. The rate of profit is then given by the directional derivative of price, which equals

$$\operatorname{grad} p' \frac{\Psi}{|\Psi|} \quad .$$

Let the unit cost of moving the commodity be $k = k(x_1, x_2)$. The net rate of profit at the locations achieved by moving the commodity in the direction of the flow field is then

$$\operatorname{grad} p' \frac{\Psi}{|\Psi|} - k \quad . \tag{38}$$

The direction in which the profit rate is maximized is easily shown to be that where Ψ is parallel to grad p, or

$$\frac{\Psi}{|\Psi|} = \frac{\operatorname{grad} p}{|\operatorname{grad} p|} \quad . \tag{39}$$

Substituting (39) into (38), we obtain the realized profit rate

$$|\operatorname{grad} p| - k \quad .$$

In equilibrium under perfect competition profits must vanish:

$$|\operatorname{grad} p| = k \quad . \tag{40}$$

Combining (39) and (40), one has

$$\text{grad } p = k \frac{\Psi}{|\Psi|} \quad , \qquad (41)$$

and this is the second condition of equilibrium. It may be shown that the two conditions (34) and (41) together with a boundary condition (35) determine the direction of the flow field $\frac{\Psi}{|\Psi|}$ uniquely in all location.

It is interesting to compare the economic equilibrium condition (34), (35) and (41) with the equilibrium conditions of physical flow in hydrodynamics.

$$\text{grad } P = h\Psi \quad . \qquad (42)$$

In hydrodynamic equilibrium P denotes pressure and flow is proportional to the gradient of pressure. Equations (42) and (34) may be combined into the single equation

$$\Delta P = \text{div grad } P = -hq \qquad (43)$$

(provided the resistance factor h is constant).

The Poisson equation (43) is related to the diffusion equation (6). It applies to the steady state when the total propensity to spend locallly and in the neighborhood is unity. The condition (37) requires that the autonomous term should vanish in the aggregate. However, economics is more difficult than physics, and the equations of economic equilibrium (34) and (41) may not be so combined. In fact, while pressure distribution $P(x_1,x_2)$ is analytic, the price distribution $p(x_1,x_2)$ is continuous but not differentiable everywhere. For example, on ridge lines of prices separating two adjacent market areas of a commodity, the flow runs off in two directions, and no well defined unique gradient or derivative exists.

Consider, however, the case of congestion, where transportation cost depends on flow. Then (42) may be a better representation of the relationship between flow, price and transportation cost than is (41). In that case potential theory based on (43) is relevant to the study of spatial economic equilibria.

INTERACTION BETWEEN PRICE AND DEMAND

So far the local excess demand function was assumed independent of price. This restriction is easily lifted. In fact let

$$\text{div } \Psi(x_1,x_2) + q(x_1,x_2,p) = 0 \qquad (44)$$

and assume

$$\frac{\partial q}{\partial p} < 0 \quad . \tag{45}$$

In order that the consistency condition (37) be satisfied it is now sufficient that there exist two prices p_1, p_2 such that

$$\int\int q(x_1, x_2, p) dx_1, dx_2 < 0 \qquad \text{for } p_1 \tag{46}$$

and

$$\int\int q(x_1, x_2, p) dx_1, dx_2 > 0 \qquad \text{for } p_2 \tag{47}$$

Stability of market equilibrium follows by an argument similar to that in the income diffusion case. For a full treatment of this topic see Beckmann-Puu (1980).

CONCLUSIONS

The purpose of this paper has been to demonstrate how a treatment of space in continuous terms will lead to the notation of flow fields, and how the study of these flow fields may be carried out with well-known tools of vector analysis. However, great care must be exercised in properly formulating the relevant economic relationships rather than following blindly the lead of mathematical physics. In the end economic relationships do differ from those of physics.

One advantage of this approach is to emphasize structure and bring it into the open: the distribution in two-dimensional space of potential functions (price or income) and of the associated flows (commodities or money). These variables can be visualized and mapped directly, and are no longer hidden under sets of single or double indices as in so much of contemporary mathematical geography.

REFERENCES

Beckmann, M., 1980. A Continuous Flow Approach to the Location of Facilities, in Giorgio Leonardi, Conference on Facility Location, IIASA (forthcoming).

Beckmann, M. and Puu, T., 1980. The Continuous Transportation Model, IIASA Report (forthcoming).

Courant, R. and John, F., 1965. Introduction to Calculus and Analysis, Wiley, Interscience, New York.

SPACE AND TIME IN GEOGRAPHY: AN AXIOMATIC APPROACH

H. Beguin

Department of Geography
Université Catholique
de Louvain

J.-F. Thisse

SPUR, Université
Catholique de Louvain

INTRODUCTION

The existing literature [a comprehensive survey and bibliography are presented in Thrift (1977) and Carlstein, et al. (1978)] clearly indicates that there is a need for a careful approach to space and time in geography. Accordingly, it seems needless to discuss the reasons for our interest in that problem, and we immediately ask the question: how should space and time be defined, compared, and integrated?

In this paper *we propose an axiomatic approach to the concepts of space and time in geography, to their comparison, and to their integration.* Space has been defined in Beguin and Thisse (1979) by measures on a metric space, and time will be defined hereafter by measures on a totally ordered set. To propose a definition of time is a matter of challenge for spatial analysts. However, a careful review of the literature suggests that some basic concepts to be used in the treatment of time could have some analogy with those used for space. We shall verify that an approach to time based upon such analogies is fully justified and leads to the above definitions. Moreover, this approach has the advantage of making the comparison and the integration of space and time much easier.

A brief warning is in order about our interpretation of time. Present geographical literature does not seem to be satisfied with just considering time as a third dimension of any located object. The main argument is that time is multidimensional; beyond unidimensional clock-time there are times associated with attributes [Thrift (1977); Carlstein, et al. (1978)]. But multidimensionality sometimes appears as a confusing concept that mixes the following two meanings: (1) time can be measured with different characteristics, viz. its duration (unidimensional time) and its attributes (multidimensional time); and (2) duration and attributes can be taken as characteristics of the attribute under study (perceived time). In the present paper multidimensionality completely excludes any perception component.

SPACE

This section summarizes the main features of the definition of space given by Beguin and Thisse (1979). Three basic components constitute the core of the definition: (1) the set of places; (2) the length-metric; and (3) the area-measure. They are completed by measures associated with simple attributes. Space is then expressed by *measures defined on a metric space.*

D.1 Let S Be a Given Set of Places.

The place, denoted s, is an undefined term. It is to be viewed as the elementary spatial unit. Set S is interpreted as the material substratum of natural and human processes. We assume that S contains at least two places.

D.2 Let d_L Be a Length-Metric Defined on S.

The places are normally separated from one another. Accordingly they can be characterized by their relative position. To express this, we use two concepts: (1) a metric; and (2) the dimension of length, denoted [L], that is a dimension endowed with the unit of length (an undefined term). Then, a length-metric d_L is a metric with values in [L]. For any ordered pair of places $(s_1.s_2)$ the distance from s_1 and s_2 is given by $d_L(s_1,s_2)$.

D.3 Let μ_A Be an Area-Measure Defined on $(S,\mathcal{B}_A)$.

A subset of places is expected to have a spatial extension. This one can be expressed by the area of the subset. Two concepts are then used: (1) a measure; and (2) the dimension of area, denoted [A], that is a dimension endowed with the unit of area (the product of the unit of length by itself). The subsets of S whose areas are defined are the elements of the Borel σ-algebra associated with (S,d_L), i.e., the smallest σ-algebra containing the topology $\mathcal{T}_L$ constructed from d_L; it is denoted by $\mathcal{B}_A$. The elements of $\mathcal{B}_B$ are called measurable and represented by B. An area-measure μ_A

is then a non-negative measure defined on $(S,\mathcal{B}_A)$ and with values in [A]. The area of $B \in \mathcal{B}_A$ is given by $\mu_A(B)$.

Singletons of S are measurable; this allows one to distinguish two types of places:

(1) *adimensional places* (also called points) when $\mu_A(\{s\}) = 0$, with $s \in S$; and,

(2) *dimensional places* when $\mu_A(\{s\}) > 0$, with $s \in S$.

C1. The restricted definition of space is given by the triplet

$$(S,d_L,\mu_A) \quad . \tag{1}$$

The logic of this definition must be emphasized. A set of places is necessary to define a length-metric. This metric is used to form the Borel σ-algebra on which the area-measure is defined.

D5. Let $(\mu_i^S;\ i \in I)$ Be a Family of Attribute-Measures Defined on $(S,\mathcal{B}_A)$.

A further characterization of space is obtained by introducing simple attributes, i.e., attributes described by a mass available over any measurable subset of places. We express a simple attribute $i \in I$ by a measure μ_i^S defined on $(S,\mathcal{B}_A)$, and with values in a dimension $[D_i]$. Then the mass of attribute i available on subset B is given by $\mu_i(B)$.

C2. An extended definition of space is obtained by the introduction of simple attributes in (1):

$$[S,d_L,\mu_A,(\mu_i^S;i \in I)]^1 \quad . \tag{2}$$

This definition integrates the basic components of space and all the elements necessary to endow space with composite attributes. One should recall that these ones are defined as functions of simple attributes and/or of basic components of space. It is also quite flexible since no restriction is imposed on the family I of simple attributes.

TIME

We propose to express time by *measures defined on a totally ordered set*. The definition is presented as follows. In the first three subsections we introduce what we consider as the three basic components of time, namely the set of moments, the total order on the set of moments, and the duration-measure. These components can receive the following intuitive interpretation: the set of moments suggests itself since time is under study; the one-way course of time justifies the use of a total order associated with the notions of anteriority and posteriority; the course of time similarly leads to the concept of duration, expressed here by a measure. As in the case of space, these three basic components are linked through their mathematical specification. We consider these components as necessary, though not sufficient, for the characterization of time. In other words they are time-specific. In a subsequent subsection a family of simple attributes is introduced to complete the definition. This will enable the extended definition to express multidimensionality time, whose characterization appears as a main purpose of the abundant literature on time in geography.

D5. Let T Be a Given Set of Moments.

The *moment* is an undefined term that is denoted by t. Intuitively the moment corresponds to the elementary temporal unit. *The temporal medium for natural and human processes is defined by a given set* T *of moments*. T is assumed to contain at least two moments. This interpretation of T is the first specificity of our model of time.

D6. Let $\leqq$ Be a Total Order Defined on T.

In a temporal context, *moments are expected to follow one another in a one-way course*. Any moment therefore must be positioned *before* or *after* any other moment. We propose to use the concept of total order to express this particular form of relative position. Such a relation allows the moments to be completely ranked inside T so that the succession of moments is fully seized.[2] The total order $\leqq$ defined on T is then to be interpreted as follows: given any pair of moments $\{t_1,t_2\}$, $t_1 < t_2$ means that t_1 *is previous to* t_2 and $t_1 = t_2$ that t_1 *and* t_2 *are simultaneous*. This relation is the second specificity of time.

Being totally ordered, now, T resembles the set of real numbers or the set of integers endowed with the natural order, denoted $(\mathbb{R},\leq)$ and $(\mathbb{N},\leq)$, respectively. Hence, to keep the mathematics simple, it would seem preferable to replace $(T,\leq)$ by $(\mathbb{R},\leq)$ or $(\mathbb{N},\leq)$. In fact this substitution is possible only if $(\mathbb{R},\leq)$ or $(\mathbb{N},\leq)$ is the image of $(T,\leq)$, and conversely; that is, only if there exists a homomorphism between the two structures.[3] But it is known that such a homomorphism does not necessarily exist when T is uncountable [Gérard-Varet, et al. (1976)]. Using $(\mathbb{R},\leq)$ instead of $(T,\leq)$ as a general concept is then justified only under some conditions, namely [see Hausdorff (1962), but also Gérard-Varet, et al. (1976) for a more recent presentation]:

(1) *Assume that* T *is uncountable. Then* $(T,\leq)$ *is homomorphic to* $(\mathbb{R},\leq)$ *if and only if* $(T,\leq)$ *is perfectly separable.*[4]

On the other hand, when T is countable the substitution is permitted since we have:

(2) *Assume that* T *is countable. Then* $(T,\leq)$ *is homomorphic to* $(\mathbb{N},\leq)$.

To construct the subsets of moments whose duration is well-defined we use as an intermediate concept the topology $\mathcal{T}_D$ associated with the total order $\leq$. Our approach therefore is similar to that followed in the definition of space where the measurable subsets of places are obtained from the topology $\mathcal{T}_L$ associated with the length-metric. The topology $\mathcal{T}_D$, called order topology, is defined by the class of subsets G of T such that:

(1) $\emptyset \in \mathcal{T}_D$ and $T \in \mathcal{T}_D$; and,

(2) $G \in \mathcal{T}_D$, with $G \neq \emptyset$ and $G \neq T$, if and only if G is the union of open intervals of $(T,\leq)$.[5]

Some equivalences between properties of the order structure $(T,\leq)$ and of the topological space $(T,\mathcal{T}_D)$ are given in Appendix A.

D7. Let μ_D Be a Duration-Measure Defined on $(T,\mathcal{B}_D)$.

In a temporal context *subsets of moments are expected to have a temporal extension*. We therefore, must be able to express their *duration*. To this effect we use a measure and the dimension of duration. This dimension is clearly the most specific one that concerns time. The *unit of duration* is taken as an undefined term (the second and last one of our definition of time). The *dimension of duration* is then defined by a dimension endowed with the unit of duration; it is denoted [D]. (In fact [T] seems to be more common, but we retain [D] in order to unify the notation.)

Similarly to the case of the area in the definition of space, the duration is not necessarily defined for all the subsets of moments. For this reason, we introduce the σ-algebra $\mathcal{B}_D$ whose elements can be characterized by their duration. This σ-algebra is associated with $(T,\leqq)$ and is defined as the smallest σ-algebra containing the topology $\mathcal{T}_D$. The elements of $\mathcal{B}_D$ are called measurable and denoted by $\overline{B}$.

A *duration-measure* is a non-negative measure defined on $(T,\mathcal{B}_D)$, and taking its values in [D]. It is denoted by μ_D. For any measurable subset of moments $\overline{B}$, $\mu_D(B)$ gives the duration of the corresponding subset. The duration-measure is the third specificity of time. Note, in passing, the relations existing between the total order defined on T and the duration-measure. These concepts are linked through $\mathcal{T}_D$, built from $\leqq$, and through $\mathcal{B}_D$, built from $\mathcal{T}_D$, on which μ_D is defined. As mentioned above, these relations are analogous to those between the length-metric and area-measure in the definition of space. We can now introduce some characterizations of time similar to those formulated for space in Beguin and Thisse (1979). When $\mu_D(T)$ is positive and finite the duration of a set of moments is said to be *finite*. When $\mu_D(T)$ is infinite and T the union of a countable family of subsets, each of which has a finite duration, the duration of the set of moments is called *σ-finite*. We also have the following important property that is analogous to Proposition 3 in Beguin and Thisse (1979).

(3) *The singletons of* T *are measurable.*

For any $t \in T$ we have $\{t\} = C[(\leftarrow,t)\cup(t,\rightarrow)]$. Given that $(\leftarrow,t)$ and $(t,\rightarrow)$ are open subsets of $\mathcal{T}_D$, $\{t\}$ belongs to $\mathcal{B}_D$ is the complementary of an open subset. This property allows us to introduce two familiar types of moments:

(a) *instantaneous moments* (also called instants) for which the duration-measure of $\{t\}$, for any $t \in T$, is equal to zero; and,

(b) *non-instantaneous moments* (also called periods) defined by $\mu_D(\{t\}) > 0$, for any $t \in T$.

Some intuitive relations can be established between the types of moment representations and the cardinal of set T [the proof is similar to that of Propositions 6 and 7 given in Beguin and Thisse (1979)].

(4) *If moments are instantaneous and if the duration of* T *is finite or σ-finite, then moments are uncountable; and*

(5) *If moments are non-instantaneous and if* T *has a finite or σ-finite duration, then the set of moments is countable.*

C3. In a conception limited to the basic components, the following restricted definition can be given: *we call time any set of at least two moments endowed with a total order and a duration-measure,* i.e., a triplet of the form

$$(T, \leqq, \mu_D) \ . \tag{3}$$

Intuitively, this definition corresponds to what is referred to in the literature as unidimensional clock-time. It has the following internal logic: the existence of at least one moment leads to the introduction of a duration-measure, while the existence of at least two moments calls for a total order on T.

D8. Let $(\mu_i^T; i \in I)$ Be a Family of Attribute-Measures Defined on $(T, \mathcal{B}_D)$.

Besides the basic components, time can also be characterized by a list of attributes. As stated in the introduction, attributes correspond to some of the characteristics mentioned in the literature advocating the multidimensional nature of time. A simple attribute, marked with $i \in I$, is an attribute that can be represented by a mass available during a certain subset of moments. The masses being additive, a non-negative measure can be used to describe their sum. Formally, a simple attribute $i \in I$ is expressed by a non-negative measure μ_i defined on $(T, \mathcal{B}_D)$ and with values in a dimension $[D_i]$. The mass of attribute i available on a subset $\overline{B}$ is given by $\mu_i(\overline{B})$

Similarly, composite attributes can be considered. They are defined by functions of simple attributes and, possibly, of basic components of time. They are thus derived, and not basic, concepts.

C4. In a more extensive conception including simple attributes and corresponding to multidimensional time, the following definition of time is proposed:

$$[T, \leqq, \mu_D, (\mu_i^T; i \in I)].^6 \tag{4}$$

This definition is flexible in that the number of attributes is not fixed. It also contains all the necessary elements for a still broader characterization of time, since composite attributes can be constructed from the components given in (4). Interestingly the comparison between unidimensional time [definition (3)] and multidimensional time [definition (4)] suggests that the two concepts are not fundamentally different.

SPACE AND TIME: A COMPARISON

A brief comparison between space and time in geography seems to be useful before suggesting an integration of the two concepts. The comparison presented in Table 1 is very suggestive.

TABLE 1

A Comparison Between Space and Time in Geography

Characteristics	Space	Time
Elements and set	s and S	t and T
Relative position	d_L on S	$\leq$ on T
Measure of extension	μ_A on $(S,\mathcal{B}_A)$	μ_D on $(T,\mathcal{B}_D)$
Attribute-measures	$(\mu_i^S; i \in I)$ on $(S,\mathcal{B}_A)$	$(\mu_i^T; i \in I)$ on $(T,\mathcal{B}_D)$

The analogy between space and time clearly holds as far as the sets of the elementary units and the measures associated with them are concerned. This amounts to observing a correspondence between place and moment on the one hand, and between area and duration on the other. Further, places and moments are both characterized by their relative position. But the relative positions inside S and inside T are expressed through two different mathematical tools, i.e., a metric and a total order.[7] This distinction, however, is less fundamental than would appear at first sight. Indeed, the analogy can be pursued further, provided some additional concepts are introduced. This can be done in two ways: (1) *from space to time*; and (2) *from time to space*. In the former we have to construct a distance on T, while in the latter a ranking over S is needed.

Given $t_1 \leq t_2$, the *time-distance from* t_1 *to* t_2 is defined by the difference of the duration-measures of the intervals $(\leftarrow,t_2)$ and $(\leftarrow,t_2)$. It is denoted $d_T(t_1,t_2)$ and is expressed in units of duration. For any $(t_1,t_2)\in\{(t_1,t_2); t_1\in T, t_2\in T \text{ and } t_1\leq t_2\} \overset{def}{=} T_<$,

$$d_T(t_1,t_2) = \mu_D[(\leftarrow,t_2)] - \mu_D[(\leftarrow,t_1)] \quad , \tag{5}$$

which amounts to

$$d_T(t_1,t_2) = \mu_D[(t_1,t_2)] \tag{6}$$

by definition of a measure. In other words, the time-distance from t_1 to t_2 is equal to the duration of the period starting at t_1 and terminating at t_2. We immediately notice that the time-distance from t_2 to t_1 is not defined since the interval (t_2,t_1) *does not* exist. In this way, we try to capture the phenomenon of *non-reversibility of time*.

So define the time-distance verifies some interesting properties:

(i) non-negativity: $d_T(t_1,t_2) \geq 0$;

(ii) partial identity: $d_T(t_1,t_2) = 0$ if $t_1 = t_2$; and,

(iii) triangle inequality: $d_T(t_1,t_3) \leq d_T(t_1,t_2) +$

$d_T(t_2,t_3)$ if $t_1 \leqq t_2 = t_3$,

$d_T(t_3,t_2)$ if $t_1 \leqq t_3 < t_2$.

Indeed, $d_T(t_1,t_2)$ is non-negative since μ_D is. Furthermore, $d_T(t_1,t_2) = \mu_D(\emptyset) = 0$ by definition of a measure. Concerning the triangle inequality, we have to consider the cases $t_1 \leqq t_2 \leqq t_3$ and $t_1 \leqq t_3 < t_2$. In the first one, we have $(t_1,t_3) = (t_1,t_2)$ (t_2,t_3) and $(t_1,t_2) \cup (t_2,t_3) = \emptyset$, so that $\mu_D[(t_1,t_2)] +$ $\mu_D[(t_2,t_3)]$ by definition of a measure. In the second case, the monotonicity property of a measure implies that $\mu_D[(t_1,t_3)] \leq$ $\mu_D[(t_1,t_2)]$ [see Halmos (1974)]. As $\mu_D[(t_1,t_2)] = \mu_D[(t_1,t_3)] +$ $\mu_D[(t_3,t_2)]$, the desired inequality is then easily deduced.

Thus we have verified that the time-distance is a weak metric on $T_<$.[8] For d_T to become a metric on T we therefore must introduce two further conditions. The first one does not seem to be very restrictive. It says that $\mu_D[(t_1,t_2)] = 0$ implies $t_1 = t_2$ for any $(t_1,t_2) \in T_<$. Intuitively this means that T is of a "single piece" with respect to the duration-measure.[9] It is then very easy to see that the axiom of identity is satisfied. The second condition is relative to the axiom of symmetry, and its soundness is much more questionable. It states that the orientation of time is not a basic attribute so that $d_T(t_1,t_2)$ and $d_T(t_2,t_1)$ can be considered simultaneously, and put equal to each other.

In other words, to be able to construct a metric on T *we must suppose that time is reversible.* Our purpose is not to engage in a discussion about the respective pertinence of the hypotheses of reversibility and non-reversibility in geography. For that the reader is referred to the existing literature. Rather, it has been to bring into light the assumption required to have a time-metric. Incidentally, we notice that in that case--and in that case only--time can be viewed as a "third" dimension of space.

The second direction aims to introduce a ranking over S. For that purpose we must assume that *there exists a unique center* s_c *of the set of places* [e.g., s_c may be a minimizer of $\int_S d_L(s,s')d\,\mu_A(s')$]. A total preorder $\lesssim$ on S can then be defined in the following way[10]:

$$s_1 \lesssim s_2 \text{ if and only if } d_L(s_c,s_1) \leq d_L(s_c,s_1) \quad . \tag{7}$$

Since different places are generally equidistant from center s_c, the condition of anti-symmetry does not hold and the relation $\lesssim$ is not a total order. Nevertheless, a total order can be built from $\lesssim$ provided the relative position of places with respect to the center is the dominant characteristic in the system of relative positions, viz. when *space is isotropic around* s_c *for the properties under analysis.*[11] In this case the places equidistant from the center can be grouped in equivalence classes, say $S(r) = \{s \in S;\ d_L(s_c,s) = r\}$ with $r > 0$, and a total order $<$ called quotient order can be defined on the set of classes $S(r)$ as follows:

$$S(r_1) < S(r_2) \text{ if and only if } r_1 < r_2 \quad . \tag{8}$$

The analysis on S then reduces itself to the analysis on the set of equivalence classes $S(r)$, a totally ordered set.

In summary, the analogy "from space to time" is valid for any problem in which the orientation of time is not a basic feature. The analogy "from time to space" is valid for any problem in which the relative position of places with respect to a center is the major relation. Finally, both attempts towards analogy resort to derived, and not to basic, concepts of the definitions.

SPACE-TIME: A PROPOSAL

As the level of the restricted definitions given by the triplets (1) and (3), *interdependence between space and time does not exist.* Other disciplines (physics, for example) may be concerned by such a possible dependence but, in geography, separate definitions seem be be fully justified. *Interaction is introduced by considering attributes.* This is because attributes can be defined as well on subsets of places as on subsets of moments. Thus, in geography space and time are to be considered together at the level of the extended definitions (2) and (4), and at that level only.

Space-time joint consideration requires us to take into account the product set $S \times T$, and the product σ-algebra $\mathcal{B}_A \otimes \mathcal{B}_D$ defined as the smallest σ-algebra containing $\mathcal{B}_A \otimes \mathcal{B}_D$. This is possible since the triplets (1) and (3) are independent. The attribute-measure μ_i associated with simple attribute $i \in I$ of space *and* time is now defined on $(S \times T, \mathcal{B}_A \otimes \mathcal{B}_D)$. For any $B^* \in \mathcal{B}_A \otimes \mathcal{B}_D$, where B^* is a subset of pairs of places and moments, $\mu_i(B^*)$ represents the mass of attribute i available on B^*.

The key question is now the meaning of $\mu_i(B^*)$ in geographical terms. The interpretation is easy to give when B^* is the product $B \times \overline{B}$ of two subsets B and $\overline{B}$ belonging to $\mathcal{B}_A$ and $\mathcal{B}_D$, respectively: $\mu_i(B^*)$ *expresses the mass of attribute i available over a subset of places* B *and during subset of moments* $\overline{B}$. It is clear in this case that the distinction between space and time is preserved: the mass $\mu_i(B^*)$ is given relatively to *fixed* and *independent* subsets of places and of moments. But a complete analysis of the space-time structure also requires an interpretation of $\mu_i(B^*)$ when B^* is not defined by a product $B \times \overline{B}$. For this, it is necessary to understand the structure of B^* in the general case; but, for all we know such a description seems to be very difficult. The reason is that *the corresponding subset of places changes over time while the corresponding subset of moments jointly changes over space.* Stated differently, the subsets of places and of moments are *changing* and *interdependent.* Until now these transformations have been neglected by geographers. This is probably because they are difficult to characterize. It is our belief, however, that the use of the interpretation of such transformation are necessary for a full exploitation of the space-time concept. Otherwise, we will go on considering attributes in space and time separately.

CONCLUSIONS

The analogy between space and time had already been pointed out by geographers and by economists. This paper has set this analogy in more formal terms. It suggests the reinterpretation of some concepts developed for space (time) in terms of time (space). The approach is very fruitful in the direction "space to time." An example is provided by the similitude between transportation and storage costs [see Samuelson (1957)]. This led to the link of location and inventory theory, and spatial price theory and intertemporal price theory. Undoubtedly other concepts could be similarly related [see Ullman (1974)]. In the direction "time to space" the approach is also promising. Such is the case, for instance, in the New Urban Economics, where optimal control theory, borrowed from dynamic optimization methods, has proved to be very useful [Mills and MacKinnon (1973)]. Note that this transposition has been made possible because a center (called Central Business District) is assumed given a priori, and because space is supposed to be isotropic around the center so that the analysis may be restricted to a straight line. However, the analogy being incomplete, exploiting the similarities between space and time is insufficient to guarantee the development of spatial and dynamic theories. A specific example is given by spatial econometrics [Paelinck and Klaassen (1979)]. The importance of the contiguity of places in spatial data precludes any total order to be built on S. Hence, classical econometrics cannot be applied and specific methods are required.

This paper has also verified the importance of measure theory for a theoretical study of time. A similar observation has been made by Beguin and Thisse (1970) for space. Nonetheless, whereas measure theory is already operational in spatial analysis, its usefulness for dynamic analysis still remains to be demonstrated. In the meantime, this should not preclude the use of other mathematical tools such as difference or differential equations [see Smith (1981)].

The space-time concept appears only in part in geography, since space and time remain separated in the existing studies. The real breakthrough will occur when an operational interpretation is given to the space-time subset B^* of places and moments taken in the product σ-algebra. Clearly, much work remains to be done in this domain.

APPENDIX A

The following properties illustrate the relations existing between the order structure $(T,\leq)$ and the topological space $(T,\mathcal{T}_D)$ [see Gérard-Varet, et al. (1976)]:

(1) *$(T,\leq)$ is perfectly separable if and only if $(T,\mathcal{T}_D)$ verifies the second axiom of countability*[12]*; and,*

(2) *$(T,\leq)$ is without gaps*[13] *and conditionally complete*[14] *if and only if $(T,\mathcal{T}_D)$ is connected.*

Proposition 1 gives the necessary and sufficient condition on $(T, \mathcal{T}_D)$ for $(T, \leqq)$ to be homomorphic with $(\mathbb{R}, \leqq)$ when T is uncountable. Proposition 2 deals with the property T being a "single piece." This expression differs in $(T, \leqq)$ and $(T, \mathcal{T}_D)$. In $(T, \leqq)$ the property says that T cannot be partitioned into two open subsets of $\mathcal{T}_D$. Though not equivalent, these two formulations nevertheless are linked, as shown by the proposition.

FOOTNOTES

1. A more compact expression of the definition of space may be proposed: (S, d_L, μ^S), where μ^S is a measure defined on the product σ-algebra $\mathcal{B}_A \otimes \mathcal{I}$, $\mathcal{I}$ being a σ-algebra on I that contains the singletons $\{i\}$. Then, the area-measure is the *projection of* μ^S on $\mathcal{B}_A$, and the attribute-measure i is the *measure* μ^S *conditional to* $\{i\}$ [see Halmos (1964) for the definition of the new mathematical concepts used here]. Interestingly, this presentation emphasizes the formal similitude between the area-measure and the attribute-measures. We, however, prefer formulation (2) because it is more intuitive.

2. An *order* $\leqq$ defined on T is a binary relation that satisfies the following axioms: $\forall (t_1, t_2, t_3) \in T$,

   (1) reflexivity: $t_1 \leqq t_1$;

   (2) anti-symmetry: $t_1 \leqq t_2$ & $t_2 \leqq t_1 \Rightarrow t_1 = t_2$; and

   (3) transitivity: $t_1 \leqq t_2$ & $t_2 \leqq t_3 \Rightarrow t_1 \leqq t_3$.

   The order is said to be *total* if and only if the following additional condition holds:

   $$\forall (t_1, t_2) \in T^2 ,$$

   (4) comparability: $t_1 \leqq t_2$ or $t_2 \leqq t_1$.

   The pair $(t, \leqq)$ is called *total order structure.*

3. A mapping f from T in $\mathbb{R}$ (or in $\mathbb{N}$) is an *homomorphism* if and only if

   $$\forall (t_1, t_2) \in T,\ t_1 \leqq t_2 \Rightarrow f(t_1) \leq f(t_2).$$

   Given that $\leqq$ is a total order, it is easily verified that

   $$f(t_1) \leq f(t_2) \Rightarrow t_1 \leqq t_2 ,$$

$$t_1 = t_2 \Leftrightarrow f(t_1) = f(t_2), \text{ and}$$

$$t_1 < t_2 \Leftrightarrow f(t_1) < f(t_2).$$

Thus, f maintains the ranks in the two structures.

4. The total order structure $(T, \leqq)$ is said to be *perfectly separable* if and only if there is a countable subset U of T such that

$$\forall (t_1, t_2) \in T^2,\ t_1 < t_2 \Rightarrow \exists\ \bar{t} \in U / t_1 \leqq \bar{t} \leqq t_2\ .$$

Roughly speaking, this characterization means that T can be "well-approximated" by a countable subset of moments.

5. An *open interval* of $(T, \leqq)$ is a subset of T defined as follows:

$$(t_1, t_2) = \{t;\ t \in T \ \&\ t_1 < t < t_2\},$$

$$(\leftarrow, t_2) = \{t;\ t \in T \ \&\ t < t_2\}, \text{ and}$$

$$(t_1, \rightarrow) = \{t;\ t \in T \ \&\ t_1 < t\}.$$

6. A remark similar to that of footnote 1 can be made for the definition of time.

7. That a total order is often sufficient to position the elements of a "one-dimensional" set is illustrated by the Weber problem with collinear points. Indeed, the order of the markets, without reference to the distance between them, is sufficient to determine the transportation cost minimizing location [see Witzgall (1965)].

8. For a discussion of the concept of weak metric see Witzgall (1965).

9. Interestingly, this characterization compares with Proposition 2 of Appendix A.

10. A *preorder* $\leqq$ defined on T is a binary relation that satisfies the axioms of reflexivity and of transitivity. It is said to be *total* if the condition of comparability is also verified.

11. We say that a space is *isotropic* around a point s_c for a given property if and only if the property is invariant on each set $S(r) = \{s \in S;\ d_L(s_c, s)r\}$.

12. The topological space $(T, \mathcal{T}_D)$ satisfies the *second axiom of countability* if and only if there exists a countable subclass of $\mathcal{T}_D$, denoted C_D, such that any open set of $\mathcal{T}_D$ is equal to the union of open sets of C_D:

$$\forall G \in \mathcal{T}_D, \exists \{C_i ; i \in I\} \subseteq C_D / G = \bigcup_{i \in I} C_i \; .$$

In other words, this definition says that the order topology can be generated from a countable subclass of open sets. Given the intuitive interpretation of the perfect separability proposed in footnote 4, Proposition 1 therefore is not surprising.

13. We say that $(T, \leqq)$ is *without gaps* if and only if for any $t_1 < t_2$ there exists $t \in T$ such that $t_1 < t < t_2$. Intuitively it is always possible to find a moment of T between any two different moments; T is therefore "full."

14. A subset V of T is called *bounded* if and only if there exist $\bar{t}$ and $\underline{t}$ in T such that $\underline{t} \leqq t \leqq \bar{t}$ for any $t \in V$.

$(T, \leqq)$ is said to *conditionally complete* if and only if any bounded subset V of T has a least upper bound and a highest lower bound in T.

REFERENCES

Beguin, H. and Thisse, J., 1979. An Axiomatic Approach to Geographical Space, Geographical Analysis, 11: 325-341.

Blaut, J, 1961. Space and Process, The Professional Geographer, 13: 42-51

Carlstein, T., Parkes, D. and Thrift, N. (eds.), 1978. 1. Making of Time; 2. Human Activity and Time Geography; 3. Time and Regional Dynamics, Arnold, London.

Gérard-Varet, L., Prévot, M. and Thisse, J., 1976. Analyse mathématique pour l'économie. Topologie, Dalloz, Paris.

Hägerstrand, T., 1973. The Domain of Human Geography, in: Directions in Geography, R. Chorley (ed.), Methuen, London.

Halmos, P., 1974. Measure Theory, Springer-Verlag, New York.

Hausdorff, P., 1962. Set Theory, Chelsea Publ. Co., New York.

Levy, J., 1967. An Extended Theorem for Location on a Network, Operational Research Quarterly, 18: 433-443.

Louveaux, F., Thisse, J.-F. and Beguin, H., 1980. Location Theory and Transportation Costs, SPUR, Louvain-la-Neuve.

Mills, E. and MacKinnon, J., 1973. Notes on the New Urban Economics, Bell Journal of Economics and Management Science, 4: 593-601.

Paelinck, J. and Klaassen, L. 1979. Spatial Econometrics, Saxon House, Founborough.

Phlips, L. and Thisse, J., 1979. Pricing, Distribution and the Supply of Storage, CORE, Louvain-la-Neuve.

Samuleson, P., 1957. Intertemporal Price Equilibrium: A Prologue to the Theory of Speculation, Weltwirtschaftliches Archiv, 181-221.

Schuler, R. and Holahan, W., 1978. Competition vs. Vertical Interaction of Transportation and Production in a Spatial Economy, Papers of the Regional Science Association, 41: 209-225.

Smith, T., 1981. Multiple Equilibria, Stability and Chaos in Master Equation Descritpions of Population Movements Driven by Externalities, in Dynamic Spatial Models, D. Griffith and R. MacKinnon (eds.), Plenum, New York, 49-66.

Smithies, A., 1939. The Maximization of Profits Over Time with Changing Costs and Demand Functions, Econometrica, 7: 312-318.

Smithies, A., 1941. Monopolistic Price Policy in a Spatial Market, Econometrica, 9: 63-73.

Swoveland, C., 1975. A Deterministic Multi-Period Production Planning Model with Piecewise Concave Production and Holding-Backorder Costs, Management Science, 21: 1007-1013.

Thrift, N., 1977. Time and Theory in Human Geography, Progress in Human Geography, 9: 65-101 and 413-457.

Tuan, Y., 1974. Space and Place: Humanistic Perspective, Progress in Geography, 6: 211-252.

Ullman, E., 1974. Space and/or Time: Opportunity for Substitution and Prediction, Transactions of the Institute of British Geographers, 63: 125-139.

Witzgall, C., 1965. Optimal Location of a Central Facility: Mathematical Models and Concepts, National Bureau of Standards, Report 8388.

---

*The authors gratefully acknowledge the help of L. Gérard-Varet, D. Peeters and T. Smith in the preparation of this paper.

TIME IN SOCIAL SCIENCES MODELS: SOME REMARKS

Bernard Marchand

Institut d'Urbanisme
Université de Paris-VIII

Most of the models built in urban and regional planning as well as other areas of social sciences, are indirectly inspired by or directly borrowed from the physical sciences. This infatuation with physics is so strong that planners trying to build models of processes changing over time speak of "dynamic model," although the true meaning of this adjective is "of or relating to physical force producing motion." Actually, physical time is quite different from mathematical time. The point I wish to make here is that time in social sciences is a much more subtle, involved and ambiguous category than the oversimplified version of it used in the so-called "hard" sciences.

As an urban planner I recently have been involved in two different tasks: (1) to build a model to try to forecast the behavior of pedestrians around a subway station; and, (2) to describe the past evolution of population and housing in Los Angeles between 1940 and 1970. Most social scientists have to act, in this way, as historians of the Past and as forecasters of the Future. Most have probably felt, as I did, an extraordinary difference between both attitudes. In the Past it seems that the evolution we observe was, if not the only possible one, at least a very likely solution among the very few which appear to have been possible. Looking to the Future, on the other hand, makes the head spin for the enormous number of possible paths an evolution can follow. This note tries to suggest some explanation for the contrast that exhibits, in our belief, the very nature of time.

STABILITY OR DISORDER: THE CONFLICTING EFFECT OF TIME

Most European cities were founded before or during the Roman Empire. Since the beginning of this era all factors that might have justified urban creation and early development have radically lost their value. The original sites of these cities, as a fortress on a hill, as a ford, as a Roman camp, as a crossroad or a frontier town, have lost their relative advantages, or in some cases have become a burden. All the British Ox-ford or the German Frank-furt have their bridges nowadays. The Man-chester or Win-chester no longer house a military camp. The "limes" cities along the Rhine are no longer frontier towns but form the urban core of Germany and Europe. In the same way, Lyon, once a contact city between the French kingdom and the German empire is today an important regional captial in France. Local sites, regional situations, urban functions have all changed; more often than not, they have turned into their opposite without seemingly hampering the irresistible growth of the settlements.

Such stability is not even a privilege of the Old World. In Los Angeles, which may well be the most fluid of all North American cities, ethnic and demographic characterizations, as well as land rent and type of dwelling (i.e., the principal aspects of population and housing) have grown in a most regular way between 1940 and 1970, following an extremely clear concentric pattern (Marchand, forthcoming). In the same metropolis, land rent spatial structures, as analyzed through the theory of regional variables and a variety of spatial auto-correlation analyses, still exhibit the spatial pattern of the old electric streetcars that in the 1920's connected the old urban centers of downtown and Hollywood to the beach resorts.

The urban planner studying the past, and for that matter the historian too, cannot help being impressed by the very small number of paths from which a spatial evolution seems to have emerged. The point here is certainly not to try to rehabilitate determinism, and to suggest history could not have taken forms other than the ones we observe; but rather, those other possible branchings of any evolution, although they did exist, and probably in very large numbers, remain very difficult, if not impossible, for us to conceive.

Investigating the past exposes us to three different traps. First, we could be led, as suggested above, to an oversimplified, mechanistic type of interpretation: nothing could have happened differently. Many Marxist scholars fall into the deterministic trap, destroying their own purpose by unwillingly showing that the revolutionary has nothing to do since "revolution" will come necessarily at its own time. A second mistake, common to Marxist and Christian doctrines is teleological: history seems to develop in such a clear direction that it is supposed to have a goal or purpose it is blindly fulfilling. The last error is more frequent but different in nature: past evolutions appear to be so strictly determined that they seem to have only one possible explanation. The researcher is too often satisfied with the detailed causes he has exhibited (i.e., the explanatory model he has devised as long as it is coherent) without realizing other equally coherent and satisfying models could be proposed. It is always possible to explain courses of actions in history, since they did really happen; the difficulty is to show with convincing arguments that among all the possible explanations the one proposed is the more likely. The difficulty, when studying the past, lies in trying to enlarge this set of possible explanatory models. Conversely, the main difficulty in modelling the future is to restrict the set of possible models until one gets a limited and workable set of possible evolutions.

I see three ways of explaining such difficulties. The first two deal with socio-economic mechanisms; they are most important but do not constitute the subject of this communication. The last one deals with the very nature of time. The apparent stability of evolutions observed in the past is certainly the result of conscious, powerful mechanisms designed to preserve the status quo and to maintain current privileged situations. The research field here is vast but has been quite extensively investigated. Effects of zoning laws, building codes, property taxes, real estate policies, illegal and unspoken but efficient racial prohibitions have been indicated and sometimes evaluated in the case of land rent and land use. Unconscious mechanisms maintaining the stability of social systems and of their spatial patterns are probably more important, still, than consciously designed policies. In my opinion, the urban neighborhood plays the main role of a physical and moral frame insuring the reproduction of the basis of social life in the city: reproduction of the working force (preserving the relations between occupational groups, housing and occupation locations), of the patrimonies (preservation of land rent), of the moral values (through the family, friends and neighbors). Both groups of mechanisms probably explain most of the stability we observe in urban structure, but it is not the purpose of this note to investigate them. My interest is presently focused on a third type of explanation to be found in the very nature of time and in the different knowledge we have about change: an individual explanation about the past, but only an aggregate one about the future.

ON TWO PROPERTIES OF TIME IN SOCIAL SCIENCES

Consider _n_ elements of space (lots, tracts, urban districts, settlements along a river) and the way their properties (such as price, land use, position in a competitive urban network) change as time passes.

The Cumulative Determination of Possible Histories

Observe the development of a Löschian central place system. At the start space is uniform, isotropic, homogeneous. When a principle of spatial organization is chosen, however, and two settlements are located, space is changed from uniform to polarized by those privileged locations. The transportation network, then, is largely determined. Once it is developed across the plain, space has become anisotropic. The following development of sectors rich in settlements, jobs, roads and activity, and of depressed regions finally produces a definite heterogeneity in space. As the system develops, space properties turn into their contraries and constrain drastically, at each step, the number of possible evolutions. Each time an element _i_ of space adopts a particular type of land use, or a particular function, the number of possible histories decreases very abruptly. The level at which this element "freezes" is capital, because as the environment changes perpetually in its properties, the type of land use adopted will change accordingly, which in turn will alter again the environment.

Let us consider another example. In Los Angeles land rent has been determined by the layout, during the 1920's, of the electric streetcar lines; land made accessible along those lines became high-priced. The exact outline of the rails was partly arbitrary, but as soon as the line was open, rent was determined and frozen along large ribbons of land. But more degrees of freedom were indirectly lost. Tracts located between the lines with poor accessibility could not be developed and put on the market at an arbitrary price. Given the supply of land at high- or medium-price, developers had to aim for a certain price bracket indirectly determined by the preceding developments. In this example, each new line determines land prices directly along the rails, and indirectly on wide portions of yet unaccessible land. Before elaborating a more precise mathematical model of the dynamical process, one intuitively feels that this model should not be linear, but rather exponential: the number of possible spatial histories decreases in a cumulative way with time. The level at which a spatial element's characteristics are defined is capital, or *relative position in history*.

Local Amenities as a Function of Time

Absolute position in history, however, is not less important because local amenities that might determine the land use, land price or function of a spatial element, vary themselves with time. Consider a situation in which a hypothetical river offers three different sites: site A is deep and narrow, very dangerous to cross, but particularly favorable for building a bridge. Site B is the opposite, shallow and wide, and easy to ford. Site C, with a steep hill, offers a naturally protected spot to build a fortress. Assume we are in Europe, where successive waves of population have open the wilderness to man. If the first important settlement happens at the peak of the Roman Empire, say at the beginning of the second century A.D., site B will be the obvious choice: the empire is well policed and thus settlers do not need extensive protection, while building a bridge, although possible with techniques of the day, is still a difficult task. A Roman road would certainly use the ford, and the settlement develop on the corresponding river bank.

The second important wave of inner colonization was triggered by the Carolingian renaissance of the IX century, which was set in a different political environment. Banditism, civil strife and general violence were the rule during feudalism; only defensive sites atop a hill ensured a settlement's survival. Site C would be the only possible one.

Finally, assume this part of the river has not been settled yet in the XIX century, and a railroad is built across the river. Then a bridge, and subsequently a center would be built at site A.

Once one of those settlements has developed, an obvious phenomenon of inertia (still largely unexplained) would give it a definite advantage over neighboring sites, so the center would most probably survive and grow till the present. Hence the problem is not only to know the *relative moment* that a spatial element gets frozen in a particular type of use or function, but also the *absolute moment* of time (i.e., the advantage of each site over the others since it is a function of time). In the successive determinations of land uses over time, with corresponding competition between spatial elements, the first property of time consists of knowing at which level in the competition a certain use is attached to a certain place. A second property is associated with the competitive process being non-commutative. These properties now make it possible to propose a mathematical model of historical evolution.

COMBINATORIAL MODELS OF HISTORICAL EVOLUTION

Recall the n spatial elements defined above. Consider the historical processes by which certain characteristics such as land use, progressively are attached to each element. Those processes are not necessarily competitive although they might take this form. One example would be the history of land development in the Los Angeles plain. Another one would be the competition between different settlements along a road or a river, during the unfolding of an urban hierarchy. Such examples are particularly simple and will be discussed hereafter, but we should keep in mind that we are building general models of historical evolution rather than spatial competition, although such process might be easier to visualize.

Let us call *chronopath* j a particular history of point j (i.e., spatial element j) from indifferentiation among n points (at level n) till the whole hierarchy (or spatial system) is built at final level [n-(n-1)] = 1. Spatial competition among the n elements might be represented as a branching process, or more conveniently, as a particular way of building parentheses in a set of n elements. For instance, consider 5 possible locations for a city along a river. One history might be: sites 1 and 2 first compete, then the winner competes with site 3; meanwhile, sites 4 and 5 compete, and both winners of the first and second subsets decide between each other which one will head the small urban system. This particular history would be represented as:

$$(((X_1 \cdot X_2)X_3)(X_4 \cdot X_5)) \quad ,$$

or in a still simpler writing:

$$(((X^2)X)X^2) \quad .$$

Therefore, the problem of evaluating the number of different possible histories among n points reduces one to computing the number of ways of building parentheses on a set of n points, given certain conditions [Comtet (1970)].

Global Knowledge: The Wedderburn Problem

Assume we lack individual knowledge in that we do not know the detailed competitive processes, but only the levels at which successive dominance is established. More precisely, we compute the number of ways of building, with parentheses, the product X^n among n points, considering two products as indistinguishable if they have the same degree. For instance.

$$(X^2X)X = (XX^2)X = X(XX^2) = X(X^2X) \quad .$$

The relation here is cummutative, which in practice means that we consider only the level at which a point enters into the competition and not the detailed account of individual competitions.

The problem is complicated, and has not yet been formalized. By recurrence, it is possible to compute the number of solutions, which increases explosively with n. It will be shown later that the resulting step-function may be approximated, as n becomes large (say, greater than 15), by a positive exponential function.

Individual Knowledge: Catalan and Schröder Problems

Assume we know the detailed competitive process, i.e., we have individual information on each confrontation between adjacent pairs of points (adjacency is no problem, here, since we can always reshuffle the points on the line). This more constrained relation is non-commutative, i.e.,

$$(X_1X_2)X_3 \neq X_1(X_2X_3) \quad .$$

Two slightly different solutions appear according to the way we assume competition. If it happens only between pairs of points, then we describe the set of possible histories among n points by different products X_n, where $X_n = \Pi X^k$ $(k = 1,2)$. Then, the number of possible solutions is given by the quite celebrated Catalan's Number

$$a_n = (n+1)^{-1} \binom{2n-2}{n-1} \quad .$$

If, considering the more general case, we accept that k in the preceding formula might be an integer larger than 2 but smaller than n, then the number of solutions is given by Shröder. Let ℓ_1 be the number of products with i elements. Note that we have

$$\ell_1 + \ell_2 + \ell_3 + \ldots + \ell_n = \ell \quad ,$$

and

$$\ell_1 + 2\ell_2 + \ldots + (n-1)\ell_{n-1} + n\ell = n \quad ,$$

with

$$\ell_n = 0 \quad \text{since} \quad \ell \geq 2 \quad .$$

Then, the number of solutions is

$$c_u = \sum_{i=1}^{n} \frac{\ell!}{\prod_{j=1}^{i} \ell_j!} \prod_j c_j^{\ell j} \quad \text{with } n \geq \ell \quad .$$

Growing Inertia and Historical Stability

As time passes by and competitive processes develop, the number of free elements to enter competition decreases. But what is the rate of this decline?

Define level i as the level at which there are still n-i points free to enter competition (or to have a certain land use or price attached to them). S_{n-1} is the number of possible histories at level i, and

$$R_i = S_{n-1}/S_{n-i+1}$$

is the rate of decrease of those numbers of solutions. In the case of the Catalan solution, we have

$$R_i = R_{i-1} = (n-1)/2(2n-3) \quad , \text{with} \quad 1 \leq i < n \quad .$$

For large n, R_i is approximately equal to 0.25. Since it is constant for large n, this step-function may be approximated by a positive exponential. Schroder's solution is much more involved. Let us use a recurrence relation

$$c_{n+1} = (n+1)^{-1}[3(2n-1)c_n-(n-2)c_{n-1}] \quad .$$

We have $R_i/R_{i-1} = (c_{i-1}/c_i)\cdot(c_{i+1}/c_i)$, which leads to

$$3(2n-1)/(n+1)R_i-(n-2)/(n+1)R_iR_{i-1}-1 = 0$$

if and only if n is large. Then $R_{i-1} = 6 - 1/R_i$. Thus we obtain a function of R_i which is $y = R_i^2 - 6R_i + 1$, with $0 < R_i, R_{i-1} < 1$, since the function c_n is monotonic.

It is easy to see that for R_{i-1} to be restricted to the closed interval [0,1], R_i must lie between 0.17 and 0.20. In the Wedderburn case, for large n the ratio R_i rapidly converges upon 0.43.

Obviously these models are very general. They are presented here as explanatory examples rather than as operational models which could be calibrated and tested. Nevertheless, they present two properties which are crucial for this argument. The progressive attribution of characteristics to spatial elements is a cumulative process that may be approximated by exponential curves. As time passes by, landscapes freeze fast and the number of possible histories left decreases drastically. This is a property common to the three models. If we have a detailed knowledge of individual evolution (i.e., if change assumes a non-commutative form), then this decrease is extremely fast. The corresponding rate is either 0.25 if points compete two by two (Catalan's process), or lies between 0.17 and 0.20 if competition occurs among groups of any number of points (but smaller, of course, than n). Without such detailed knowledge of individual evolutions, meaning we know only the level at which points enter the competition, without being able to identify individually those points (i.e., if we only have a global or aggregated knowledge), then the number of possible historical paths open to a set of spatial elements in a landscape decreases much more slowly, at an approximate rate of 0.43. Let us put it in simpler words. If I know individual behaviors, the number of possibilities as time passes by and as the landscape becomes progressively frozen decreases by 3/4 or 4/5 each time a spatial element's land use, for instance, is determined. In this case spatial order develops very fast, hence the impression the geographer might get is that a strongly determined process is operating. Conversely, if our knowledge is only global (i.e., about aggregates of points) without any possibilities of distinguishing individual spatial elements, the number of possibilities is reduced by 57% each time a particular point receives a determined characteristic.

THE HISTORIAN VERSUS THE FORECASTER: CHANGE AS A DILECTICAL PROCESS

It is now possible to suggest an explanation to the paradox presented at the beginning of this essay. How is it that when the geographer looks to the past and tries to explain the history of a landscape, everything seems to have been determined in such a way that a different evolution appears inconceivable? Meanwhile, when he looks to the future and tries to forecast even a simple evolution, he cannot help getting lost immediately in the immense number of possible paths open in front of him?

Time Is Dissymmetrical

Consider the case where n is the number of free spatial elements not fixed yet (i.e., to which a given land use, for instance, has not been yet allocated). The two variables t and n vary in opposite directions. The number of possible historical paths open in a spatial system decreases almost exponentially with time. When he acts as a historian, looking to the past to explain it, the urban planner sees the low tail of the quasi-exponential curve. Time seems to very rapidly narrow down the number of possible branchings any evolution could have taken, and this combinatorial trick makes history look well determined. The closer in time the period analyzed, the smaller the number of different paths evolution could have followed: explaining the recent past is easy. Conversely, the planner trying to forecast the future looks in the direction of time through the larger tail of the curve. Practically anything seems possible, at least at the beginning, and anticipation appears as a hopeless task. Of course it is the same time-curve, but being dissymmetrical it poses a very different problem when we try to follow it in opposite directions.

Global Knowledge Versus Individual Knowledge

If the step-functions corresponding to the preceding models can be approximated, as soon as n is larger than 10 or 15, by an exponential one, their gradients (i.e., the slope of those quasi-exponentials) differ. In all practical cases, predictions can be made only about aggregates. We do not know, as a rule, how every individual element of space will be used and priced, or how it will be settled. We must swap individual information for a global one, at a higher scale. This is the main goal of models. By concentrating a lot of individual but minute information we may hope to produce a larger quantity of information, but at an aggregated level. A model does not create information, but merely transforms it, like a machine would transform energy, and with a similar loss of information through "friction" [Marchand (1972)].

Since we have more information (and of better quality at the individual level) about the past than about the future, we can only predict the level of entry of elements in an historical process, like in the Wedderburn model, whereas knowing the individual behavior of these elements in the past allows us to use a Catalan or Schröder type of model. In other words, in most if not all cases, we have only aggregated, global forecasts, while it is logically possible, with sufficient data available, to describe the past evolution of each individual tract in a city, or each potential settlement along a river.

Remember now that the slopes (to use a term loosely that is not appropriate for step-functions) vary with the type of knowledge. In the Wedderburn case, with only global knowledge the number of possible branchings decreases only about one half with time, whereas in the Catalan or Schröder cases, with detailed knowledge about individuals, it decreases approximately by four-fifths. Here again, time takes a different aspect if we look to the past or to the future. For the historian it freezes very fast and the puzzling number of paths remaining open decreases drastically as each element in turn adopts a certain characteristic. For the forecaster, on the other hand, this puzzling number decreases very slowly and leaves too many questions open for too long a period.

Knowledge about the past and the future is markedly different in quantity and in quality. Here we presented a few of the differences, enough we hope to make people think about the nature and the role of time, and wonder about representing it as a mere variable, among others, in a model. This (philosophical) problem has been considered, although superficially, by Catell [in Harris (1961)], who writes that if time is only a variable among others, it would be conceivable to extract it, say through principal component analysis from a set of variables. But then what would all the other variables mean without the time framework? Can we talk of land rent or land use in a world where time no longer exists? If time is a basic component to each and all variables, without which they would not exist anymore; then how do we study it? This is where mathematics and philosophy join as the necessary background for the planner.

REFERENCES

Comtet, L., 1970. Analyse Combinatoire. PUF, Paris, 2 volumes.

Harris, C. (ed.), 1961. Problems in Measuring Change, University of Wisconsin Press, Madison.

Marchand, B., 1972. Geography and Information Theory, Geographical Analysis, 4: 234-257.

Marchand, B. (forthcoming). The Emergence of the City of Dreams: Los Angeles, 1940-1970, Pion, London.

SPATIAL DEMOGRAPHIC CHANGE

Predicting the spatial distribution of human population has recently become one of the most intensively studied areas in geography, planning, sociology and economics. The demand for housing and consumer goods as well as many public services is fundamentally dependent on where people live and work and the age and social structure of an area's residents. Predicting and describing the consequences of different fertility, mortality, and migration rates and identifying the processes by which these rates themselves undergo change is the concern of the papers in this section.

Smith uses a master equation framework to study interregional population movements. He reviews some of the problems associated with constructing transition probabilities, particularly those which are spatially non-stationary. A family of models based on elimination by aspects in which information flows constrain the choice sets is presented. Three examples are given of the analysis of master equations applied to population distributions.

Information flows and non-stationary transition rates also play an important role in MacKinnon and Rogerson's analysis of interregional migration. A feedback mechanism that builds up the impressions people have of various regions on the basis of interpersonal and media information flows is postulated and analyzed. Ways in which fluctuations in vacancy rates and population levels can be influenced by selective advertising and stimulating new job vacancies are examined. Suggestions for introducing a complementary micro analysis using job search theory are also advanced.

Huff provides a review of alternative methods for small area forecasts of population. Theoretical elegance often must be compromised when appropriate data and knowledge of relevant relationships are not available. Practical issues such as the sparse matrix problem are identified and discussed. Models that are simple enough to be implemented but complex enough to account for non-trivial population change are emphasized.

It is desirable that macro models be consistent with and ultimately based upon a set of micro, or behavioral relationships. Clark discusses the antecedents of the present housing search models. The results of an empirical study of search in the Los Angeles housing market are analyzed in terms of the dimensions of search and search strategies. Several existing hypotheses from the empirical search literature are retested. The dimensions of search are used as a basis for translating the cumulative-inertia stress model of mobility into a housing search model.

The final paper of this set examines the pervasive process that most nations are currently undergoing--the aging of their populations. The demographic transition from a relatively young population to a relatively older population structure is examined statistically and cartographically. Sonis postulates an ergodic hypothesis whereby current spatial variation in aging provides a mirror of the temporal stages through which any one country is likely to pass.

TRANSITION PROBABILITIES AND BEHAVIOR IN MASTER EQUATION DESCRIPTIONS OF POPULATION MOVEMENTS

Terence R. Smith

Department of Geography
University of California/Santa Barbara

INTRODUCTION

A task of particular challenge to the social scientist is to model the behavior of spatially distributed populations that are characterized by flows between various subsets of locations. The so-called "master equation" frequently provides a natural starting point for the analysis of such systems. In the case of a social system represented by continuous variables, a fairly general form for the master equation would be

$$\frac{\partial n(c,s,x,t)}{\partial t} = \iiint_{R_O} n(c,s,x,t)p_O(c,c',s,s',x,x',t)dc'ds'dx'$$
$$+ \iiint_{R_I} n(c',s',x',t)p_I(c',c,s',s,x',x,t)dc'ds'dx'$$
$$+ B(c,s,x,t) + D(c,s,x,t) \qquad (1)$$

where n(c,s,x,t) is a density function describing the distribution of individuals in the system over a set of classes c, a set of states s, a set of locations x, and at time t. In general, the arguments c,s,x may be vectors. Since p_O (c,c',s,s',x,x',t) is a probability density function describing transitions from configuration (c,s,x) to configuration (c',s',x') at time t and p_I (c',c,s',s,x',x,t) is the probability density function describing transitions from (c',s',x') to (c,s,x), it follows that the first term on the RHS describes the rate of flow of individuals from (c,s,x) to all feasible configurations, while the second term on the RHS describes the rate of flow of individuals to (c,s,x) from all feasible configurations. The terms B(c,s,x,t) and

$D(c,s,x,t)$ represent pure birth and death processes, respectively.

There are several cogent reasons for employing a master equation formulation. First, the basic concept underlying equation (1) is a simple conversation, or accounting, principle. It should be noted, however, that although the accounting principle holds under all conditions, the master equation is restricted to providing a representation of system behavior in the mean. It provides no insight into random fluctuations. A second reason for using the formulation is its flexibility and generality. For example, one may choose to make any of the variables c,s,x,t either discrete or continuous--the function $n(c,s,x,t)$ need not be interpreted in terms of a density of individuals--while the ranges of possible behaviors embodied in master equations is almost unlimited. A third and very important reason is that the flows between locations are not subject to the extreme "localness" assumptions that are implicit in some of the other approaches to modelling spatial systems, such as the approach employing partial differential equations. The assumptions of spatial localization implicit in the usual applications of partial differential equations may be quite unsuitable in representations of distributed social systems.

Several stages typically are involved in analyzing a problem in terms of the master equation. First, one must decide on the appropriate set of variables and whether they should be continuous or discrete. A second and critical stage involves the introduction of assumptions concerning individual behavior by way of the transition probability densities. Third, one must decide upon the appropriate initial and boundary conditions, while the fourth stage is concerned with the analysis of the behavior implicit in the model. Such analysis may involve questions concerning the existence, uniqueness and characterization of steady state solutions, or conditions for stability of steady state solutions in the linear regime.

The aim of the present paper is to exemplify the application of master equation techniques to distributed social systems in which there are flows of individuals. The first section addresses the problem of constructing appropriate transition probabilities. In this context, a new set of spatial interaction models is introduced. A second section presents examples of the analysis of master equations applied in a spatial context.

TRANSITION PROBABILITIES

The transition probability densities contain the key behavioral assumptions in a master equation. The manner in which the probabilities are modelled depends on both the nature of the system under study and goals of the analysis. At a strategic level one may choose functional forms for the probabilities that satisfy certain criteria. For example, these forms may be chosen as representations of individual decision making, and they may also be constructed to incorporate some of the feedbacks that occur between individual and aggregate behavior.

The approach taken to specify transition probabilities in this essay is that they represent the outcome of individual decision making. Hence, in modelling the probabilities it is useful to introduce concepts relating to preference structures, information flows and mental processing that are important in modelling decision making processes. How closely one is able to model individual decision making processes in constructing transition probabilities evidently depends upon the problem under consideration.

A general but strategically useful classification of the transition probabilities is dependent on their degree of temporal and spatial stationarity. A simple classification is shown in Figure 1. Although cases I and IV of spatially

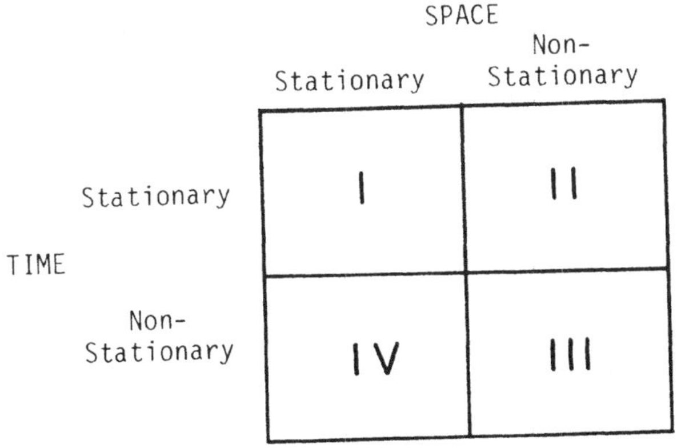

Fig. 1. Classification of transition probabilities.

stationary probabilities have limited applicability to many problems of interest, they sometimes may be used to obtain insights into aggregate system behavior. Although the present essay focusses on cases II and III of spatially non-stationary probabilities, a few simple examples of spatially stationary probabilities are considered.

Spatially Stationary Transition Probabilities

There are several classes of problems in which it is sometimes advantageous to employ an assumption of spatially stationary transition probabilities. Examples of such problems include the distribution of towns by size and the numbers of people who know some message in a given spatial system. It is frequently possible to analyze such problems using a master equation, although this is rarely done in an explicit manner in the social sciences. For example, one may approximate the growth of towns of different sizes by assuming that some towns grow at the expense of other towns through a net movement of individuals from the less preferred to the more preferred towns. If one assumes that the relative attractivity of towns is a function of their size, then a master equation describing the process may be written as

$$\begin{aligned}\frac{dn}{dt}(s,t) &= -\int n(s)p_O(s,s')ds' \\ &\quad +\int n(s-s')p_I(s-s',s')ds',\end{aligned} \tag{2}$$

where s is the size of the town. It is evident that the process of accretionary growth changes the <u>local</u> spatial distribution of town sizes. As a result, one would expect the transition probabilities to be spatially non-stationary. Hence equation (2) at best can be only an approximation.

Similar comments apply when master equations with spatially stationary probabilities are applied to spatial systems in which messages are being exchanged by individuals. For instance, consider a system in which messages can only be learned and not forgotten and in which the total population has a fixed size N. The appropriate master equation is

$$\begin{aligned}\frac{dn(s,t)}{dt} &= -n(s,t).\ o + n\ (\bar{s},t).\ p(\bar{s},s,t) \\ &= [N - n(s,t)].\ p(\bar{s},s,t) \quad ,\end{aligned} \tag{3}$$

where s represents the state of knowing and $\bar{s}$ the state of not knowing a given message. In particular, if the transition probability $p(\bar{s},s,t)$ is proportional to the number of knowers, one obtains the logistic equation.

Spatially Non-Stationary Probabilities

Although the assumption of spatially stationary transition probabilities may prove of value in a few limited cases, as noted below, it will generally lead to an inadequate representation of a system. In turning to consider representations with non-stationary probabilities, the discussion is limited to systems in which space is represented as a set of discrete locations and in which individuals belong to discrete classes. The master equation without births or deaths now takes the form

$$\frac{dn_1^k}{dt} = -\sum_j^{\ell} n_\ell^k p_{\ell j}^k + \sum_j^{\ell} n_j^k p_{j\ell}^k \qquad k = 1,k \qquad \ell = 1,L$$

$$= -n_\ell^k + \sum_j^{\ell} n_j^k p_{j\ell}^k \quad ; \qquad k = 1,2,\ldots,K; \ \ell = 1,2,\ldots,L, \tag{4}$$

where $\ell = 1,2,\ldots,L$ represents the set of locations, $k = 1,2,\ldots,K$ represents the set of classes, and $p_{j\ell}^k$ is the probability that an individual of class k will relocate from location j to location ℓ in a unit interval of time. The second form of the equation follows from the condition that

$$\sum_{\ell=1}^{L} p_{j\ell}^k = 1 \, , \tag{5}$$

which implies that the system is closed.

It is a natural first step to assume that the probabilities are stationary in time. Migration systems with such probabilities have been considered by a number of authors including Cordey-Hayes (1975), Liaw (1975) and Rogers (1975). In general, it is clear that serious problems may arise from the assumption of time-independent probabilities if one is describing the behavior of a system over time. Cordey-Hayes (1975) has illustrated the magnitude of the errors in prediction that may arise as a result of using such an assumption. When the transition probabilities are viewed as a representation of the outcome of individual decision making, it is clear that most of the complexities of decision making processes are lost in assuming temporal stationarity.

A natural procedure for introducing temporal non-stationarity is to induce a functional dependence on time dependent variables, such as the population distribution. When the probabilities are interpreted as the outcome of individual decision making, one may implement such a procedure by first expressing the probabilities in terms of utilities, and then expressing the utilities in terms of time-dependent arguments.

Probabilistic choice models provide one convenient procedure for expressing the probabilities in terms of utilities. If $u_j(\ell)$ is the utility that a randomly chosen individual would achieve on moving from j to ℓ, then

$$p_{j\ell} = \text{Prob}[\tilde{u}_j(\ell) > \tilde{u}_j(k),\ k \neq \ell] \ . \tag{6}$$

If we write

$$\tilde{u}_j(\ell) = A_\ell f_{j\ell} + \tilde{\varepsilon}_{j\ell} \ , \tag{7}$$

where A_ℓ is a representative utility function characterizing preferences for destination ℓ, $f_{j\ell} = f_{\ell j}$ is some interaction term and $\tilde{\varepsilon}_{j\ell}$ is a random component of utility, one obtains a large class of transition probabilities. In particular, the restriction that $\tilde{\varepsilon}_{j\ell}$ be i.i.d. Gumbel leads to the singly constrained spatial interaction (SPIN) model,

$$P_{j\ell} = \frac{A_\ell f_{j\ell}}{\sum_k A_k f_{jk}} \ , \tag{8}$$

which, as noted in Smith and Clayton (1978), is a particular case of the Luce choice model [Luce (1969)].

Despite their appealing simplicity, the transition probabilities (8) are associated with several difficulties. A theoretical difficulty is that (8) is only derivable from (6) under strong assumptions on the random utility component $\tilde{\varepsilon}_{j\ell}$. A second problem, discussed by Luce (1977) and Tversky and Sattath (1979), among others, is related to the fact that the choice probabilities do not reflect the degree of similarity between alternatives, whereas empirical evidence supports the existence of such a dependence. A related problem, discussed in greater detail below, is that the SPIN model for transition probabilities can be viewed as a very special case of two models of much greater generality.

A fourth class of difficulties relates to empirical problems encountered in applying models of the form (8). For example, the spatial separability inherent in (8) implies that population movements described by (8) should be transitive in the mean for both probabilities and numbers. In symbolic notation

$$p_{ij} > p_{ji}, p_{jk} > p_{kj} \Rightarrow p_{ik} > p_{ki}, \text{ and}$$
$$N_i p_{ij} > N_j p_{j\ell}, N_j p_{jk} > N_k p_{kj} \Rightarrow N_i p_{ik} > N_k p_{ki} \quad , \tag{9}$$

where N_i is the number of individuals leaving destination i. U.S. migration data at various scales of spatial resolution do not possess the transitivity property, and one finds relatively large numbers of intransitivities in flow numbers and a somewhat lesser number of intransitivities in the computer probabilities.

Goodchild and Smith (1980) attempted to reconcile model and data by modifying the basic SPIN model (8). They found that neither the absolute degree of intransitivity nor the relative proportion of numbers and probability intransitivities in the data could be explained as the result of random variations that arise because the transition probabilities in this case are the probabilities of a multinomial process for which each migration pattern is a particular realization. They also were unable to explain the observed patterns of intransitivity in terms of the superimposition of different migration streams, each obeying its own SPIN probabilities. Finally, Goodchild and Smith were unable to simulate the transitivity pattern of U.S. migration streams by perturbing the interaction term f_{ij} into asymmetry, although the results were in the right direction.

The preceding difficulties with the transition probabilities (8) are inherent in the functional form of the probabilities, and independent of any application in the context of a master equation. When probabilities of the SPIN model are employed in the master equation (4) another difficulty of a conceptual nature arises. If one assumes conditions of costless relocation (implying that distance has no effect on relocation probabilities), and if the utilities of alternative locations depend on the distribution of population (by way of rents and externalities, for example), then one would expect at least some of the stationary states of the system (4) to be spatial equilibria. By stationary state is meant the condition

$$\frac{dn_\ell^k}{dt} = 0 \; , \; \forall \ell \; , \tag{10}$$

and by spatial equilibrium is meant the condition (10) coupled with

$$A_\ell^k = A^k \quad , \quad \forall \ell \ . \tag{11}$$

The assumption behind this expectation is that individuals will continue to move from areas of lower utility to areas of higher utility under costless relocation, until a condition of spatially uniform utilities is reached.

Applying the condition (10) of a steady state to (4) implies that

$$n_\ell^k = \frac{N\, A_\ell^k}{\sum_j A_j^k} \ . \tag{12}$$

Since costless relocation implies that

$$f_{j\ell}^k = \text{Constant}, \quad \forall j, \ell \quad , \tag{13}$$

it follows that condition (11) of spatial equilibrium implies a uniform distribution of population. This appears to be an unnatural restriction on system behavior, since one would expect the possibility of spatial equilibrium with agglomeration in systems with spatial externalities. Hence, if one is interested in studying systems with agglomerative spatial equilibrium, then it is necessary to modify the simple probabilities of the Luce model.

Since it is generally true that spatial equilibrium occurs only under conditions of uniform population distributions when relocation is costless, and when the probabilities depend only on the utilities of the alternatives (A_ℓ), one must introduce other arguments into the specifications of the probabilities. For example, one may introduce the population distribution into the general form of the probabilities

$$p_{j\ell}^k = f^k[A_1^k, \ldots, A_n^k \ (n_1^1, \ldots, n_n^1, \ldots, n_1^k, \ldots, n_n^k)] \ . \tag{14}$$

One justification for the entry of the population distribution into (14) is that it may be used to represent information flow in the system. Smith and Papageorgiou (1980) used this device in constructing probabilities for a master equation (4) that was employed in studying the stability of population distributions subject to spatial externalities. The analysis was restricted to the case of both costless relocation and two classes of individuals.

In order to specify further the functional form of the transition probabilities, Smith and Papageorgiou assumed that the probability of an individual relocating from j to ℓ is proportional to the level of utility attainable at ℓ and to the level of information flowing from ℓ. This flow of information was assumed proportional to the numbers of individuals at ℓ. Hence,

$$p^k_{j\ell} \propto n^k_\ell A^k_\ell \quad , \tag{15}$$

for costless relocation. The constant of proportionality may be evaluated using condition (5). It is easy to show that every steady state of the master equation with probabilities (15) is a spatial equilibrium, and it is in fact possible to produce an agglomerative spatial equilibrium.

It should be noted that the introduction of the population distribution into (14) induces non-temporal stationarity. Further, non-stationary effects can be induced, as noted above, by allowing the utilities to depend on time-dependent arguments. Smith and Papageorgiou specified the utilities of the destinations using

$$A^k_\ell = A^k[y^i_\ell - r^k(n^1_\ell, n^2_\ell), E^k_\ell] \quad , \tag{16}$$

where y^k_ℓ is the income that would accrue at location ℓ and $r^k(n^1_\ell, n^2_\ell)$ is the housing cost at ℓ. The housing cost is assumed to depend on the demand for housing that is assumed to be inelastic. E^k_ℓ represents an externality experienced by individuals of class k at place ℓ, and is represented as

$$E^k_\ell = \sum_m \sum_j f^{mk}_{j\ell} n^m_j \quad , \tag{17}$$

where $f^{mk}_{j\ell}$ denotes the effect of an individual of class m at location j on an individual of class k at location ℓ, and depends only on distance. Hence, the externality (17) represents the aggregate effects of the population distribution at any given location.

The Luce model of choice, and the variants such as that examined by Smith and Papageorgiou, may be used in master equation formulations to provide useful insights into certain aspects of system behavior. In order to exemplify this contention, some of the results from Smith and Papageorgiou's study are described briefly below.

The difficulties that were noted previously in connection with such models, however, suggest that alternative models may provide more suitable transition probabilities. A first set of models may be obtained by using alternative probability distributions for $\tilde{\varepsilon}_{j\ell}$ in (6). For example, if $\varepsilon_{j\ell} \approx$ MVN, then one obtains the general class of Probit models. While such models are free of some of the difficulties inherent in the Luce probabilities, they generally have no tractable mathematical form and computationally are difficult to handle. From the point of view of analysis using master equations, alternative forms of transition probabilities may be preferable.

One attractive alternative is based on the conceptually simple idea of endowing alternative locations with certain set membership properties, and applying the law of total probabilities. Using such an approach the probability of an individual making a transition from location j to location ℓ (given a feasible set S of alternative locations) may be written

$$P_j(\ell|s) = \sum_{\alpha\varepsilon A} \text{Prob}_j(\ell|S_\alpha)\ \text{Prob}_j(\alpha) \ , \tag{18}$$

where α is an index indicating various set membership properties, and $S_\alpha \subset S$ is the subset of alternatives characterized by property α. $\text{Prob}_j(\alpha)$ is the probability that property α is chosen at origin j as the selection criterion.

Model (18) is the "elimination by aspects" (EBA) model of Tversky (1972). In the present context the α's may be interpreted as characteristics of the destinations. Associated with each destination ℓ is a set of characteristics

$$\ell' = (\alpha, \beta, \gamma, \ldots) \ . \tag{19}$$

Tversky's model assumes the existence of a scale function $u(\alpha)$ defined on the set of characteristics such that the probability of a given α being the choice criterion is given by its relative importance, or

$$\text{Prob}_j(\alpha) = \frac{u_j(\alpha)}{\sum_{\beta\,\varepsilon S_1} u_j(\beta)} \ , \tag{20}$$

which is the choice model applied to characteristics. It is now clear that if there is a one-to-one correspondence between characteristics and destinations ($\ell' = \alpha_\ell$), then (18) becomes

$$\text{Prob}_j(\ell|S) = \frac{u_j(\alpha_k)}{\sum_{k\varepsilon S_1} u_j(\alpha_k)} \quad , \tag{21}$$

which is the Luce choice model. The manner in which this model arises as a special case of the EBA model illustrates why the degree of similarity between alternative destinations fails to affect the transition probabilities.

Advantages of the EBA family of transition probability models include their having relatively simple analytic forms, their taking into account of similarities between alternative destinations, and their only moderately stochastically transitivity. On the other hand, they generally require large quantities of empirical data for purposes of calibration. Tversky has constructed special cases of the EBA model in which the characteristics have a tree structure. These special cases require less data for calibration and may actually represent the mental processing that individuals go through in reaching decisions.

Probability transition models having the structure (18) do not appear to have been applied in spatial problems, although there has been an awareness of the need for models of constrained choice [Burnett and Hanson (1979)]. A natural approach to constraining the choice sets implicit in the transition probabilities involves the introduction of spatial information flows. It is clear from many studies of migration and relocation that both preferences and information are important determinants in the selection of destinations. In particular, the absence of information about some of the choice alternatives may severely constrain the set of locations considered.

A model that explicitly incorporates preferences and information flows has been recently constructed and investigated by Smith and Slater (1980). This model is designed to represent migration probabilities, and is based on the assumption that individuals must possess a message about a job vacancy before relocating. In order to produce a relatively simple model, it is assumed that an individual possesses either one vacancy message or no job vacancy message from each of n feasible locations. Hence, there are 2^n feasible choice sets that may confront any decision maker. As in the basic representation (18) one must specify two sets of probabilities. The first set relates to 2^n subsets of locations from which messages are received, while the second set contains the conditional probability of choosing a given destination from the subset of destinations from which vacancy messages have been received.

The probability of receiving a vacancy message from location ℓ while at location j is represented by $r_j s_\ell \exp(-\beta d_{j\ell})$, where $0 \leq r_j \leq \ell$ is a measure of message receptivity at location j, $0 \leq S \leq 1$ is a measure of message emissivity at location ℓ, and $\exp(\beta d_{j\ell})$ is a distance attenuation factor. If it is assumed that the event of receiving a message from one location is independent of the event of receiving a message from any other location, the probability of receiving a message from a subset S' of the S destinations is

$$\prod_{\ell \in S_1} r_j s_\ell \exp(-\beta d_{j\ell}) \prod_{k \notin S_1} [1 - r_j s_k \exp(-\beta d_{jk})] \quad .$$

It is assumed that the probability of receiving a message from the location at which the decision maker currently resides is unity.

The conditional probability of choosing destination ℓ from a subset S_1 of locations from which vacancy messages have been received is assumed to be given in the SPIN model

$$\text{Prob}_j(\ell | S_1) = \frac{A_\ell \exp(-\gamma d_{j\ell})}{\sum_{k \in S} A_k \exp(-\gamma d_{k\ell})} \quad . \tag{22}$$

Hence, using the general EBA model (18), the transition probabilities are given by the expression (for the case $j \neq \ell$)

$$P_j(\ell | S) = \sum_{S_1 \epsilon 2^S} \frac{A_\ell \exp(-\gamma d_{j\ell})}{\sum_k A_{k\ell} \exp(-\gamma d_{jk})} \prod_{m \epsilon S_1} r_j s_m \exp(-\beta d_{jm}) \tag{23}$$

$$\cdot \prod_{n \epsilon S_1} [1 - r_j s_n \exp(-\beta d_{jn})] \quad ,$$

where 2^S represents the power set of S. For the case $j = \ell$, an additional term is required to account for the possibility of no job vacancy messages.

There are several points of interest concerning the transition probabilities (23). First, the model incorporates two separate distance attenuation factors, one reflecting the attenuation of information flow with distance and the other reflecting the decrease of attractivity with distance from a given origin. A second point is that it is possible to simplify the probabilities by truncating the series after various numbers of terms. For example, if each $p_j d_\ell \exp(-\beta d_{j\ell})$ is sufficiently small, then all terms relating to the possibility of receiving two or more job messages may be ignored (this may be called the one-message model). If $p_j d_\ell \exp(-\beta d_{j\ell})$ is somewhat larger, but still not too large, terms relating to the possibility of three or more messages may be ignored (this may be called the two-message model). On the other hand, if each $p_j d_\ell \exp(-\beta d_{j\ell})$ is sufficiently close to unity, then all terms with less than n-1 job messages may be dropped. In particular, in the limit as each $p_j d_\ell \exp(-\gamma d_{j\ell})$ becomes unity (a message is received from each location with certainty) one obtains

$$p_j(\ell|S) = \frac{A_\ell \exp(-\gamma d_{j\ell})}{\sum_{k\varepsilon S} A_k \exp(-\gamma d_{jk})} , \qquad (24)$$

which is just the singly constrained SPIN model. Hence the SPIN model is a special case of model (23) under the conditions that job vacancy messages are obtained from each feasible location. For the case of interstate migration it would seem that these job message models are more applicable to representing employment related interstate migration, since the probability of a prospective migrant receiving job messages from each state would appear to be extremely low.

An interesting property of model (23) is that it is capable of generating intransitive behavior in the mean. In fact, it can be shown that the one message model is capable of producing intransitivities in the flow numbers, but transitive behavior in the flow probabilities. In this regard the model is not inconsistent with migration data.

Smith and Slater (1980) have calibrated the model using time series interstate migration data from the Continuous Work History Study. They calibrated the one message, two messages and SPIN models using non-linear least squares. Both the one-message and the two-message models gave significantly better results than the SPIN model in describing the destinations of out-migration from the serveral states considered.

Several lines of further research suggest themselves in relation to the construction of transition probabilities. First, models of the forms (18) and (23) should be explored in greater detail, both from a theoretical and an empirical point of view. Second, research efforts should be directed to the construction of new transition probabilities that capture properties of individual decision making, and the nature of information flows not

inherent to examine the behavior of master equations under different assumptions concerning the transition probabilities. While Liaw (1975) has examined the behavior of the master equation (4) under the assumption of temporally stationary transition probabilities, and Smith and Papageorgiou (1980) have examined the behavior using the probabilities (15), there is a need to extend these analyses to other models of transition models, and to obtain results on the comparative behavior of such systems.

SOME RESULTS FROM THE ANALYSIS OF MASTER EQUATIONS

The remainder of the essay is a brief description of the analysis of three master equations. This description is aimed at exemplifying techniques for analyzing such equations and the results that are obtainable using these techniques.

An Example with Spatially Stationary Probabilities

Although models with spatially stationary transition probabilities are limited in their applicability to the analysis of spatial systems, occasionally they may provide useful insights into such systems. It was suggested above that models of the form (2) might provide some insight into the question of city size distributions. Although such an equation does not appear to have been analyzed in relation to this particular problem, it has been analyzed in other contexts. For example, Hayashi and Nakagawa (1975) used a specific form of (2) to examine the process of agglomeration of interstellar dust grains.

The transition probabilities to describe the processes of grain-grain collisions and agglomeration were obtained using physical principles. The resulting master equation was analyzed using a combination of analytical and numerical techniques. For example, the form of the distribution of particles by size was obtained for very small and for very large particles using similarity solutions. In this manner the rate at which the distribution approached zero was obtained for both cases. They were also able to show with the use of numerical techniques that in the limit of long time the distribution of particles by size approached a form resembling a log-normal distribution. This is an interesting result, and suggests that analogous results may be obtained for the case of distributions of cities by size.

Two Examples with Spatially Non-Stationary Probabilities

Smith and Papageorgiou (1980) analyzed the master equation (4) in relation to a number of questions. Using the general form of the transition probabilities (14) they first examined the existence and uniqueness of steady state solutions to (4), under the assumption that the representative utilities for destinations A_i^k and the function $f^k(*)$ are continuous functions of the population numbers n_i^k. The main questions of the analysis concerned the form of stability of the steady state solutions. In order to examine these questions it proved convenient to specify the form of the transition probabilities in greater detail. In particular, Smith and Papageorgiou (1980) adopted the form (15), for which every steady state is a spatial equilibrium, with the utilities A_i^k given by equations (16) and (17). Two cases were examined in relation to the form of the spatial equilibrium solutions and their ability, the first case being a uniform distribution of income

$$y_i^k = y^k \quad \forall i \quad , \tag{24}$$

and the second case involving small but spatially arbitrary perturbations away from the uniform income distribution. In the first case, the uniform population distribution provides a spatial equilibrium. Smith and Papageorgiou derived necessary and sufficient conditions for the stability of these solutions. In particular it was found that the spatial form of the spatial externality diffusion operator $f_{j\ell}^{km}$ of (17) entered the stability conditions only by way of its aggregate form

$$F^{km} = \sum_j f_{j\ell}^{km} \quad . \tag{25}$$

This result implies that stability only depends on the aggregate level of spatial externality affecting each location, and not on the spatial origins of the externalities. It was also found that the same stability results held for both one-dimensional and two-dimensional spatial systems.

In the case of small perturbations away from a uniform income distribution non-uniform spatial equilibria exist, which Smith and Papageorgiou (1980) were able to express in terms of linear combinations of the spatial perturbations in income. Again, explicit, necessary and sufficient conditions for stability were found that possessed the two properties that characterized the conditions for stability of uniform distributions of population. In addition, however, a new condition of interest held. If the perturbations to income resulted from a redistribution of income in a system with an initially uniform spatial distribution of income, then the conditions for stability were the same in both systems. On the other hand, perturbations to a uniform income distribution that resulted in net additions or subtractions of income from the aggregate system caused a change in the stability conditions. This implies that under certain conditions such systems may be stabilized or destabilized with small injections of aggregate income.

The full non-linear behavior of the system investigated by Smith and Papageorgiou (1980) proves extremely difficult to investigate analytically. A very special case of model (4), however, has been extensively investigated in the non-linear regime, and reveals the complex behaviors possible even in simple master equation formulations. If model (4) is limited to discrete time, two locations, and one class of individuals, one obtains the system

$$n_{1t+1} = n_{1t} - n_{1t}p_{12} + (N - n_{1t})p_{21} \quad . \tag{27}$$

If one further specifies the probabilities as

$$\begin{aligned} p_{12} &= \alpha(N - n_{1t}) \ , \text{ and} \\ p_{21} &= \beta n_{1t} \ , \end{aligned} \tag{28}$$

one obtains the logistic equation in finite difference form:

$$m_{t+1} = \phi m_t(1 - m_t) \ , \tag{29}$$

where

$$\begin{aligned} \phi &\equiv 1 + (\beta - \alpha)N \ , \text{ and} \\ m_t &\equiv \frac{\beta - \alpha}{\phi} n_t \quad . \end{aligned} \tag{30}$$

May (1976) has summarized the results of a large number of investigations of the non-linear behavior of this equation for parameter values in the range $1 \leq \phi \leq 4$. Over this range of values there are stable periodic solutions for all possible periodicities. An important point about these solutions is that although the equation is deterministic, it would prove almost impossible to differentiate finite segments of some of the solutions from a stochastic process if one were observing a system governed by (29). There are also parameter values that are points of accumulation for different solutions. At such points of accumulation the behavior of the solutions is truly chaotic. Overall, as the parameter ϕ increases from one to four an uncountable number of bifurcations occurs, at which the behavior of solutions changes in the qualitative manner. Similar chaotic behavior can be obtained from master equations with continous time dependence (4) if the number of locations considered is at least three.

CONCLUSION

The master equation formulation provides a natural and flexible framework for examining many questions that relate to spatial systems in which flows of various kinds are occurring. The critical problem employing this formulation is in the specification of the transition probabilities. In many cases of interest these probabilities represent the outcomes of individual decision making and information flows. A good deal more research is required in specifying appropriate models for transition probabilities and in examining the behavior of master equations under various assumptions concerning the transition probabilities.

REFERENCES

Burnett, P. and Hanson, S., 1980. A Rationale for an Alternative Mathematical Approach to Movement as Complex Human Behavior, forthcoming in Transportation Research Record.

Cordey-Hayes, M., 1975. Migration and the Dynamics of Multiregional Population Systems, Environment and Planning 7: 793-814.

Goodchild, M. and Smith, T., 1980. Intransitivity, the Spatial Interaction Model and U.S. Migration Streams, Environment and Planning A, 12: 1131-1144.

Hayashi, A. and Nakagawa, Y., 1975. Size Distribution of Grains Growing by Thermal Grain-Grain Collison, Progress in Theoretical Physics, 54: 93-103.

Liaw, K., 1975. A Discrete Time Dynamic Analysis of International Population Systems, Geographical Analysis, 7: 227-244.

Luce, R., 1959. Individual Choice Behavior: A Theoretical Analysis, Wiley, New York.

Luce, R., 1977. The Choice Axiom after Twenty Years, Journal of Mathematical Psychology, 15: 215-233.

May, R., 1976. Simple Mathematical Models with Very Complicated Dynamics, Nature, 261: 459-467.

Rogers, A., 1975. Introduction to Multiregional Mathematical Demography, Wiley, New York.

Smith, T. and Clayton, C., 1978. Transitivity, Spatially Separable Utility Functions and U.S. Migration Streams, 1935-1970, Environment and Planning A, 10: 399-414.

Smith, T. and Papageorgiou, G., Spatial Externalities and the Stability of Interacting Populations Near the Center of a Large Area, unpublished manuscript.

Smith, T. and Slater, P., A New Set of Interaction Models with Choice Sets Constrained by Information Flows, Applied to U.S. Interstate Migration, unpublished manuscript.

Tversky, A., 1972. Elimination by Aspects, A Theory of Choice, Psychological Review, 79: 281-299.

Tversky, A. and Sattath, S., 1979. Preference Trees, Psychological Review, 6: 542-573.

PERSPECTIVES ON INFORMATION SENSITIVE MODELS OF MIGRATION

Ross D. MacKinnon and Peter A. Rogerson

Department of Geography
State University of New York at Buffalo

Employment opportunities and information about them are crucial in being able to predict and understand the flows of migrants between regions. Although climatic and other factors relating to amenities are increasingly cited as major influences affecting the rapid growth and decline of regions, economic opportunities, in particular job vacancies, would seem to continue to be a major contributor to differential regional population growth. But it is argued here and elsewhere that perceptions or personal impressions about job opportunities are as important as the existence of the jobs themselves. A study of the channels that are used to disseminate and receive information is important in understanding how cyclical growth and decline might occur, how some regions might attempt to remedy misperceptions that are impeding growth, and how the amplitude, frequency and mean of vacancy rates might be dampened. Further, under what conditions do initial conditions (a head start, for whatever reason, or the geographical position of places) strongly condition long run behavior? What are the relative merits of two competing strategies--to increase mobility in part by enhancing the information flow about job opportunities that arise from the "natural" workings of the economy of "artificially" stimulating growth that would allow more people to obtain jobs without having to migrate?

The major underlying premise of this paper is that information (and the lack of it) plays a major role in the choice of a job (and a residence) and therefore, in the aggregate, on migration flows, population distributions, vacancy and unemployment rates. There are at least two broad classes of information of relevance to the migration processes studied here--specific information about a particular job vacancy and more general information about the economic climate of a region. With respect to opportunity-specific information, it is unlikely that an individual will be aware of more than a very few specific job opportunities. On the other hand, an individual's general impression of the spatial distribution of economic opportunities arises from a multitude of information flows from friends, relatives and associates, purposive advertising, and reports in the mass media. In the models to be discussed here, we assume that these general impressions have as an objective basis actual job vacancies that exist in each region. This is our "objective reality"--an index of economic health, at least from the point of view of a prospective job seeker. This objective world is modified depending on the way the channels by which a person receives his information bias or filter the true picture. This filtering process, it is hypothesized, is dependent in part on a person's location relative to job opportunities and the rest of the population.

Several variants of a model that incoporates these features have been postulated and their behavior analyzed.

Before introducing substantive information dependent postulates it is useful to present some standard accounting equations that ensure that every individual in the population occupies one and only one of the permissible states in the system. Such a components of change model is standard in the demographic literature:

$$N_i(t+1) = N_i(t) + B_i(t) - D_i(t) + \sum_j M_{ji}(t) - \sum_j M_{ij}(t), \quad (1)$$

where $N_i(t)$ is the population in region i at time t,
$B_i(t)$ is the number of people born in region i during the time interval beginning at t,
$D_i(t)$ is the number of people who die in region i during the time interval beginning at t, and
$M_{ij}(t)$ is the number of people in i who move to j during the time interval beginning at t.

Frequently, demographers divide through equation (1) by $N_i(t)$ to generate a set of behavior equations, that in matrix form may be expressed as:

$$\underline{N}(t+1) = (\underline{I} + \underline{\beta} + \underline{M})\underline{N}(t) \quad , \quad (2)$$

where $\underline{I}$ is the identity matrix,
$\underline{\beta}$ is a diagonal matrix of net birth rates by region, and
$\underline{M}$ is a matrix of migration rates

(M_{ij} = the relative frequency of movement from region i to j within any time period of specified length).

This model may be highly disaggregated by age, sex, socio-economic status, and the such, and incorporated within a cohort-survival framework [Wilson and Rees (1976), Rogers (1975)]. Alternatively, it may be simplified by ignoring births and deaths and re-stating it as a simple Markov chain:

$$\underline{N}(t+1) = \underline{MN}(t) \quad , \tag{3}$$

where M_{ij} is the probability that someone will reside in j at the end of the time period given he began the time period in i, and
$\underline{N}(t)$ is the vector of expected values of regional populations.

Critical theoretical and practical issues center around the estimation of the matrix $\underline{M}$, in either (2) or (3). The statistical estimation problem of $\underline{M}$ as relative frequencies does not concern us here. Although making migration probabilities or rates consistent with an accounting for every individual in the population is necessary, it begs the question of why these relative frequencies take on the values we observe. Moreover, if this model is to be used for prediction purposes we must either assume stationarity ($\underline{M}$ constant) or develop a relationship that allows $\underline{M}$ to vary as the system evolves through time.

Before specifying such relationships, we now state a model analogous to (3) from Evtushenko and MacKinnon (1976) that relates migration flows to the spatial distribution of economic opportunities:

$$\underline{N}(t+1) = \underline{N}(t) + (\underline{I} - \underline{\tilde{M}})\underline{V}(t) \tag{4a}$$

$$\underline{V}(t) = \underline{\tilde{M}V}(t-1) + \underline{V}_z(t) + \underline{V}_u(t) \quad , \tag{4b}$$

where $\underline{V}(t)$ is a vector of the number of job vacancies existing in each region at the beginning of time period t,
$\tilde{M}_{ij}$ is the probability that a vacancy in j will be taken up by someone formerly resident in i (thus assuming full employment $\tilde{M}_{ij}$ is the probability a vacancy in j will generate a vacancy in i in the subsequent time period),
$\underline{V}_z(t)$ is a vector of exogenously created job vacancies, arising perhaps from secular economic growth or decline and perhaps predictable by another model of the economy or alternatively from random, unpredictable shocks, and

$\underline{V}_u(t)$ is a vector of government induced new jobs either government jobs or jobs resulting from government incentive programs.

$\underline{V}_z(t)$ and $\underline{V}_u(t)$ are variables that link the model to exogenous economic and political factors. Although they are not exploited in the research reported here, it should be emphasized that the modular structure of this model allows it to accommodate more realistic and complex structures than those described in this paper, including control theory and other optimization approaches [Evtushenko and MacKinnon (1976), Tan (1980)].

In spite of the potentially richer structure of (4) than (3), both ignore the problem of non-stationary migration rates (i.e., time varying $\tilde{\underline{M}}$ and $\underline{M}$, respectively). This is the second major feature in addition to information, sensitivity of our modelling strategy--to allow the rules of change to vary through time in response to changing spatial patterns of population, jobs, and information flows. We are thus providing a partial response to Sheppard's (1980) "plea for greater consideration of changing inter-urban spatial interaction patterns in our theories of urban system development." "The evolution of spatial configurations is a function of interaction patterns which themselves are generated by past configurations. There is an interaction feedback model with a theory of interaction dependent on a theory of configurational change and vice-versa" [Sheppard (1980), p. 5].

Although we are fundamentally concerned with information sensitive models of migration, it is instructive to analyze the "perfect information" case. To do this, we reformulate equation (4), using scalar notation and an intervening opportunities model of spatial interaction.

Again, ignoring births and death, the basic accounting equation is similar to (1):

$$N_i(t+1) = N_i(t) + \sum_j M_{ji}(t) - \sum_j M_{ij}(t) \tag{5}$$

$$M_{ij}(t) = N_i(t)\left[(1-\delta)^{\sum_{k<j} V_k(t)} - (1-\delta)^{\sum_{k\leq j} V_k(t)}\right], \quad i \neq j \tag{6a}$$

$$M_{ii}(t) = N_i(t)\left[1 - (1-\delta)^{V_i(t)}\right] \tag{6b}$$

$$V_j(t) = V_j(t-1) + \sum_i M_{ji}(t-1) - \sum_i M_{ij}(t-1) \tag{7a}$$

$$V_j(t) \geq 0 \quad , \tag{7b}$$

where δ is the probability that an individual will migrate to occupy or apply for any particular vacancy.

Equations (5) and (7) are accounting equations whereas (6) postulates a behavioral relationship between migration, on the one hand, and relative location of the migrant vis-a-vis job opportunities on the other.

The properties of even this simple model can be investigated analytically only if we use the well-known binomial approximation whereby

$$(1-\delta)^{\sum_{k\leq j} V_k(t)} \approx [1 - \delta \sum_{k\leq j} V_k(t)] \quad . \tag{8}$$

For this approximation to be even conceivable, it obviously must be true that

$$\delta \sum_{k\leq j} V_k(t) < 1 \quad .$$

This approximation is not good unless $\delta \sum_{k\leq j} V_k(t)$ is *much* less than unity. If the approximation *is* used to permit analytical tractability, it follows that

$$M_{ij}(t) = \delta N_i(t) V_j(t) \quad . \tag{9}$$

Equation (9) states that migration is proportional to the product of the population at i and the opportunities at j, in effect a simple gravity model *without any distance effect.* Thus by using this approximation, we are effectively saying that distance is unimportant. We then have precise and complete analytical results for the model composed of (5), (9), and (7) instead of (5), (6), and (7)--a *very* special case where information is perfect and distance is irrelevant. In this case, the following results have been derived [MacKinnon and Rogerson (1980)], where $\bar{N}$ and $\bar{V}$ are total system population and vacancies respectively:

(a) If $0 < \delta \ (\bar{N} + \bar{V}) \leq 1$, the behavior of the system monotonically approaches the equilibrium described in (d) below;

(b) If $1 < \delta \ (\overline{N} + \overline{V}) \leq 2$, the behavior of the system is characterized by oscillatory behavior, but the oscillations are dampened such that the equilibrium described in (d) below is approached;

(c) If $(\overline{N} + \overline{V}) > 2$, the system is characterized by explosive oscillations such that no equilibrium results;

(d) The equilibrium, if it exists, is such that $\frac{V_i(t)}{N_i(t)} = \frac{\overline{V}}{\overline{N}} \ \forall \ i$, i.e., "vacancy rates" are everywhere constant.

Schematically, the trajectories towards equilibrium may be represented by Figure 1 where the regional population and vacancy levels adjust themselves so that at equilibrium vacancy levels for all regions lie on the line

$$N_i = \frac{\overline{N}}{\overline{V}} V_i \quad .$$

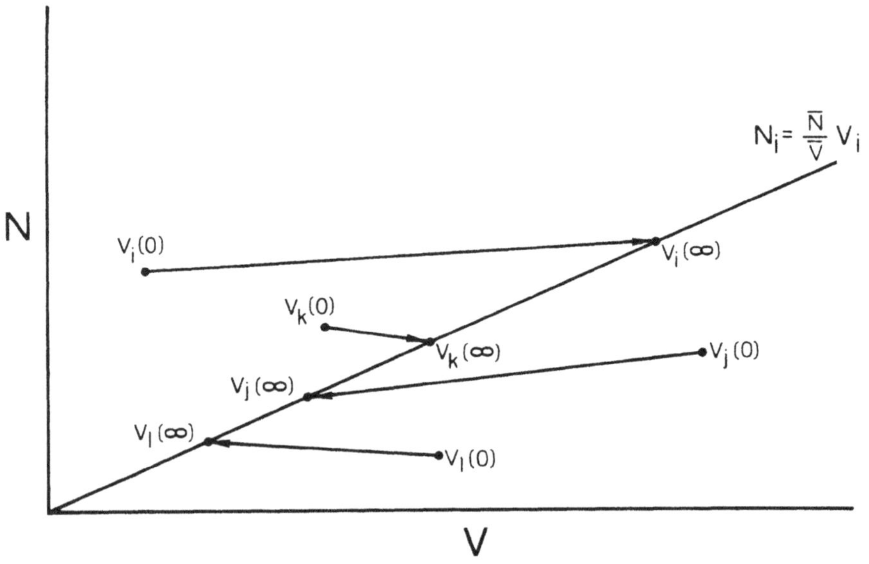

Fig. 1. Schematic Representation of Regional Vacancy/Population Trajectories to Equilibrium Under Assumption of Perfect Information and Distance Irrelevance.

The results for the case where information is still perfect, but distance *is* important are analogous but significantly different. Vacancy levels approach the line $\frac{\bar{N}}{\bar{V}} V_i$, but never reach it, even in the limit. That is, there is some spatial variation in the vacancy rate that arises from the effect of distance itself. Distance inhibits the process from reaching an "equitable" equilibrium.

What happens when this same model is augmented to make it sensitive to interpersonal information flows? It is hypothesized that in the first instance people communicate with people they know, and, with respect to interregional migration, people are assumed to know other people in their region of previous residence.

Thus in addition to (5) and (7), the first information sensitive model, continuing to use the intervening opportunities postulate is developed by having people react to perceived opportunities that are themselves generated by interpersonal communication and combined with past perceptions. This highly non-linear model is described in detail and extensively analyzed in MacKinnon and Rogerson (1980). Briefly, three categories of behavior are possible:

(i) With sufficiently small values of communication rate and information retention rate parameters, migration will eventually dampen to zero. People will either forget vacancy distributions too quickly or not receive enough new information so that people do not migrate in as great numbers as previously. This in turn has a feedback effect in the next time period since recent migrants are the only source of information; thus less information means even fewer will migrate in the subsequent time period.

(ii) With sufficiently large values of these same two parameters, migration flows will increase explosively as people react to "too much" or highly dated information about job vacancies that may in fact not exist anymore.

(iii) Only with specific linear combinations of these two parameters does the system behave such that a constant level of migration occurs. Under these very special circumstances vacancy levels oscillate in elliptical and apparently irregular orbits--orbits that are apparently centered around the $\bar{V}/\bar{N}$ equilibrium condition for the no distance, perfect information case (see Figure 2). Terry Smith has suggested this behavior is rather similar to chaos described by May (1976) and others.

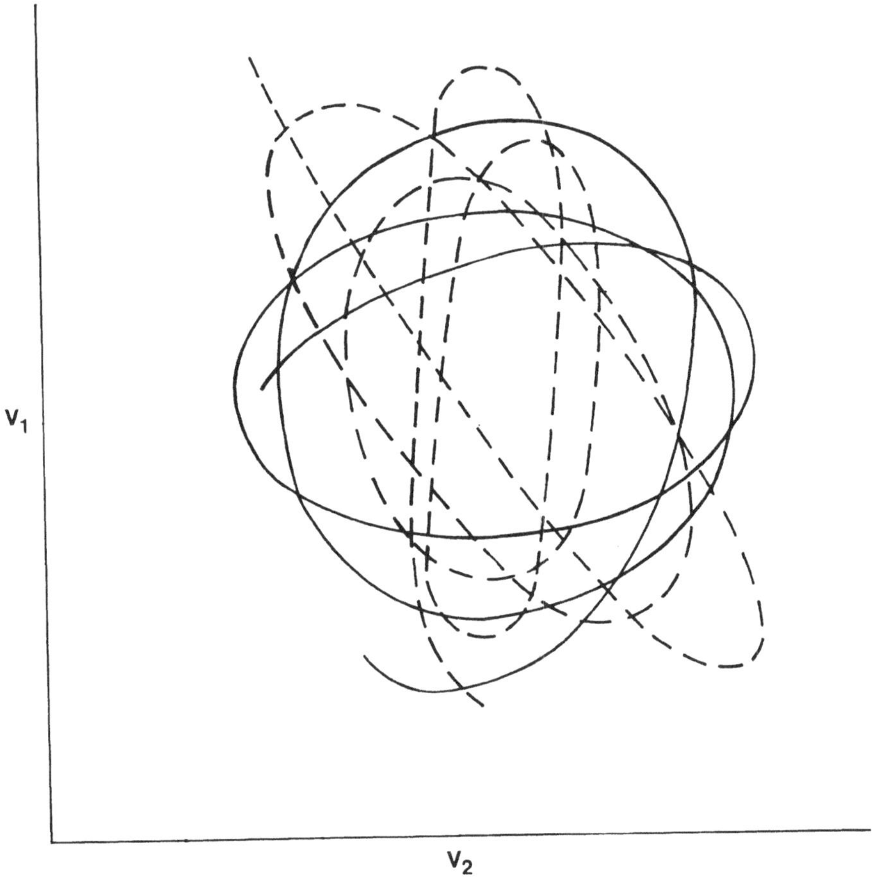

Fig. 2. Trajectories of Vacancies with Interpersonal Information Flows, Intervening Opportunities Model of Spatial Interaction.

Thus with this version of the model, there is a single equilibrium (the trivial null case). Stable, but oscillating, behavior may occur in other circumstances, but those circumstances are extremely particular and most small deviations of parameter values will cause the system to explode or collapse. The steady state behavior is rather interesting, but the knife edge conditions associated with it are unfortunate. Moreover, it is clear that other information flows (e.g., advertising, the mass media) should also be included. It would appear that incorporating these other information flows may prevent migration from dampening to zero even for low values of the parameters associated with interpersonal communication rates and information retention.

Subsequent versions of the model do provide for both background and control information flows. *Background* information is a steady distance dependent stream of information about job opportunities that appears in newspapers, trade journals, radio and television reports. By law or custom, these information channels are stationary--the amount of information changes with actual job vacancies but the vacancies together with a gravity type information dissemination function determine the flow of information of this type. That is, flow *rates* are assumed to be constant. *Selective or control* information is a result of a conscious choice on the part of the employer or regional manager who wishes to fill his vacancies. In principle this information will be directed to areas that would be more likely to satisfy the employer's labor demands--i.e., probably to areas of low vacancy rates or high unemployment.

These different types of information flow are incorporated into a single model that is presented and analyzed in Rogerson and MacKinnon (1980). Background information does stabilize the system in the sense that wider areas of the parameter space become relevant if stationary information channels are introduced. Selective information flows using plausible, satisficing decision rules have the effect of decreasing interregional variances in vacancy rates, whereas intertemporal variances for every region increase as employers and industrial development agencies jockey with one another to gain advantage--an advantage that is quickly dissipated by the actions of other employers in the system. A control theoretic approach could be useful in identifying an optimal combination of intertemporal and interregional variances.

The models outlined above are very much rooted in social-demographic accounting frameworks. They are enriched, however, by the fact that they are tied to specific causal relationships associated with spatial interaction and information flow postulates. Secular economic growth and decline is not incorporated into the analysis. The introduction of exogenously determined vacancies $\underline{V}_z(t)$ and government induced vacancies $\underline{V}_u(t)$ would, however, allow the systematic spatial repercussions of such growth to be traced through the system. In addition, random fluctuations in parameters or variables, superimposed on the causal structure postulated with perhaps certain thresholds introduced could lead to models of the self organization type outlined by Allen and Sanglier (1979, 1981).

In addition to introducing formal control frameworks and specifying a mechanism for secular growth and decline, it would also be beneficial to develop a set of micro, behavioral models consistent with the macro frameworks thus far developed. Although it may be argued that people move or consider moves in response to aggregate perceptions regarding relative economic climates of regions, they eventually must interact with specific potential employers, perhaps via a labor exchange agency. Whether and whence a person moves is a complex question that any single model can only partially answer. The models described above assume people respond to gross perceived numbers of vacancies. It would seem obvious that the perceived skills mix or job opportunities would be relevant to the destination choice as would the mismatch of job skills and skills requirements of the job currently occupied. Thus a person who is overqualified for his job is likely to enter the "mover pool" just as is the unemployed person.

A model currently under development deals with a migrant as he enters a region searching for a job. Whether or not he is successful in obtaining a job is dependent on his own skills, the skill requirements of the job vacancies, and the random matching and mismatching of job applicants with job opportunities. Tracing typical cohorts of unemployed people, at given locations, with specified skills levels, through the migration and job search process may generate insights regarding constraints imposed on the equilibrating process by spatial structure and information filters of various types, as well as suggesting how the efficacy of job and employee search strategies may be improved. Similarly, if a mismatch between job skills and skill requirements exists, causing stress, how long will it take and in what way does an idealized system adjust to reduce this mismatch for the system as a whole. and for people within particular subregions? Do hard core pockets of unemployment and job vacancies persist in certain regions and what strategies are available for employers, employees, and governments to reduce these anomalies?

REFERENCES

Allen, P. and Sanglier, M., 1979. A Dynamic Model of Growth in a Central Place System. Geographical Analysis 11: 256-272.

Allen, P. and Sanglier, M., 1981. A Dynamic Model of a Central Place System - II. Geographical Analysis 13 (in press).

Evtushenko, Y. and MacKinnon, R., 1976. Nonlinear Programming Approaches to National Settlement System Planning, Environment and Planning A 7: 781-792.

MacKinnon, R. and Rogerson, P., 1980. Vacancy Chains, Information Filters, and Interregional Migration. Environment and Planning A 12: 649-658.

May, R., 1976. Simple Mathematical Models with Very Complicated Dynamics. Nature 261: 459-467.

Rees, P. and Wilson, A., 1976. Spatial Population Analysis, Wiley, New York

Rogers, A., 1975. Introduction to Multiregional Mathematical Demography, Wiley, New York.

Rogerson, P. and MacKinnon, R., 1980. An Interregional Migration Model with Source and Interaction Information. Unpublished manuscrript, Geography Department, State University of New York at Buffalo.

Sheppard, E., 1980. Spatial Interaction in Dynamic Urban Systems. Working Paper WP-80-103. International Institute for Applied Systems Analysis, Laxenburg, Austria.

Tan, K., 1980. Solution Strategies for National Settlement System Planning Models. Geographical Analysis 12: 68-79.

POPULATION DYNAMICS AND POPULATION PROJECTION TECHNIQUES FOR LOCAL AREAS

James O. Huff

Department of Geography
University of Illinois

INTRODUCTION

Increasing pressure is being brought to bear on local planning authorities to provide accurate estimates of current and future changes in the size, composition, and distribution of the population within their jurisdictions. The projection techniques commonly employed by local planners are simple extrapolation or ratio methods which are easy to use and require no specialized data [Greenberg, et al. (1978); Isserman (1977); Pittenger (1976)]. These methods, although adequate for certain planning purposes, leave a great deal to be desired as policy instruments since policy relevant variables rarely if ever are included as bases for the projections. On the theoretical side, the projections describe expected outcomes of a dynamic population system. However, when we investigate the kind of system which would generate population distributions consistent with those projected by trend line methods or step-down methods, we discover that the system either violates basic demographic principles at some point or is extremely naive. The recognition of these weaknesses in existing projection methods has stimulated an increasing interest in methods that are derived from models of local population dynamics containing causal variables, some of which can be either directly or indirectly influenced by local policy decisions. Two such methods are discussed in this paper.

The science or the art of making small-area population projections is still in its infancy and, as might be expected in these circumstances, there is no generally accepted "best" method. In fact the methods which have been applied within actual planning context are often _ad hoc_ in the sense that they have been tailored to specific planning needs and their final form often has been dictated by the nature of the data that could be generated from readily accessible sources.

The methods, if they are to be effective, should be _ad hoc_ in that they should be tailored to the context and the questions of interest to the planner/client/policy maker, and they should be designed to make maximal use of existing data or they will not be used. This is not to say that we should stop working on robust models capable of answering a diverse set of questions, or that we should cease to explore certain classes of models because they require data that are not readily available. Rather, we should never lose sight of the fact that projection methods are first and foremost practical tools. The best measure of the method ultimately is how well it does the specific job required by the planner.

The growing number of _ad hoc_ methods tends to obscure the fact that there are a series of basic issues that each method either explicitly or implicitly resolves. The form that the method ultimately takes depends upon the resolution of the following issues or questions:

(a) Can we predict population changes directly or must we break down population change into its component parts and predict birth, death, and migration effects separately? The distinction here is between components and non-components methods.

(b) Must we explicitly consider the possibility that population change (or change in a population related variable) in one part of the region may influence the magnitude and direction of change in another part of the region? If the answer to the question is "yes," then the method is an interaction model that specifies between-area interdependencies.

(c) What time-dependent variables influence the future population size (and composition) of an area? If population variables are deemed to be the only important variables in the change process, then we have a pure demographic projection model. When other control and explanatory variables are introduced we move into the realm of augmented demographic models. In the extreme, population projections are by-products of models designed primarily for the purpose of projecting the values of population related variables such as employment.

The answers to the above questions form the basis for the three-way classification of population projection methods presented in Figure 1.

Inter-area relationships \ Demographic processes	non-component pure	non-component augmented	component pure	component augmented
not explicit	Isserman(1977)	Greenberg(1978)	Pittenger(1976)	Ledent(1978)
interaction methods		Lowry(1964) Barber(1977)	Rees and Wilson(1977) Rogers(1973)	Huff(1979) Wilson(1974)

Fig. 1. A Classification of Population Projection Methods

The methods discussed in this paper are components methods, which means that population change is decomposed into its component parts: birth, death, and migration. The basic demographic equation underlying any components method is of the following form:

$$\Delta P_i(\hat{t}) = B_i(\hat{t}) - D_i(\hat{t}) + M_{\cdot i}(\hat{t}) - M_{i\cdot}(\hat{t}) \quad , \qquad (1)$$

where $\Delta P_i(\hat{t})$ is the change in the population size of area i over the time interval (t,t+1), and the number of births, deaths, inmigrants and outmigrants during (t,t+1) are given by the four terms on the right hand size of (1).

When the population is disaggregated into cohorts this basic equation is complicated by the possibility that a member of cohort j at time t may become a member of cohort k by time t+1. The one year age cohort equation is the simplest version of the general cohort model, and will be the only cohort model discussed in this paper. The basic age cohort equations are

$${}_oP_i(t+1) = B_i(\hat{t}) - {}_oD_i(\hat{t}) + {}_oM_{\cdot i}(\hat{t}) - {}_oM_{i\cdot}(\hat{t}) \quad ,$$

and

$${}_{k}P_i(t+1) = {}_{k-1}P_i(t) - {}_{k}D_i(\hat{t}) + {}_{k}M_{\cdot i}(\hat{t}) - {}_{k}M_{\cdot i}(\hat{t}) \quad .$$

The age cohort k consists of all those people who are (or would have been) age k at time t+1.

The main appeal of the components approach to small area population projection is that the effects of changing in- and outmigration are specifically identified. As we move down to the local level an increasing amount of the observed change in the size and composition of the population is caused by migration or residential mobility. As a consequence, the key to a good local projection method will be the migration component of the method. Although net migration models such as those proposed by Pittenger (1976) and Ledent (1978) have shown considerable promise, as have the step-down or ratio-trend models of migration, the migration models discussed in this paper are origin-destination accounting models [Rees and Wilson (1977)].

The two origin-destination models discussed in this paper are based upon the assumption that significant interaction effects exist and must therefore be built into the structure of the population method. The first example is a semi-Markov model of the form

$$M_{ij}(\hat{t}) = M_{i\cdot}(\hat{t})\ \omega_{ij} \quad ,$$

where

$$M_{i\cdot}(\hat{t}) = P_i(t)a_i(t) \quad .$$

The second model is an origin-constrained interaction model in which

$$M_{ij}(\hat{t}) = M_{i\cdot}(\hat{t})\ b_j(t)\ \omega_{ij}/\sum_k b_k(t)\ \omega_{ik} \quad ,$$

where

$$M_{i\cdot}(\hat{t}) = P_i(t)a_i(t) \quad .$$

A CONSTANT TRANSITION MATRIX METHOD

Origin-destination migration data consisting of the number of people moving from one area in a region to another are normally stored and portrayed as an (N+1) × (N+1) matrix, $\underline{M}^*$, with N areas inside the region and one area outside containing the rest of the world. As is the case with most components methods, we will work with rates rather than totals--the key rates in this instance being the origin-destination transition rates

$${}_k t_{ij}(t) = {}_k m_{ij}(\hat{t}) / {}_k M_{i\cdot}(\hat{t}) \ .$$

The simplest model involving transition rates is the familiar first-order Markov model, which is based upon the assumption that the rates do not change over time such that

$${}_k t_{ij}(t) = {}_k t_{ij}$$

Population projection methods based upon origin-destination migration predictions are very appealing from the theoretical standpoint; but, to date they have rarely been used as a basis for local population projection primarily because of prohibitive data requirements. In some instances the data problem is a practical or an institutional issue hinging on the cost and/or the legality of generating and maintaining an extensive and highly individualized data base. In other instances data requirements may even exceed the limits of an ideal data set for the planning region, as would tend to be the case when parameter estimates are required for infrequent events experienced by relatively small target populations. Origin-destination components methods will be adopted more widely only if: (1) we can demonstrate that they are superior to other methods in the sense that they provide better projections and more insights into the dynamics leading to observed change; and (2) we can show that the methods are flexible enough to accommodate non-optimal data bases.

Origin-destination methods attempt to predict directly the number of migrants moving between area i and area j during (t,t+1), $M_{ij}(\hat{t})$, as a function of origin characteristics, $a_i(t)$, destination characteristics, $b_j(t)$, and relationships between origin and destination, $\omega_{ij}(t)$, such that

$$M_{ij}(\hat{t}) = f[a_i(t), b_j(t), \omega_{ij}(t)] \ .$$

Given the expected flow of migrants from area i to area j, the total number of migrants leaving i during (t,t+1) is

$$M_{i\cdot}(\hat{t}) = \sum_j M_{ij}(\hat{t}) \quad \text{for all destinations j,}$$

and the total number entering i is

$$M_{\cdot i}(\hat{t}) = \sum_j M_{ji}(\hat{t}) \quad \text{for all origins j.}$$

The method is described by a Markov chain if the age-specific mobility rates for the population residing in area i are assumed to be constant over time. The model which we shall explore in this section is a semi-Markov model, which means that the transition matrix remains constant over time but the mobility rates are allowed to vary such that

$${}_kM_{i\cdot}(\hat{t}) = {}_ka_i(t)\,{}_kP_i(\hat{t}) \quad ,$$

and

$${}_kM_{\cdot i}(\hat{t}) = \sum_j {}_kt_{ji}\,{}_kM_{j\cdot}(\hat{t}) \quad .$$

The Markov chain model has been explored as a means of projecting interregional and intermetropolitan population changes with apparently satisfactory results [Rees and Wilson (1977)], although significant non-stationarities appear to exist at the intra-urban level [Huff and Clark (1978)]. The main practical difficulty in applying the Markov model in the projection of minor civil division populations is small size of the base population for many small towns and rural areas (assuming for the moment that origin-destination data were available at this scale). The observed matrix of inter-area moves, $\underline{M}$*, tends to be extremely sparse, implying severe estimation problems. On the theoretical side, there are several reasons for suspecting that the migration rate, a_i, for an area will not remain constant over time. If we do not have cohort specific flow data, then the gross migration rate will tend to change if the age composition of the area changes, since the variation in age specific migration rates is considerable.

If we solve the heterogeneity problem by controlling for age composition, it is still quite possible that the migration rate could change as a function of changes in the volume of inmigrants to the area. The argument in this instance has a behavioral basis, and has been termed the cumulative inertia hypohtesis or the "mover-stayer" dichotomy. The idea is that people who have moved recently have a greater probability of moving again than is the case for people who have remained in the same place for a long period of time. This argument has been used to explain the apparently contradictory (from a labor mobility perspective) observation that metropolitan areas with high "rates" of inmigration also tend to have high rates of outmigration, and areas with low "rates" of inmigration also have low rates of outmigration.

The following projection method suggests possible solutions to both the sparse matrix problem and the non-stationarity problem. The method is discussed with reference to a specific planning problem--the projection of age cohorts for the counties in the U.S. on the basis of origin-destination data that are not broken down by age; but, the basic techniques employed are relevant to most small area projection methods based on origin-destination data.

A Collapsed Transition Matrix

The proposed solution to the sparse matrix problem is a projection method that is mid-way between a step-down components approach and a fully specified origin-destination method. The complete flow matrix will be selectively collapsed around each area so as to increase the number of migrants in each cell of the new matrix.

Although we could design a separate collapsing scheme for each area, a more practical solution is to set up a hierarchical regionalization scheme. If B is the set $\{b_i\}$ of local planning areas (counties for example) then C is the set of disjoint subsets $\{c_j\}$ of B (SEAs for example), D is the set of disjoint subsets $\{d_k\}$ of C (states for example), and so on up the hierarchy of regions. Next we define the complement of c_j contained in d_k as

$$\overline{C}^j = (C - c_j) \cap d_k \quad \text{, and similarly}$$

$$\overline{D}^k = (D - d_k) \cap e_\ell \quad \text{, and}$$

$$\overline{E}^\ell = E - e_\ell \ ,$$

where E is the set of regions at the top of the regionalization hierarchy (census divisions for example). A simplified example of the setting is given in Figure 2 where local area b_i is contained in c_j which is contained in d_k.

For each element, c_j, of C we now propose to define a partial transition matrix, $\underline{T}_j$, that is a $Q_j \times N_j$ matrix, where Q_j is the number of areas or regions contributing migrants to areas contained in c_j under the collapsing scheme proposed below and N_j is the number of areas in c_j. Throughout the discussion the

Fig. 2. A Nested Hierarchical Structure

areas in question are assumed to be contained in c_j, which is contained in d_k, which is contained in e_ℓ and whose numbers are N_j, N_k, and N_ℓ respectively and $Q_j = N_j + (N_k-1) + (N_\ell-1)$ in the following example.

The first N_j rows of $\underline{T}_j$ contain the transition probabilities estimated by the number of people moving from one small area b_i contained in c_j to another area in c_j, divided by the total number of migrants leaving b_i. The next N_k-1 rows contain the probabilities that a migrant from $\overline{C}^j$ moves to a small area b_i contained in c_j, and the last $N_\ell-1$ rows are the probabilities that a migrant from $\overline{D}^k$ moves to b_i (the rest of the world is included as a region in the highest level of the hierarchy). If $M(j\cdot,t)$ is the $1 \times Q_j$ vector of migrants from areas in c_j and the regions in $\overline{D}^k$ and $\overline{E}^\ell$ and $M(\cdot j,t)$ is the $1 \times N_j$ vector of inmigrants to areas in c_j then

$$M(j\cdot,t)T_j = M(\cdot j,t) \quad .$$

After going through the above procedure you may justifiably ask why we went to all this trouble. Are the estimation problems really all that bad in a straight origin-destination method, and if they are, can this collapsing method significantly improve the situation? The answer to both questions can best be illustrated by a simple example. There are approximately 5000 counties in the U.S., which means that we must attempt to estimate 25,000,000 parameters in the transition matrix if we wish to use a straight origin-destination method. The values in the vast majority of the cells of $\underline{M}$* will either be zero or so small that estimated transition rates are almost meaningless. On the other hand, let us assume that there are 10 census divisions each containing 5 states which each contain 10 SEAs with 10 counties in each SEA. We must estimate 500 32-by-10 matrices, which means that the total number of transition parameters is 160,000--a large number but manageable--and we can be reasonably certain that most of the cells in the new $\underline{M}^*$ will have sufficient numbers to make reliable estimates of the transition probabilities.

Another attractive feature of this collapsing method from an applied standpoint is that the whole system need not be projected forward on a county by county basis if we are only interested in projecting populations for counties within a particular SEA or state, for example. All we need in order to project the population of a specific county $b_i \varepsilon c_j \varepsilon d_k$ are the migration data required for the estimation of $\underline{T}_j$, the migration rates used in the estimation of $M(j\cdot,t)$, and exogenous estimates of the population in $\bar{C}^j$, $\bar{D}^k$, and $\bar{E}^\ell$ at time t. If reliable estimates for $\bar{C}^j$ do not exist, this level of the regionalization scheme could be "dropped out," or the elements in $\bar{C}^j$ could be projected using the same limited projection strategy as was proposed for the local areas in c_j.

The Cohorts Problem

The second problem to be solved is the cohorts problem that will be an issue whenever we either do not have direct access to cohort specific migration data, or the cohort by origin by destination data matrix is so sparse as to preclude direct estimation of the transition rates. We shall begin by assuming that the following data are available:

(1) The age specific population for each area at the beginning and the end of the time interval (0,1) [i.e., ${}_kP_i(0)$ and ${}_kP_i(1)$];

(2) age specific fertility and mortality rates for each area (often stepped down from estimates at higher levels of aggregation);

(3) the number of people in the age k cohort who either entered the area, ${}_kM_{\cdot i}(0)$, or left the area, ${}_kM_{i\cdot}(0)$, during (0,1) (given inmigration, then outmigration can be determined, and vice-versa); and,

(4) the total number of people moving from area i to area j during (0,1) as summarized by the transition matrix $\underline{T}^*$ where

$$t^*_{ij} = m^*_{ij}/M^*_{i\cdot}$$

(note that the origin-destination flows are not broken down into age cohorts).

At the outset our best estimate of the age-specific transition matrix, ${}_k\underline{T}$ would appear to be $\underline{T}^*$ such that

$${}_kt_{ij} = t^*_{ij} \quad .$$

We can check this estimate against the known number of cohort k inmigrants to area j by comparing ${}_kM^*_{\cdot j}$ with ${}_k\hat{M}_{\cdot j}$, keeping in mind

$${}_k\hat{M}_{\cdot j} = \sum_i t^*_{ij}\ {}_kM^*_{i\cdot} \quad .$$

The observed and the initial estimate of the cohort k migrants to area j will be similar only if the age k cohort is behaving like the "average" members of the total population. In general the best estimate of the true age specific transition probabilities, given the data at hand, will be

$${}_k\hat{\underline{T}} = \underline{BT^*}\ \overline{\underline{B}} \quad ,$$

where $\underline{B}$ and $\overline{\underline{B}}$ constitute a bi-proportional operator [Bacharach (1970)] that "minimizes the differences" between ${}_k\hat{t}_{ij}$ and $\hat{t}^*_{ij}$ while ensuring that

$$\sum_j {}_k\hat{t}_{ij} = 1 \quad \text{for all } i \quad ,$$

and

$$\sum_i {}_k\hat{t}_{ij}\ {}_kM^*_{i\cdot} = {}_kM^*_{\cdot j} \quad .$$

The age specific transition matrices may now be used in conjunction with age specific outmigration rates, ${}_ka_i(t)$, to project the cohort k population living in area j at time t. We also avoid the heterogeneity problem alluded to earlier since the age composition of the population is projected forward through time with accompanying age-specific migration rates.

Duration of Stay Effects

Thus far attention has been directed primarily toward the problems associated with the estimation of the transition matrix in the Markov model. We will now turn to a discussion of the migration rates associated with each age-area cohort. The Markov chain model assumes that these rates are constant over time, and it is suggested that this assumption should be modified in view of empirical and theoretical evidence indicating that the probability of moving is dependent upon duration of stay [Ginsberg (1971)]. If we assume that duration of stay does affect the probability of moving, then it has been argued that places experiencing rapid inmigration should also experience rapid outmigration--since by definition new inmigrants have short durations of stay and should therefore be more prone to move again. If P^i is the number of people who last moved i time periods ago and α_i is the associated mobility rate (the cohort and area subscripts will be implicit), then it is assumed that

$$M(\hat{t}) = \alpha(t)P(t) = \alpha_1 P^1(t) +\ldots \alpha_i P^i(t) +\ldots \alpha_n P^n(t) \quad . \quad (2)$$

The number of people in $P^1(t)$ is

$$P(t+1) = M(\hat{t}) + IN(\hat{t}) - OUT(\hat{t}) \quad , \quad (3)$$

where $M(\hat{t})$ is the total number of migrants generated by the area, and $IN(\hat{t})$ and $OUT(\hat{t})$ are the number of migrants entering and leaving the area during (t,t+1). The number of people in $P^i(t)$ is

$$P^i(t) = (1-\alpha_{i-1})P^{i-1}(t-1) \quad , \text{ and} \quad (4)$$

the number of people leaving the area, $OUT(\hat{t})$, is assumed to be proportional to the total number of movers such that

$$OUT(\hat{t}) = \beta a(t)P(t) \quad (5)$$

(this condition on outmigration must hold if we have an embedded transition matrix containing a main diagonal). If we let $\Delta M(t)$ be $M(\hat{t}) - M(\hat{t}-1)$ and $\Delta IN(t) = IN(\hat{t}) - IN(\hat{t}-1)$, then equations (2) through (5) imply that

$$\begin{aligned} \Delta M(t) &= \alpha_1[P^1(t) - P^1(t-1)] +\ldots+ \alpha_n[P^n(t) - P^n(t-1)] \\ &= \alpha_1[(1-\beta)\Delta M(t) + \Delta IN(t)] + \alpha_2(1-\alpha_1)[(1-\beta)\Delta M(t-1) \\ &\quad + \Delta IN(t-1)] +\ldots \quad . \end{aligned} \quad (6)$$

The result is a difference equation with parameters that are interpretable, and if the α_i's cannot be obtained from survey data then they can be estimated using longitudinal data on ${}_kM(t)$ and ${}_kIN(t)$ across similar areas. The time series need not be particularly long since the later terms in (6) tend to zero fairly rapidly for the most mobile sectors of the population. With this final modification the projection method is completely specified. The parameter estimation problem is resolved through the use of bi-proportional operators to project age specific flows from total population flows, and the use of a hierarchical regionalization scheme that substantially reduces the size of the transition matrices, thereby reducing the number of parameters to be estimated. The final procedure relates changes in the migration rate to changes in the number of inmigrants to an area, thereby changing the model from a simple Markov chain to a semi-Markov process with constant transition probabilities.

ORIGIN CONSTRAINED INTERACTION METHODS

We have assumed in the previous section that the characteristics of the destination that tend to affect the flow of migrants into that locale do not change over time. If these characteristics were subject to change the transition probabilities would not be constant. Changes in the availablility of housing in different parts of the city have significant effects upon the observed mobility pattern, for example. One of the simplest models capable of describing migration processes that are sensitive to changes in destination characteristics is the familiar origin constrained interaction model. The interaction model serves as the basis for the methods described in this section, and it takes the following general form:

$$M_{i\cdot}(\hat{t}) = P_i(t)a_i(t) \quad ,$$

and

$$M_{ij}(\hat{t}) = M_i(\hat{t})b_j(t)\omega_{ij}/\sum_k b_k(t)\omega_{ik} \quad .$$

The first method described in this section is a modified spatial interaction model in which ω_{ij} is a function of the distance between areas i and j. The two other methods are vacancy chain models in which the mobility patterns and the distribution of housing vacancies are jointly determined.

A Spatial Interaction Model

The standard form of a spatial interaction model is as follows;

$$M_{i\cdot}(\hat{t}) = \rho_i P_i(t) \quad ,$$

and (7)

$$M_{ij}(\hat{t}) = M_{i\cdot}(\hat{t}) D_j(t) c_{ij} / \sum_k D_k(t) c_{ik} \quad ,$$

where $D_j(t)$ is some measure of the attractiveness of destination j, and c_{ij} is a function of the distance s_{ij} between areas i and j. The main advantage that this model has over many other origin-destination migration models is the small number of parameters requiring estimation. The price that we pay for simplicity is that all variation in the interaction term is summarized by the distances between places. Because this particular version of the general interaction model explicitly focuses on distance effects, it has received a great deal of attention in the geographic literature. Because the model is well known I do not propose to discuss the standard entropy maximizing methods normally employed in the estimation of the model's parameters; rather, I would like to explore a different estimation method which appears to have certain advantages over the entropy maximizing approach.

Begin by assuming that $D_j(t)$ and c_{ij} in (7) have specific functional forms:

$$D_j(t) = \exp[a_1 x_{1j}(t) + a_2 x_{2j}(t) + \ldots] \quad , \tag{8}$$

and

$$c_{ij} = \exp(\lambda s_{ij}) \quad .$$

Instead of estimating $M_{ij}(t)$ directly as a function of the independent variables $\{x_j\}$, consider the ratios (age cohorts are implicit)

$$r(i,j,k) = M_{ij}(t)/M_{ik}(t) = D_j(t)c_{ij}/D_k(t)c_{ik} \quad . \tag{9}$$

If these ratios can be estimated then the actual rates can also be estimated, since we know that $\sum_j M_{ij}(t) = \sum_j r(i,j,k)M_{ij}(t) = M_{i\cdot}(t)$ for any k. Since $M_{i\cdot}(t)$ is given, $M_{ik}(t)$ is determined as is $M_{ij}(t)$ for all j.

It follows from (8) and (9) that

$$\ln r(i,j,k) = \ln M_{ij}(t) - \ln M_{ik}(t) = \tag{10}$$
$$\lambda(s_{ij}-s_{ik}) + a_1[x_{1j}(t)-x_{1k}(t)] + a_2[x_{2j}(t)-x_{2k}(t)] + \ldots \quad .$$

Standard estimation techniques may now be applied to (10), giving direct estimates of λ for example (which may or may not vary with a change in the origin, i). The time dependent variables on the RHS of (10) in fact may be endogenous to the model--the age k population residing in j at time t, for example--or they may be exogenous--the expected number of job vacancies in industry k and area j during (t,t+1), for example. Other destination characteristics besides the distance to the origin in question may also be time invariant--good climate/bad climate as a dichotomous variable, for example.

The main advantage that the ratio model has over single destination regression models is that the competitive effects of other potential destinations are explicitly incorporated within the structure of the model. Its appeal over standard origin constrained spatial interaction models is that the "destination effects" may be expressed as a function of several variables, and the estimation of the parameters is direct rather than via an iterative procedure (requiring additional conditions on c_{ij} in the case of entropy maximizing methods).

Simultaneous Household and Housing Vacancy Projection Methods

I will conclude this discussion of population projection methods with two methods based on a model of residential mobility within the city that are modified vacancy chain models [White (1971)]. The basic idea is that the observed change in the population size and composition within small areas of the city is primarily a function of residential mobility. Rather than attempting to model the results of residential mobility, it is argued that the mobility process itself should be modelled. Residential mobility is viewed as a process by which households select a new residence on the basis of the attributes of the dwelling and the households' housing requirements (in conjunction with a housing budget constraint). The aggregate outcome of individual selection processes is indeed a pattern of moves between areas of the city, but it is essential to recognize that dwellings (with situational attributes that may affect the selection process) are being selected, and not areas per se. The implication is that the mobility model should be designed to allocate households to vacancies in a particular class of dwellings. Once we know the number of households moving into and out of a given dwelling class, households can be assigned to areas on the basis of the proportion of the vacancies in that dwelling class located in that particular area.

Given this change in perspective from a focus on areas to a focus on dwelling types, we begin by partitioning the housing stock into J distinct dwelling classes--on the basis of housing characteristics found to be important in the choice process, such that $H_j(t)$ is the number of dwelling units in the j-th housing class at time t. Households are similarly partitioned on the basis of the type of dwelling currently occupied, and household characteristics (such as family size and income) that are found to play a significant role in the mobility process, such that ${}_kP_i(t)$ is the number of households in the k-th sub-population living in type i housing at time t.

A Housing Choice Projection Method

The model presented in this section is discussed in Huff (1979), and is only briefly summarized below. The key to the model is the assumption that households either directly or indirectly compare alternatives against their present dwelling, and that the selection of a new residence is based upon this comparison. The probability of moving to a new residence in housing class j is shown to be solely a function of the number of vacancies, $V_j(t)$, in the various housing classes, and a set of parameters $\{{}_ka_{i\ell}\}$ indicating the relative attractiveness of these vacancies.

If the unit time period is small enough (a matter of weeks, for example) to preclude the possibility that a large number of multiple moves will occur during any unit of time, then the expected number of type k households moving from housing class i to a new residence in housing class j during (t,t+1) is

$$ {}_kM_{ij}(\hat{t}) = {}_kP_i(t)V_j(\hat{t})\,{}_ka_{ij}/{}_kW_i(t) \quad , $$

where

$$ {}_kW_i(\hat{t}) = \sum_{\ell} V_\ell(\hat{t})\,{}_ka_{i\ell} + {}_ka_{io} \quad . $$

The total number of type k households leaving residences in housing class i is therefore

$$ {}_kM_{i\cdot}(\hat{t}) = {}_kP_i(t)[1 - {}_ka_{io}/{}_kW_i(\hat{t})] \quad , $$

and the total number of type k households entering residences in housing class i is

$$ {}_kM_{\cdot i}(\hat{t}) = [\sum_{\ell} {}_kP_\ell(t)\,{}_ka_{\ell i}/{}_kW_\ell(\hat{t})] \quad . $$

With a knowledge of the number and type of households entering and leaving dwellings in different housing classes, the size and composition of in- and outmigration and for any small area of the city may be determined. If $[{}_kP_i(t)|A]$ represents the number of households in area A that are members of ${}_kP_i(t)$, and $[V_j(\hat{t})|A]$ represents the number of vacancies in H_j located in area A, then the number of type k households leaving class i residences located in A is

$$ {}_kM_{i\cdot}(\hat{t})|A = {}_kM_{i\cdot}(\hat{t})\{[{}_kP_i(t)|A]/{}_kP_i(t)\} \quad , $$

and the total number of type k households leaving residences in area A during (t,t+1) is

$$ {}_kM_{A\cdot}(\hat{t}) = \sum_i {}_kM_{i\cdot}(\hat{t})|A $$

$$ = \sum_i [{}_kP_i(t)|A][1-{}_ka_{io}/{}_kW_i(\hat{t})] \quad . $$

The number of type k households entering class j residences located in A is similarly designated as

$$ {}_kM_{\cdot j}(\hat{t})|A = {}_kM_{\cdot j}(\hat{t})\{[V_j(\hat{t})|A]/V_j(\hat{t})\} $$

since by assumption every vacancy included in $V_j(\hat{t})$ has an identical probability, ${}_k\alpha_{ij}(t)$, of being filled by a member of ${}_kP_i(t)$, for all k and i. The total number of type k households entering residences in area A during (t,t+1) is

$$ {}_kM_{\cdot A}(\hat{t}) = \sum_j {}_kM_{\cdot j}(\hat{t})|A $$

$$ = \sum_j [V_j(\hat{t})|A][\sum_\ell {}_kP_\ell(t)\,{}_ka_{\ell j}/{}_kW\,(\hat{t})] \quad . $$

Before the above model can be used as a forecasting device, the parameters, ${}_ka_{ij}$, and the vacancies, $V_j(\hat{t})|A$, occurring in each area A for each housing class j during (t,t+1) must be estimated. As a first step in the estimation of ${}_ka_{ij}$ the attractiveness of the present residence, ${}_ka_{io}$, is arbitrarily set equal to 1 for all k and i. Since the model is concerned with the *relative* attractiveness of each vacancy, there is no loss of generality in establishing an arbitrary base of comparison. If ${}_k\bar{M}_{ij}(\hat{t})$ is the observed number of households in ${}_kP_i(t)$ that move to a dwelling in $H_j(t)$ during (t,t+1), ${}_k\bar{M}_{io}(\hat{t})$ is the observed number of non-movers in ${}_kP_i(t)$, and $\bar{V}_j(\hat{t})$ is the observed number of vacancies in $H_j(t)$ occurring during (t,t+1), then

$${}_k\overline{M}_{ij}(\hat{t})/{}_kP_i(t) = \overline{V}_j(\hat{t})[{}_ka_{ij}/{}_ka_{io}]{}_k\overline{M}_{io}(\hat{t})/{}_kP_i(t) \quad .$$

Since ${}_ka_{io} = 1$ by assumption,

$${}_ka_{ij} = {}_k\overline{M}_{ij}(\hat{t})/[\overline{V}_j(\hat{t}){}_k\overline{M}_{io}(\hat{t})] \quad \text{for all k, i and j.}$$

The estimation of $V_j(\hat{t})|A$ is less straight forward since the number of vacancies occurring in housing class j during (t,t+1) is directly related to the number of households leaving dwellings in class j during (t,t+1), which implies that the number of vacancies in an area and the number of households leaving the areas are jointly determined. As can be seen from the discussion of the model, the simultaneity problem arises because ${}_kM_{j\cdot}(\hat{t})$ is a function of $V_\ell(\hat{t})$ for all ℓ. In the following estimation procedure for $V_j(\hat{t})|A$ each variable is indexed on A, the small area in question; but, to simplify notation the areal unit designation is omitted.

If for any area A the effects of decrements and additions to H_j in A are exogenously determined and the instantaneous number of vacancies in A at time t is $V_j(t-1) - \sum_k {}_kM_{\cdot j}(t-1)$, then the number of vacancies in A during (t,t+1) that are members of H_j is

$$V_j(\hat{t}) = V_j(\hat{t}-1) - \sum_k {}_kM_{\cdot j}(\hat{t}-1) + \sum_k {}_kM_{j\cdot}(\hat{t}) - r_jR_j(\hat{t}) + n_jN_j(\hat{t}) , \qquad (11)$$

where

$${}_kM_{j\cdot}(\hat{t}) = {}_kP_j(t)[1 - {}_ka_{jo}/\sum_\ell V_\ell(\hat{t}){}_ka_{j\ell} + {}_ka_{jo}] \quad ,$$

and r_j is the proportion of the decrements, $R_j(\hat{t})$, to the housing stock that were vacant, and n_j is the proportion of the additions, $N_j(t)$, to the housing stock that were vacant.

To solve for $V_j(\hat{t})$ one must solve the system of N simultaneous equations implied by the substitution of the expression for ${}_kM_{j\cdot}(\hat{t})$ into (11). Alternatively, $V_j(\hat{t})$ may be solved iteratively. Taking ${}_kM_j^{(1)}(\hat{t})$ to be first approximation to the number of type k households leaving class j housing in A where

$${}_kM_j^{(1)}(\hat{t}) = {}_kP_j(t)[1 - {}_kq_{jo}(t-1)] \quad ,$$

then the first approximation to $V_j(\hat{t})$, $V_j^{(1)}(\hat{t})$, is found by substituting ${}_kM_{j\cdot}^{(1)}(\hat{t})$ for ${}_kM_{j\cdot}(\hat{t})$ in equation (11). In the n-th iteration $V_\ell^{(n-1)}(\hat{t})$ is substituted for $V_\ell^{(n-1)}(\hat{t})$ such that

$$ {}_kM_{j\cdot}^{(n)}(\hat{t}) = {}_kP_i(t)[1 - {}_ka_{io}/{}_kW_i^{(n-1)}(\hat{t})] \quad , $$

and

$$ V_j^{(n)}(\hat{t}) = V_j(\hat{t}-1) - \sum_k {}_kM_{\cdot j}(\hat{t}-1) + \sum_k {}_kM_{j\cdot}^{(n)}(\hat{t}) - r_jR_j(\hat{t}) + n_jN_j(\hat{t}) \; . $$

With the estimation of the number of vacancies occurring in each housing class for each area A, the residential change model is completely specified in terms of observed or predicted values from the previous time period as input into $V_j(\hat{t})$, ${}_kP_j(t)$ and three sets of exogenous variables: increments and decrements to each housing class, $N_j(\hat{t})$ and $R_j(\hat{t})$ for each area, and the number of type k inmigrants entering the study area during (t-1,t), ${}_kM_{N+1}(t)$. Given these estimates and the number of type k households entering and leaving each small area of the city, equations (1) and (2) may be predicted for each time period. The model provides a simultaneous projection of the number and type of households living in an area, and the number of vacancies within the area's housing stock. It also provides a mechanism for predicting changes in the cohort specific mobility rates as a function of changes in the vacancy set (the cohorts are household types in this instance). Finally, the model is compatible with a widely used individual choice model, thereby providing a link to the more micro studies of residential search and choice.

On the debit side, the model presumes that individuals are engaged in an unconstrained selection process. Each household is treated in isolation or independently of every other household that is searching for a new residence during the same time period. When excess capacity exists in all housing classes this assumption is not a problem; but, when the market is "tight" the vacancy set experienced by the household during its search may be quite different in composition from the set of all vacancies generated during a given period of time. On the other hand, each additional vacancy in a given housing class is assumed to increase the probability of selecting a dwelling in that class by a constant increment. What is more likely is that an additional vacancy in a housing class with a low instantaneous vacancy rate would have a significantly greater impact upon the probability of selecting a dwelling in that class than would be the case for an additional vacancy in a class with a high instantaneous vacancy rate. If the vacancy rate in the stock reaches very high levels within a given area, the externalities may be such that an additional vacancy may even tend to reduce the probability of selecting a dwelling in that housing class concentrated in the high vacancy areas.

A Constrained Choice Model

The mobility model presented below is designed to redress the unconstrained choice problem mentioned above. The model is based upon the assumption that the housing market is imperfect in the sense that certain classes of dwellings are in short supply and other classes have an excess number of dwelling units, relative to demand. Furthermore, these imperfections are assumed to be chronic, meaning that the imbalances are persistent and are not easily resolved through existing clearing mechanisms.

Let ${}_kN_{ij}(\hat{t})$ be the expected number of type k households living in $H_i(t)$ that would move to a new residence in $H_j(t)$ during (t,t+1) if given a choice. It is assumed that housing preferences are invariant over the planning horizon such that

$$ {}_kN_{ij}(\hat{t}) = \omega_{ij}\ {}_kP_i(t) \quad , \tag{12} $$

where $\sum_j {}_k\omega_{ij} = 1 - {}_k\omega_{io}$, ${}_k\omega_{io}$ being the preference for the current dwelling. If every household were able to achieve its first choice, thus satisfying (12) above, the mobility process could be described as a simple Markov chain with $\{\omega_{ij}\}$ as the matrix of transition probabilities. When a significant number of households are unable to find a dwelling in the preferred house class, the Markov model will break down as the magnitude of the flow varies not only as a function of ${}_kP_i(t)$ but also as a function of the number of vacancies in surplus demand housing classes. Under these conditions it is assumed that the flow of type k households from housing class i to housing class j, where $j \varepsilon D$ is the set of surplus demand classes, is

$$ {}_kM_{ij}(\hat{t}) = V_j(\hat{t})[{}_k\omega_{ij}\ {}_kP_i(t)/\sum_k \sum_\ell {}_k\omega_{\ell j}\ {}_kP_\ell(t)],\ j \varepsilon D \quad . \tag{13} $$

Equation (13) is the familiar destination constrained interaction model.

If a household preferring a dwelling in a surplus demand class is unable to find a dwelling in D, then it is assumed that the household selects a dwelling in S, the set of surplus supply areas, with probability ${}_k\delta_{ij}$, $j \varepsilon S$. The household's current dwelling is also assumed to be a member of S such that ${}_k\delta_{io}$ is the probability that the household stops searching before it finds an acceptable alternative. If we now make the logical assumption that ${}_k\delta_{ij}$ is proportional to ${}_k\omega_{ij}$, the number of type k households moving from class i to class j, $j\varepsilon S$, consists of those whose first preference was j plus those who moved into j only after failing to find a dwelling in D, such that

$$\underset{j\varepsilon S}{M_{ij}(\hat{t})} = \omega_{ij}\, P_i(t) + \delta_{ij}[N_{iD}(\hat{t}) - M_{iD}(\hat{t})] \quad , \tag{14}$$

and

$$\delta_{ij} = \omega_{ij}/W_{is} \quad ,$$

where

$$N_{iD}(\hat{t}) = \sum_j N_{ij}(\hat{t}),\ j\varepsilon D \quad ,$$

$$M_{iD}(\hat{t}) = \sum_j M_{ij}(\hat{t}),\ i\varepsilon D \quad ,$$

and

$$W_{is} = \sum_k \omega_{ik},\ k\varepsilon S$$

(the household subscript is implicit).

If we expand the second term on the RHS of (14),

$$\delta_{ij}[N_{iD}(\hat{t}) - M_{iD}(\hat{t})] = \delta_{ij}[P_i(t)-M_{iD}(\hat{t})] - \delta_{ij}[P_1(t)-N_{iD}(\hat{t})] \ . \tag{15}$$

Upon expanding the second term on the RHS of (15) we have

$$\delta_{ij}[P_i(t) - N_{iD}(\hat{t})] = \delta_{ij}[P_i(t) - (1-W_{is})P_i(t)]$$

$$= \delta_{ij}\, W_{is}\, P_i(t) \quad . \tag{16}$$

$$= \omega_{ij}\, P_i(t) \quad .$$

Substituting (16) back into (15) and then into (14) we have

$$\underset{j\varepsilon S}{M_{ij}(\hat{t})} = [\omega_{ij}/W_{is}][P_i(\hat{t}) - M_{iD}(\hat{t})] \quad . \tag{17}$$

The parameter estimation is done in two stages: surplus demand area parameters are estimated first, and surplus demand parameters afterward. Both stages presume that we have information on the number of type k households leaving class i dwellings and entering class j dwellings during an interval of time (0,1), as represented by a 3-dimensional mobility matrix $\underline{M}^*$ (the household subscript is implicit in the following discussion).

From (13), it follows that

$$\omega_{ij}/\omega_{\ell j} = [P(0)/P_i(0)]\ (M^*_{ij}/M^*_{\ell j}),\quad j\varepsilon D\ ,$$

which implies that

$$M_{ij}(\hat{t}) = V_j(\hat{t})P_i(t)[M^*_{ij}/P_i(0)]/\sum_k [M^*_{kj}/P_k(0)]P_k(t)\quad . \tag{18}$$

On the surplus supply side it follows from (17) that

$$\omega_{ij}/W_{is} = [P_i(0) - M^*_{iD}]/M^*_{ij},\quad j\varepsilon s\ ,$$

which implies that

$$M_{ij}(\hat{t}) = M^*_{ij}[P_i(t) - M_{iD}(\hat{t})]/[P_i(0) - M^*_{iD}],\ j\varepsilon s\ , \tag{19}$$

where $M_{iD}(\hat{t})$ is determined from (18).

A housing class moves from a surplus supply class to a surplus demand class when

$$\sum_i \omega_{ij}\ P_i(t) > V_j(\hat{t})\quad , \tag{20}$$

and back into a surplus supply class when the inequality in (20) is reversed.

In order to convert the household mobility models discussed in the two previous sections into population projection methods, we must resolve two important problems. Existing households, like individuals, give birth to new households, cease to exist, and change internally. We must establish cohort specific "birth rates," "death rates," and inter-cohort transition rates, where the cohorts in this instance are household types. The second problem is that we would like age specific population projections whereas the present model deals only with households. The transformation from households to age specific population requires that we have estimates of the probability, ${}_kq_{\ell j}$, that an age cohort k person is a member of a type ℓ household residing in dwelling class j, such that the number of age cohort k people living in class j dwellings is ${}_kQ_j(t)$, and ${}_kQ_j(t) = {}_kq_{\ell j}\ {}_\ell P_j(t)$.

CONCLUDING REMARKS

The population projection methods discussed in this paper are components methods that explicitly incorporate between area interdependencies as integral to the migration component of the projection method. The methods are differentiated on the basis of the assumed invariance in the migration process, and the time dependent variables that influence the future population size and age composition of each local area.

The parameters in any projection method reflect either implicit or explicit assumptions concerning the invariances in the population change process. From a practitioner's point of view the key to the selection of one method over another is often found in an analysis of the parameters imbedded in the method, since parameters must be estimated and estimation requires data. If the prospective user does not have access to the type of data needed to estimate a given parameter, then either another method must be found or the parameter may be estimated indirectly using available surrogates for the missing data.

On the theoretical side, we would like to have projection techniques that are consistent with existing theory. An analysis of the parameters in existing projection methods can lead to statements concerning the necessary conditions for parameter stability. These conditions will take the form of assumptions about the population change process, and then can be checked for internal consistency and can be compared against existing theory.

The question of which time dependent variables to incoporate in the projection method is potentially the most interesting question from a dynamic systems point of view. Changes in the population size and composition of an area obviously are entwined and enmeshed with a large number of other dynamic processes going on simultaneously at the local, regional, national and international level. If we attempt to model these complexities and interdependencies we soon find ourselves in the realm of the large urban and regional simulation models. The logic behind this tendency toward increasing complexity is compelling from a theoretical point of view, but the farther we go down this road the farther we get from anything that could be used as a practical population projection method, given the current state of the art in urban and regional modelling.

In practice we want a projection model that requires a minimum of information on other population related variables. If the future values of these variables are exogenously specified (e.g., changes in basic employment at some future time), then the accuracy of future population estimates will necessarily depend upon the accuracy of the exogenous estimates. If the values of the population related variables are endogenous to the model, then any misspecification of the functional relationship between population and other endogenous variables may tend to concatenate population projection errors. On the other hand, projection models that include no policy relevant control variables are passive planning tools that provide little insight into the future consequences of prospective planning decisions, and hence may indirectly affect the size and composition of local area populations.

The projection methods discussed in this paper represent an attempt to find a middle ground between nonoperational complex systems models and extrapolation methods currently used by local planners. The methods are designed to provide insight into the dynamics of population change at the local level while ensuring that the methods can be applied within a local planning context.

REFERENCES

Bacharach, M., 1970. Biproportional Matrices and Input-Output Change, Cambridge, University Press.

Barber, G.M., 1977. Urban Population Distribution Planning, Annals, Association of American Geographers 67: 239-245.

Ginsberg, R.B., 1971. Semi-Markov Processes and Mobility, Journal of Mathematical Sociology 1: 233-262.

Greenberg, M., Krueckeberg, D. and Michaelson, C., 1978. Local Population and Employment Projection Techniques, Center for Urban Policy Research, New Brunswick.

Huff, J., 1979. Residential Mobility Patterns and Population Redistribution within the City, Geographical Analysis 11: 133-148.

Huff, J. and Clark, W.A.V., 1978. The Role of Stationarity in Markov and Opportunity Models of Intra-Urban Migration, in: Population Mobility and Residential Change, W.A.V. Clark and E.G. Moore (eds.) Northwestern University, Evanston.

Isserman, A., 1977. The Accuracy of Population Projections for Sub-County Areas, Journal of American Institute of Planners 43: 247-259.

Ledent, J., 1978. Regional Multiplier Analysis: A Demometric Approach, Environment and Planning A 10: 537-560.

Lowry, I.S., 1964. A Model of Metropolis. The Rand Corporation, Santa Monica.

Morrison, P., 1971. Chronic Movers and the Future Redistribution of Population, Demography 8: 171-184.

Pittenger, D., 1976. Projecting State and Local Populations, Ballinger, Cambridge, Mass.

Rees, P.H. and Wilson, A.G., 1977. Spatial Population Analysis, Wiley, New York.

Rogers, A., 1973. The Mathematics of Multiregional Demographic Growth, Environment and Planning 5: 3-29.

White, H., 1971. Multipliers, Vacancy Chains, and Filtering in Housing, Journal of American Institute of Planners, 37: 88-94.

Wilson, A., 1974. Urban and Regional Models in Geography and Planning, Wiley, London.

ON MODELLING SEARCH BEHAVIOR*

W.A.V. Clark

Department of Geography
U.C.L.A., Los Angeles

INTRODUCTION

In the past decade there has been a significant increase in the research literature dealing with search behavior. Research on job search, shopping search, and search in the housing market has become an important component of the study of social processes in economics, geography, and psychology. The initial interest in simple translations of job search models to the processes of search in the housing market has been replaced by attempts to develop models specifically designed to deal with the particular temporal and spatial dimensions of the processes of housing choice and selection. Some of this interest has been stimulated by advances in the modelling of job search, but many of the housing search investigations have been equally stimulated by the data resources generated from programs formulated by the Department of Housing and Urban Development (HUD) to stimulate home ownership, to provide access to the housing market for low income populations, and to remove barriers to the housing market, particularly for minority households. Large data bases such as the Annual Housing Survey and the Experimental Housing Allowance programs (EHAP) provided resources and a context in which both policy questions could be posed and analytic models of mobility and search could be developed.

The result of this sequence of events has been both increased analytic sophistication and greater data richness. Although the data are rich in socio-demographic characteristics, economic information and cross-sectional data on mobility, they are still lacking in sequential information on mobility and search. In many cases as well, the empirical models of search are often reformulations of models of the probability of moving and reflect only simple analyses of the categorical decision to search or not. They are not models of the search process, of search effort and search strategy. Despite these caveats, the research of the past half dozen years by geographers and economists has stimulated a considerable analytic literature on consumer search in the housing market. These disequilibrium models of search have added a formal dimension to earlier descriptive investigations.

The aims of this paper are to review the disequilibrium models of search, to present some additional empirical results of search in a large housing market (but within the strucutre of dimensions of search effort and search strategy), and to speculate on a translation of the cumulative inertia-cumulative stress model of mobility specifically for search in the housing market.

THE ANTECEDENTS OF HOUSING SEARCH MODELS

The genesis of the present focus on jobs and residential search lies in the strategic questions stimulated by the Second World War and which were focused on the location of targets and other military operations [Enslow (1966); Silk (1971)]. However, it was not until some years after the war that these geometric concerns (the location of points) were replaced with an interest in the *behaviors* associated with search and choice. While searching physical space for some tangible object is still of interest in computer science, the main focus in search investigations is on questions of individual decision making and with the role of information in the decision making process. The processes by which choices are made, including the process of search prior to choice, are of increasing interest to a wide range of social scientists. In psychology, research has been concerned with a wide range of issues related to individual decision making, and in economics the study of behavior under conditions of incomplete knowledge or information has been important. Attempts to formalize search processes, particularly with respect to jobs by the economists, and the formalization of judgment processes in psychology, have set the scene for much of the recent work on search in the housing market. The housing market search literature is in its infancy, and is still dependent upon both the original economic formalizations and the emphasis on behavioral processes by psychologists. Excellent reviews of the economic approach to search by Flowerdew (1976), of the role of information in migration by Goodman (1980), and of decision making by Slovic and Lichtenstein (1971) and Slovic, Fischoff, and Lichtenstein (1977) obviate the need for a detailed discussion of these antecedents to the present modelling strategies.

Stigler (1962) originally suggested that search might be modelled as a process of sampling from a distribution of vacancies, and in economics most of the literature has been focused on job vacancies and job search. McCall (1970) expanded the notion of sampling from a distribution of vacancies to allow for sequential sampling, and thus sequential search. These original papers have been expanded in a variety of ways, including the introduction of imperfect information and other questions of uncertainty with respect to optimal search procedures. The description of search behavior as sequential decision making under uncertainty has been emphasized by Kohn and Shavell (1974), Rothschild (1974), and Ioannides (1975). In these studies the notion that consumers learn from the probability distribution of prices while they search from it was introduced, and hence learning is emphasized as well as decision making as a part of the search process. Once the process of learning is introduced, it is possible to discuss the role of prior information and the updating of prior beliefs as search occurs. If the searcher is informed about the distribution of interest, say the distribution of housing vacancies, he can search optimally. With ignorance, he must postpone his decisions and gather information.

Even though economists have contributed a great deal in terms of the formalization of search models and have constructed some relatively elaborate structures of the search process, almost all of the work has been conducted in terms of the utility maximization framework, and has focused on general statements with respect to such characterizations as aggregate unemployment. They have not generally focused on the behavioral components of search in which there are specific behavioral rules which market participants are presumed to follow.

In psychology and in consumer research in management, a major focus has been on the role of information as it influences the process of decision making. Indeed, most of the questions in psychology are not concerned directly with questions of search and search processes but rather are focused on questions of choice. Even though choice implies an evaluation process which includes the process of search, there has not been a specific concern with search in the psychology literature. Two broad reviews by Slovic and Lichtenstein (1971) and Slovic, Fischoff, and Lichtenstein (1977) structure the analysis of information and information use by whether or not it is organized under (1) a regression approach or an analysis of variance methodology, or (2) a Bayesian approach. The latter approach emphasizes the role of subjective or personal probabilities and the revision of opinion in the light of new information. Almost all of the psychological studies are focused on the way in which information modifies the beliefs of individuals or households and causes changes in their actions. The way in which information is acquired and processed is critical in the final decision.

The emphasis on information acquisition and information processing has become especially important in the management and marketing fields. While the focus in psychology has been on the decision and composition rules with respect to the ways in which individuals put information together, and the relative weighting that individuals give to the stimulus parameters (i.e., to the variables that are involved in information development), the marketing literature has been more empirically focused on specific questions of experimental interest with respect to brand-choice decision making. Marketing research has emphasized the amount and methods of information acquisition, the effects of information retention on decision strategies and the relationship of memory and stored information to the type of decision process (Bettman, 1976). Researchers contributing to the market literature have made considerable progress by the use of simple experimental designs in the analysis of information processing, but as yet the information analysis in experimental marketing situations has remained an end in itself and has not been incorporated within a larger theoretical context.

Although at the present time the economic literature seems to have had a greater influence on the development of residential search models, the importance of information is increasingly recognized in both economic and geographic models, and investigations which are currently ongoing have attempted to evaluate the impact of information on residential decision making [Clark and Smith (1979)].

MODELLING HOUSING SEARCH

Only recently has there been a concerted effort to investigate the search behavior of households who relocate within cities. Earlier investigations of search behavior, including those by Hempel (1970), Brown and Holmes (1971) and Barrett (1973), were largely descriptive investigations of the results of search. Articles which have been published more recently have been concerned with developing both theoretical structures and empirical analyses of the search behavior of mobile households. Although the analytic literature on search in the housing market is recent, there are quite distinct approaches and themes within it. The substantive research can be organized into five broad categories.

First, there are a number of studies of a disequilibrium model of housing adjustment. The initial disequilibrium model (Goodman, 1976) has been elaborated by Hanushek and Quigley (1978a,1972b) and tested by Weinberg, et al. (1979) and Cronin (1979a). This model is basically concerned with the decision to move, and only in the broadest sense are these studies concerned with search. The focus is specifically on the benefits of moving. A second group of articles has applied the basic disequilibrium framework to the decision to search [Weinberg, et al. (1977); and Cronin (1979b)]. A third focus is directed to specific aspects of spatial search. One approach utilizes a disequilibrium approach [Smith, et al., 1979)] and another is concerned with specific spatial models of search [Huff (1980)]. A fourth approach to search in the housing market includes the analysis of constraints and the role of institutions (usually real estate brokers), and is represented by studies by Palm (1976), Yinger (1978), Courant (1978), and Smith and Mertz (1980). Finally, some recent studies have attempted to analyze the specific role of information in search behavior [Clark and Smith (1979); MacLennan (1979); Smith and Clark (1980)]. The discussion in this paper will be limited to the disequilibrium versions of the search model.

Basic Disequilibrium Models

The disequilibrium model and its variations were developed as alternatives to the more traditional sociological and geographic models of residential mobility. The proponents of the disequilibrium approach emphasize the link between mobility and economic decision making based on the benefits and costs of moving. The models have three characteristics in common. They apply to individual households, the household is assumed to make a decision on mobility based on the economic benefits and costs of moving, and a logit or a probit model is used to link the measures of benefits and costs to the rate of moving. However, empirical analysis indicates that these economic models generally do not have good predictive power. Even where they are significant, variations in the economic factor do not lead to large changes in the rates of mobility.

In an initial specification of the disequilibrium model, Goodman (1976) expressed the probability that a household will move at least once during a given time period as a linear function of the benefits and costs of moving. Housing consumption is expressed as a vector of four factors: expenditure (E), size (S), quality (L) and workplace accessibility (A). The deviations of current consumption of these factors from those of optimal consumption (E*,S*,A*,L*) are assumed to be the potential benefits of moving. Following traditional economic theory, the optimal consumption of housing services, hence also expenditure, is derived from maximizing household utility subject to a budget constraint, and is computed from an estimated demand function for housing. The costs of moving include the transaction cost and the psychological cost of leaving the present unit and neighborhood. Out-of-pocket moving cost is assumed constant for all households and hence is ignored in the analysis. The transaction and psychological costs are proxied by a vector $\underline{Z}$ which includes measures of tenure, year of last move, age of household head, and household life cycle stage.

The probability that a household will move at least once during the study period 1969-1971 is then expressed as

$$P(M) = f(E^*/E, S-S^*, A-A^*, L-L^*, \underline{Z}) \quad , \tag{1}$$

where

$$P(M) = \begin{cases} 1 & \text{if a household moves at least once between 1969 and 1971,} \\ 0 & \text{otherwise,} \end{cases}$$

and f is linear. All independent variables are transformed into categorical variables with arbitrarily defined intervals, and the dependent variable is estimated by multiple classification analysis.

In the Hanushek and Quigley (1978a,1978b) formulation household mobility (probability of move) is assumed to be a probit function of the difference between actual housing consumption at time t and equilibrium consumption demanded at time t+1, divided by equilibrium consumption demanded at time t:

$$M_t = f\left[\frac{H^d_{t+1} - H_t}{H^d_t}\right] \quad , \tag{2}$$

where

M_t = probability of a move between t and t+1,

H_t = actual consumption of housing services, measured by monthly rent, adjusted for contract terms and length of tenure, and

H^d_t = equilibrium housing consumption demanded at time t.

All transaction costs (search and moving costs) are assumed to be randomly distributed, independent of the demand disequilibrium, and are therefore ignored in the model. Housing demand is assumed to be a linear function of a household's socio-economic characteristics, including household income, household size, and race and age of the household head.

The basic model is modified in order to separate the effects of (1) change in equilibrium housing demand between t and t+1, and (2) current disequilibrium in consumption:

$$M_t = f' \left[\frac{H^d_{t+1} - H^d_t}{H^d_t} + \frac{H^d_t - H_t}{H^d_t} \right] , \tag{3}$$

or, defining new variables for change in equilibrium demand, D_E, and current disequilibrium, D_o,

$$M_t = f'(D_E, D_o) \quad . \tag{4}$$

The model is also further disaggregated by separating the contributions of positive disequilibria from negative disequilibria; i.e.,

$$M_t = f''(D^+_E, D^-_E, D^+_o, D^-_o) \quad , \tag{5}$$

where f, f' and f" are all assumed to be probit functions.

Extensive tests of the basic disequilibrium model were undertaken by Hanushek and Quigley (1978b), Weinberg, et al. (1977), Weinberg, Friedman and Mayo (1979) and Cronin (1979a). The data used in these tests were derived from the Housing Allowance Demand Experiment (HADE) for either or both of Pittsburgh and Phoenix, and for varying time intervals depending on the date of the study. As Hanushek and Quigley (1978b, p. 90) note the results "provide modest support for the underlying models". But in Weinberg, Friedman and Mayo (1979), which is designed to specifically test for the effect of a rent rebate on a household's residential mobility, the model fits poorly with coefficients of determination varying between .05 and .09. Overall, the impact of housing allowances on residential mobility is very small.

Cronin (1979a) investigates the basic disequilibrium model as defined by Hanushek and Quigley, but with a different approach to estimating the equilibrium demand function. In this alternative approach to the benefits of moving, he uses income equivalent variation (IEV) in lieu of a simple difference between current non-optimal level of expenditure and optimal level of housing expenditure. Income equivalent variation is the amount of additional income necessary to make the household as well off with its current consumption of housing as it would be with its equilibrium consumption of housing. The models are quite similar in their predictive power to the original Hanushek and Quigley models.

Disequilibrium Approaches to Search

All of the models which were estimated for the probability of moving have been adapted to the probability of search. In Hanushek and Quigley (1978b) the models are re-estimated for the trichotomous case--move, search without moving, no search/move--but although there is some evidence that the disequilibrium motivates search, the simple nature of the models precludes any real understanding of search behavior. Weinberg (1977) and Cronin (1979b) also adapt their disequilibrium models to the specific problem of search. Weinberg estimates a logit model of probability of search. Variables included in the model are primarily socio-economic characteristics of the household, including previous mobility, race, response to questionnaires on housing unit and neighborhood satisfaction, and a dummy variable separating experimental from control households. The variables with the most significant coefficients include: previous mobility (number of moves within previous three years), age of household head, dissatisfaction with housing unit and dissatisfaction with neighborhood. The coefficient of determination is 0.154 for Pittsburgh and 0.165 for Phoenix. The subjective statement of housing unit dissatisfaction is compared with two measures of housing inadequacy: overcrowding and physical inadequacy. The data indicate that not all who are living in inadequate housing express dissatisfaction, and not all dissatisfied households decide to search (approximately 20% of dissatisfied households do not search). The minimum housing requirements established for the recipients of the Housing Gap allowance appear to discourage some households from searching, presumably because of the difficulty involved in locating a unit which meets the standards. Among families receiving Percent of Rent allowance, there is sometimes a misunderstanding that moving may jeopardize the allowance.

The Cronin analysis of search examines both the usefulness of an economic measure of the benefits of moving in explaining the rate and characteristics of search, and the effect of race on search. The probability that a household will undertake residential search within a twelve-month period is assumed to be a logit function of benefits, costs, psychological factors (including expressed dissatisfaction with the existing housing unit, and social bond to the neighborhood), past mobility, and race.

The model used is a modified version of the Stone-Geary, IEV model developed in Cronin (1979a). While both economic and race variables significantly affect search behavior, the mean values of some of the variables related to search are strikingly different for minority and non-minority households. Average IEV for minority is $524/yr., whereas it is $156/yr. for non-minority. The average number of search days is over twice as many for minority as non-minority. The mean minority household size is 4, while the mean non-minority household size is 3. On the average, minority households appear to have greater potential benefits from moving, but also to face greater costs of search, than the non-minority households. The overall probability of move does not differ very much between the races.

Minority and non-minority households, however, follow significantly different strategies during search. Significantly more non-minority households use newspapers than minority households, and minority households tend to use vacancy signs more often than the non-minority. Significantly fewer minority households use their own automobile for search, relying more on taxis and public transportation than do the non-minority households. Cronin examines a number of indicators of search effort, such as number of days searched, number of neighborhoods searched, and number of dwelling units searched. He estimates separate multiple regressions using the indicators of search effort as dependent variables and economic variables and household characteristics as independent variables. The results indicate that IEV affects search behavior in an expected way. The greater the IEV, the shorter the search period and the more intense the search. Cronin also finds that race differences persist even after other variables are controlled. The data seem to support Courant's (1978) observation that minority households may be faced with higher search costs than non-minority households.

A Spatial Disequilibrium Model

In contrast with the disequilibrium models, the locational stress model explicitly incorporates space as a component of the search process. Within a modified disequilibrium framework the model addresses both where the individual will look and when he will stop looking. The key concept, "locational stress," is defined as the difference between the expected utility of search (for a house) in a neighborhood and the utility which could be obtained by moving to the current best alternative (house) among all neighborhoods. In order to compute the locational stress, the model assumes a household utility function, conditioned on the consumption of housing services at a given price, and a subjective probability distribution of housing prices and characteristics available to a neighborhood.

The household is assumed to follow a fixed algorithm in searching through the market. The household computes the locational stress for each neighborhood in the city and begins search in the neighborhood with the highest stress. The household locates in the target neighborhood a candidate unit which yields a higher level of utility than the previous best alternative (including the currently occupied unit), and the candidate unit becomes the new best alternative. The procedure iterates until locational stress is less than or equal to zero for all neighborhoods.

Formally, a household may have a present housing situation H_0 (giving rise to utility V^o), and the household may have a housing option $H_B = (X_B, M_B)$ giving rise to a utility $V^B > V^0$ (where X is the characteristics of the home and M the associated costs). Let $E_t^i(V^B)$ be the expected utility of search at time t in region i given that the best option currently available is H_B. With the assumption that the household attempts to maximize its expected utility, the decision as to whether a household searches or not at any period and in any region is determined by comparing the expected utility of search with the best utility which is available with certainty (V^B):

$$\Psi^i = E_t^i(V^B) - V^B , \qquad (6)$$

(where V^B may equal V^0). If $\Psi^i > 0$ for any i, search will occur [Smith, et al. (1979)].

A simplified version of the expected utility of further search is given by:

$$\begin{aligned} E_t^i(V^B) = {} & \rho\left(V^0 \int_{-\infty}^{V^0} dG(\nu) + \int_{V^0}^{\infty} \hat{\nu}\, dG(\nu)\right) \\ & + (1-\rho)\left(V^B \int_{-\infty}^{V^B} dG(\nu) + \int_{V^B}^{\infty} \hat{\nu}\, dG(\nu)\right) \end{aligned} , \qquad (7)$$

where $\hat{\nu}$ is the utility when transactions costs are incurred and $G(\nu)$ is the probability distribution function representing the household's beliefs concerning utility levels [Smith, et al. (1979), p. 13].

It is clear that the "locational stress" and the disequilibrium models share a common assumption that households search and move in order to maximize their utility. However, the two models are directed toward different aspects of residential search. The locational stress model specifically incorporates the spatial component of search, and to that extent is a model of search strategy. In contrast, the disequilibrium model is a model of housing demand and, by implication, of household mobility. As a result, the two models differ significantly in a number of areas.

First, the disequilibrium model implicitly assumes that all households have perfect knowledge of the housing market, while the locational stress model assumes that households have knowledge only of the distribution of prices and characteristics in each neighborhood. Second, the disequilibrium model typically assumes that housing price and characteristics are perfectly correlated, while the locational stress model assumes only a joint distribution of prices and characteristics. Third, using traditional economic analysis, the disequilibrium model derives a demand for housing as a function of income, prices, and household characteristics. In the locational stress model, the demand for housing depends on income, prices, and the supply distribution. A household is assumed to maximize a conditional utility defined by the consumption of non-housing goods, given a level of housing expenditure for some set of housing characteristics. The household makes a decision to undertake search based not on deviation from some optimum level of consumption, but on the distribution of expected utilities obtainable from the available supply. Thus, if the probability of locating an appropriate unit is small, the household may not undertake search, even when the level of current consumption is not optimum. The model assumes that the housing market is not perfectly competitive and that the price of housing services may not reflect the adjustment of demand and supply. Consequently, the model does not contain a clear definition of household demand for housing services. Finally, the disequilibrium model assumes fixed search and moving costs for all households of given characteristics, transforming the search and moving costs into a transaction cost. The locational stress model includes a variable search cost, which acts as one of the constraints on search.

EMPIRICAL CONTRIBUTIONS TO UNDERSTANDING HOUSING SEARCH

The models outlined in the previous section represent the development and tests of two distinct approaches to search in the housing market. Paralleling these theoretical analyses have been several largely empirical investigations of home buyer behavior. In particular, studies by Hempel (1970) and Barrett (1973) emphasize behavioral processes in the housing market at the micro level (individual participants), including buyer preferences, information collection, and searching behavior. Hempel (1970) organizes the buying process into five behavioral components, which include: (1) problem recognition (the household disequilibrium), (2) search behavior, (3) choice behavior, (4) purchase behavior and (5) post-purchase behavior. In this breakdown, disequilibrium, active search, and choice behavior are separately identified as components of the composite activity of home buying. This contrasts with the models which have considered disequilibrium and search as coincident.

The analyses by both Hempel and Barrett are primarily descriptive and tabular. There is a great deal of information on the components of search and search behavior. This material is reviewed briefly as a basis for later empirical tests in this paper. Hempel points out that home buying involves a large expenditure, is infrequent (so that it is difficult to generalize about the processes), and that there are little data generally available on house purchase. Commonly there is also a great deal of secrecy involved in the transactions. The process also involves information seeking to solve a particular problem. For the search process itself Hempel considers the length of time expended in search, the amounts and kinds of information collected, and the spatial or geographical range over which search takes place. He presents detailed information on the expenditures involved, the information collected, and the amount of time search. He notes that it is common to use more than one agent, that older households search longer, as do larger families, while higher income families search for shorter times. Households with lower education levels also search longer. Unfortunately, there is very little spatially specific data collected in the Hempel study. The data on information sources have been reported and evaluated elsewhere, but they emphasize newspapers, real estate brokers, walking or driving around, friends and relatives, and for sale signs [Clark and Smith (1979)].

Important from a modelling perspective is the suggestion that the home buying process seems not to be related to a unique market but is generalizable from market to market. But Hempel (1970, p. 156) does indicate that at least some of the variability in local housing markets may be related to information flows.

In a more recent paper, Hempel (1977) develops a model of the "transition period influences" (life cycle changes) on the home buying processes and related consumption systems. This is a somewhat elaborate attempt to analyze home purchasing behavior in a conceptual context. Utilizing correlations between the duration of search activity and a variety of life-space changes and residential mobility measurements, he is able to indicate some significant but not particularly strong relationships between family size and duration of search activity (negative), family income and duration of search activity (negative), and length of time in prior residence and search activity (negative).

Barrett (1973) also provides a set of quantitative observations on the search process. He analyzes information sources used, the length of search, the number of houses searched, and the areas of the city searched by the respondents. He tests a number of null hypotheses: (1) there is no relationship between length of search and number of houses searched (rejected); (2) there is no relationship between the length of time and the area of search (rejected); (3) there is no relationship between the price paid for the house and the length of the search (accepted); (4) there is no relationship between the price paid and the size of the search area (accepted); and, (5) real estate agents are not the major information channel used by home searchers (rejected). Using an index of search behavior, he concludes that few houses were considered seriously and most respondents have a relatively small search cluster. The conclusions of his study were to emphasize the shortness of search and the relatively limited number of houses seen in a small area. He concludes that search behavior is not a thorough process and that behavior (in search for a house) is not consistent with the obligations of the decision.

DIMENSIONS OF SEARCH

Two problems characterize the housing search literature and attempts to model search behavior at this point. First, the theoretical developments and their translation into simple models of housing search have not been related to the empirical survey evidence generated by studies of realtors and house buying behavior. And, second, the models of search behavior have been largely extensions of simple logit models of the decision to search rather than of the search process. The dimensional structures outlined in this section of the paper do not completely overcome this problem. However, the organization of dimensions of search allows multivariate tests of empirical data, a canonical analysis of the dimensions of search and the development of inputs for a dynamic, albeit speculative, model of search strategy.

In considering the elements involved in search we can suggest, on the one hand, that there are a set of (a) household characteristics, and (b) housing attributes relating both to the previous unit and the desired unit. On the other hand, there are measures of (c) the kinds of search strategies, and (d) the search efforts that households will utilize in achieving a new

TABLE 1

Dimensions of Housing Search

Household Characteristics

Age
Household size
Education
Income
Length of residence in Los Angeles
Out of town buyer
Previous tenure

Search Effort

Number of houses searched
Number of areas searched
Looked outside San Fernando Valley
Length of search
Spatial extent of search

Housing Attributes (current)

Size of current unit
Construction quality (rating)
Neighborhood (rating)
Price of current unit
Tract unit (yes - no)

Search Strategy

Number of days to bid
Number of realtors used
Used newspapers a lot
Used realtors a lot
Drove around a lot
Visited open houses a lot

location. Characteristics such as holdehold size, age, and income are standard household characteristics and have been utilized in the disequilibrium approaches to search by economists. Desired housing attributes also influence search effort and search strategy. Measures of current (purchased) housing such as price, whether the unit is in a new tract and desired neighborhood criteria are also likely to influence search intensity.

Search itself involves both effort and strategy. Search effort is involved with the temporal and spatial dimension of realizing a vacancy for purchase. Search effort is reflected in measures of the number of houses that are searched, the number of areas that are searched, and the length of the search. Whether the total spatial extent of the search is an element of effort or strategy is not yet clear. For the present, it is incorporated as an element of search effort. Strategy is clearly related to informational sources and risk incentives. The number of days to bid is a measure of the level of risk that a household is willing to take in its search process. The other variables are measures of the usefulness of information and whether or not the household relies on realtors, newspapers, or more individual activities.

The result of the interaction of search effort and search strategy is the new housing unit. One of the most difficult questions in analyzing search is the outcome of search. One measure of this might be the satisfaction with the current unit. It can be assessed for its satisfaction and re-related to initial household characteristics and housing attributes. An alternate approach to analyzing the outcomes of search is to focus on measures of success of search. A recent paper by McCarthy (1979) discusses this issue (for renters) with respect to rent discounts and rent premiums. In any event, the question of outcomes or the results of search is as yet unspecified.

The advantage of this simple dimensional structure is to incorporate elements of the disequilibrium approach and to recognize the interconnections of search strategy and search effort in terms of risk, information, and an implied temporal process (even though it will be analyzed from a cross-sectional framework). This structure also suggests a number of simple questions which can be analyzed via multivariate means tests and canonical correlation analysis. At a later time, it may be possible to relate these simple tests to housing preferences and housing utilities.

Multivariate Tests of the Dimensions of Search

The data for the present study are part of a larger analysis of residential search and buying behavior in the Los Angeles area [Smith and Clark (1980)]. As part of that study 120 retrospective interviews were completed with recent home buyers. These questionnaires were used to collect data on information sources and search behavior patterns, including houses inspected, houses on which bids are offered, and houses purchased. To the extent that these are retrospective one time interviews, they do have the difficulty that there is some lapse of memory and not all houses which were searched will be included. To minimize this problem all questionnaires were completed within one month of purchase, and the questionnaires were structured to "think back" over the search process. These data can be used to compare the raw statistics in Hempel (1970) and Barrett (1973) and other studies which have collected descriptive observations on a variety of search characteristics, housing attributes, and household characteristics. The means and variations for the raw variables are reported in Table 2.

As a preliminary analysis of the dimensional structure and to test some elementary descriptive statements of the relationships between search, information use, and purchase behavior, univariate and multivariate analyses of means were used. Hypotheses on the length of search (greater than or less than one week, Table 3),

TABLE 2: VARIABLE STATISTICS BEFORE TRANSFORMATION

Variable	Mean	Variance
Household Characteristics		
Age of head	37.14	108.68
Household size	2.72	1.68
Education (years)	14.92	6.29
Income (00)	472.67	47745.14
Length of residence	16.18	181.42
Out of town buyer (yes = 1, no = 0)	.20	.16
Previous tenure (own = 1, Rent = 0)	.68	.22
Housing Attributes		
Size of current unit (square feet)	2087.29	509223.69
Construction quality of present unit (1-5)	3.51	.87
Neighborhood quality of present unit (1-5)	4.03	.61
Price of current unit (000)	144.91	5750.87
Tract unit (yes = 1)	.23	.16
Search Effort		
Number of houses	15.34	272.34
Number of areas	2.75	2.55
Out of main area (yes/no)	.44	.25
Length of search (weeks)	3.70	37.53
Area of search cluster (miles)	3.18	14.48
Search Strategy		
Days to bid	17.63	6173.24
Number of realtors used	1.36	1.19
Used newspapers a lot (yes = 1)	.26	.19
Used Realtors a lot (yes = 1)	.63	.23
Drove around a lot (yes = 1)	.43	.25
Visited open houses a lot (yes = 1)	.18	.15

TABLE 3: MULTIVARIATE ANALYSIS OF MEANS - LENGTH OF SEARCH LESS THAN/GREATER THAN ONE WEEK

Variable	t value	p value
Household Characteristics		
Age of head		
Household size	2.47	.015
Education (years)		
Income		
Length of residence		
Mahalanobis D square .49 .083		
Housing Attributes		
Size of current unit		
Construction quality of present unit		
Neighborhood quality of present unit	2.36	.020
Price of current unit		
Tract unit		
Owned previous unit		
Mahalanobis D square .50 .049		
Search Effort/Search Strategy		
Number of houses	-5.64	>.000
Number of areas	-3.23	.002
Out of main area		
Area of cluster search	-.397	>.000
Days to bid	-1.96	.053
Out of town		
Number of realtors		
Newpapers a lot		
Realtors a lot		
Drove around a lot		
Open houses a lot		
Mahalanobis D square 2.57 p>.000		

Only significant (.05) individual means listed.

the area of the search cluster (less than one mile or greater than one mile, Table 4), whether or not the buyer was from out of town (Table 5), whether or not individuals used realtors (Table 6), and whether the home buyer purchased a new tract-house or a non-tract house (Table 7) were tested. The variables to be tested are organized within the dimensions of household characteristics, housing attributes, search effort, and search strategy.

The tests of the five characteristics of search by the four groups of independent variables are reported with Mahalanobis's D^2 and individual t values. Only the significant *individual* means are listed. For the length of search, less than or greater than one week, it is quite clear that the major influence on short versus long search is closely related to the other elements of search effort--houses searched and the number of areas searched. Naturally, the longer the search, the more houses and the more areas visited. Information does not seem to be a major factor in generating a short or long period of search. In terms of the household characteristics and housing attributes, only household size is individually significant, and in this case for a larger household there is a tendency towards shorter periods of search (this is in contradistinction to Hempel). In terms of housing attributes, neighborhood quality seems to be a critical dimension.

Variables included as elements of search effort and search strategy are important in distinguishing between limited radius and more extensive search. Some specific aspects of information are also important. The use of realtors generates a much larger cluster, and using realtors a lot and a large number of realtors generates a larger search cluster. Older and larger households generate smaller search clusters.

In terms of the use or non-use of realtors all of the dimensions of the household characteristics or housing attributes appear to be important. Within search effort and search strategy the number of areas and the number of realtors seem to be critical in relationship to behavior of out of town buyers. A larger number of areas are searched if the household is from out of town, and (expectedly) there is a greater reliance on realtors.

In terms of the use of non-use of realtors all of the dimensions are significant in terms of their ability to discriminate between the use or non-use of realtors. Amongst household characteristics the length of residence is a critical dimension in the use of realtors--the longer the length of residence, the smaller the use of realtors. Within housing attributes the construction quality of the present unit appears to be higher with the use of realtors; but, at the same time, the price of the new unit is also higher with the use of realtors. With respect to search effort, the

TABLE 4: MULTIVARIATE ANALYSIS OF MEANS - AREA OF SEARCH CLUSTER LESS THAN 1 MILE/GREATER THAN 1 MILE IN DIAMETER

Variables	t value	p value
Household Characteristics/Housing Attributes		
Age of head	2.40	.018
Household size	1.93	.057
Education (years)		
Income		
Length of residence		
Mahalanobis D square .517 p = .060		
Size of current unit		
Construction quality of present unit		
Neighborhood quality of present unit		
Price of current unit		
Tract unit	3.81	>.000
Owned previous unit		
Mahalanobis D square 1.22 p = >.000		
Search Effort/Search Strategy		
Number of houses	-4.30	>.000
Number of areas	-3.36	.002
Out of main area		
Length of search		
Days to bid		
Out of town		
Number of realtors	-3.43	.001
Newspapers a lot		
Realtors a lot	-2.49	.014
Drove around a lot		
Open houses a lot		
Mahalanobis D square 1.34 p = .003		

Only significant (.05) individual means listed.

TABLE 5: MULTIVARIATE ANALYSIS OF MEANS - OUT OF TOWN OR NOT

Variables	t value	p value
Household Characteristics/Housing Attributes		
Age of head		
Household size		
Education (years)		
Income		
Length of residence (not included)		
Size of current unit		
Construction quality of present unit		
Neighborhood quality of present unit		
Price of current unit		
Tract unit		
Owned previous unit		
Mahalanobis D square .37 p = .278		
Search Effort/Search Strategy		
Number of houses		
Number of areas	-2.35	.020
Out of main area		
Length of search		
Days to bid		
Out of town		
Number of realtors	-2.30	.023
Newspapers a lot		
Realtors a lot		
Drove around a lot		
Open houses a lot		
Mahalanobis D square 1.358 p = .046		

Only significant (.05) individual means listed.

TABLE 6: MULTIVARIATE ANALYSIS OF MEANS - USED REALTORS/DID NOT USE REALTORS

Variables	t value	p value
Household Characteristics		
Age of head		
Household size		
Education (years)		
Income		
Length of residence	2.55	.012
Mahalanobis D square .84 p = .046		
Housing Attributes		
Size of current unit		
Construction quality of present unit	-2.47	.015
Neighborhood quality of present unit		
Price of current unit	-2.43	.017
Tract unit		
Owned previous unit		
Mahalanobis D square .81 p = .015		
Search Effort/Search Strategy		
Number of houses	-3.29	.001
Number of areas	-2.88	.005
Out of main area		
Length of search		
Area of search cluster	-4.06	>.000
Days to bid		
Out of town		
Newspapers a lot		
Realtors a lot	-4.77	>.000
Drove around a lot		
Open houses a lot		
Mahalanobis D square 3.631 p>.000		

Only significant (.05) individual means listed.

TABLE 7: MULTIVARIATE ANALYSIS OF MEANS - TRACT VERSUS NON-TRACT PURCHASE*

Variables	t value	p value
Household Characteristics		
Age of head		
Household size		
Education (years)	3.44	.001
Income	2.14	.035
Length of residence	-2.97	.004
Mahalanobis D square 1.31 p = .001		
Housing Attributes		
Size of current unit		
Construction quality of present unit	3.64	>.000
Neighborhood quality of present unit	-2.39	.019
Price of current unit	2.52	.013
Tract unit		
Owned previous unit		
Mahlanobis D square 1.86 p>.000		
Search Effort/Search Strategy		
Number of houses	2.99	.01
Number of areas	3.31	.001
Out of main area		
Length of search		
Area of search cluster	3.80	>.000
Days to bid	-2.81	.006
Out of town	2.40	.019
Number of realtors	6.20	>.000
Newspapers a lot		
Realtors a lot	7.01	>.000
Drove around a lot		
Open houses a lot		
Mahalanobis D square 5.51 p>.000		

*Only significant (.05) individual means listed.

number of houses and the number of areas are both larger with the use of realtors. The area of the search cluster is also significantly larger. All dimensions are significant discriminators between tract and non-tract housing purchases.

The tests confirm several of Hempel's and Barrett's observations and add several others. More importantly, the tests indicate the usefulness of a more general dimensional structure. Rather than an item-by-item analysis, we are able to show the general contribution of the groups of variables and the individual contribution of variables within those dimensions. The relationship amongst the dimensions themselves is explored in the next section.

Canonical Relationships

The relationships amongst the dimensions--between household characteristics and housing attributes on the one hand and search effort and search strategy on the other--are reported in Tables 8-11. The relationship between household characteristics and housing attributes as a set, and search effort and search strategy as a set, yields two dimensions with significant canonical correlations of .80 and .68. The first dimension reflects a relationship between non-tract housing purchase of smaller units by younger households who use realtors a great deal (and a multiple number of realtors) and who took a relatively short time to search. The second dimension reflects tract purchases of more expensive units by higher income, older households and by individuals who relied less on realtors in their search strategy.

When we analyze the relationship between household characteristics and housing attributes, and search effort and search strategy, individually, the relationships are somewhat weaker, with two dimensions in the first instance and one in the second. These dimensions are similar to those outlined for the full analysis of housing attributes/household characteristics and search. For search effort alone, the first dimension reflects the relationship between smaller households who buy less expensive units and of people who look at a large number of less expensive houses in a large number of areas. The second dimension reflects larger households paying more money for their unit and who look at a smaller number of units. Very few units are outside of the main area of search. The relationship between household characteristics and housing attributes, with search strategy alone, yields one dimension which reflects the focus on non-tract purchases by younger households who use realtors a great deal. The search strategy dimension clearly reflects the use of realtors.

TABLE 8: CANONICAL RELATIONSHIPS
HOUSEHOLD CHARACTERISTICS & HOUSING ATTRIBUTES
VS.
SEARCH EFFORT & SEARCH STRATEGY

Variables:	Canonical Variable Loadings 1	2
Household Characteristics/Housing Attributes		
Age of head	-.322	.206
Household size	-.176	.490
Education (years)	.409	.158
Income	.218	.400
Length of residence	-.407	-.016
Owned previous unit	-.398	-.191
Out of town buyer	-.001	.209
Size of current unit	-.000	.641
Construction quality of present unit	.301	.206
Neighborhood quality of present unit	-.354	.339
Price of current unit	.052	.701
Tract unit	-.919	-.243
Search Effort/Search Strategy		
Number of houses	.486	.318
Number of areas	.482	-.096
Out of main area	.079	-.001
Length of search	.176	-.373
Area of Search Cluster	.646	-.215
Days to bid	-.423	.059
Number of realtors	.756	.176
Used newspapers a lot	.159	-.105
Used realtors a lot	.692	.281
Drove around a lot	-.120	.172
Visited open houses a lot	.180	.085
Canonical Correlation	.802	.688
Bartlett's test	>.001	.026

TABLE 9: CANONICAL RELATIONSHIPS
HOUSEHOLD CHARACTERISTICS & HOUSING ATTRIBUTES
VS.
SEARCH EFFORT

Variables:	Canonical Variable Loadings 1	2
Household Characteristics/Housing Attributes		
Age of head	-.416	.154
Household size	-.277	.503
Education (years)	.211	.093
Income	-.023	.564
Length of residence	-.316	-.141
Owned previous unit	-.251	-.188
Out of town buyer	.191	.274
Size of current unit	-.003	.592
Construction quality of present unit	.211	.211
Neighborhood quality of present unit	-.365	.422
Price of current unit	-.103	.702
Tract unit	-.762	-.400
Search Effort		
Number of houses	.607	.416
Number of areas	.648	.081
Out of main area	-.092	-.003
Length of search	.285	-.567
Area of search cluster	.938	-.157
Canonical Correlation	.677	.544
Bartlett's test	>.001	.026

TABLE 10: CANONICAL COEFFICIENTS
HOUSEHOLD CHARACTERISTICS AND HOUSING ATTRIBUTES
VS.
SEARCH STRATEGY

Variables:	Canonical Coefficients
Household Characteristics/Housing Attributes	
Age of head	-.255
Household size	-.079
Education (years)	.431
Income	.293
Length of residence	-.429
Owned previous unit	-.394
Out of town buyer	.003
Size of current unit	.002
Construction quality of present unit	.293
Neighborhood quality of present unit	-.230
Price of current unit	.088
Tract unit	-.920
Search Strategy	
Days to bid	-.422
Number of Realtors	.753
Used newspapers a lot	.181
Used realtors a lot	.749
Drove around a lot	-.149
Visited open houses a lot	.144
Canonical Correlation	.771
Bartlett's test	>.001

TABLE 11: CANONICAL RELATIONSHIPS
SEARCH EFFORT AND SEARCH STRATEGY

Variables:	Canonical Variable Loadings 1	2
Search Effort		
Number of houses	-.404	.666
Number of areas	-.388	.466
Out of main area	-.039	.554
Length of search	.500	.779
Area of search cluster	-.503	.437
Search Strategy		
Days to bid	.795	.449
Number of Realtors	-.732	.646
Used newspapers a lot	-.020	.266
Used realtors a lot	-.276	.175
Drove around a lot	.024	.359
Visited open houses a lot	.027	.346
Canonical Correlation	.566	.461
Bartlett's test	>.001	.016

The most interesting of the relationships are those between search strategy and search effort. The analysis yields two dimensions which are moderately strongly associated. The first canonical variant can be described as a measure of low intensity search with relatively long periods of search where few houses or areas are examined. This is associated with a low utilization of realtors. Households tend to be from within Los Angeles and take a long time between inspection and bid. The second canonical variant we can describe as high intensity search. It describes relatively long periods of search that is accompanied by viewing large numbers of houses and areas as well as searching outside the main area of search. This dimension is associated with a high utilization of realtors.

The canonical analysis has confirmed the plausibility of the dimensions outlined and analyzed in this section. The study confirms the household age--size component of household characteristics, the length of residence component, and the usefulness of overall dimensions of search effort and search strategy, with particular reference to the role of realtors in the strategy dimension. Now the problem is to identify a model which allows for age-size and length of residence elements of the decision to search, and at the same time can generate processes of search consistent with "the intensity" dimensions. The preliminary observations in the next section are designed to accomplish this task--to develop a model which generates strategies consistent with some of the dimensions we have outlined.

A STRESS-INERTIA MODEL OF SEARCH

As we have noted earlier almost all the models of search are not dynamic or at best they have implicit temporal and spatial components. In the disequilibrium models [Hanushek and Quigley (1978a; 1978b)] search is the outcome of a difference between actual and optimal situations. In the locational stress model [Smith, <u>et al</u>. (1979)] search is also the outcome of a difference with recursiveness built into the model. But, while the model specifically addresses the question of when search will stop, it does not tell us anything about the process of search. The adaptation of the stress-inertia model of mobility to search yields cases of search behavior which incorporate the processes that are involved in search, and that are consistent with the dimensions that have been established in the empirical analysis.

In the initial formulation of the stress-inertia model, an individual household at time (t) has a resistance to moving $R(t) \geq 0$ and a residential stress (or dissatisfaction) $S(t) \geq 0$, which is a positive pressure to move [Huff and Clark (1979)]:

$$p(t) = \begin{cases} kS(t) - R(t) \text{ if } S(t) > R(t) \\ 0 \text{ if } S(t) \leq R(t) \end{cases} , \tag{8}$$

where

$$R(t) = R(t-1) + \hat{\rho}[\tilde{R}-R(t-1)] \quad 0< \hat{\rho} <1,\ 0<\tilde{R} , \tag{9}$$

and

$$S(t) = (1-\hat{\sigma})S(t-1)+\hat{s} \quad 0< \hat{\sigma} < 1 ,\ 0\leq\hat{s} ; \tag{10}$$

or in a continuous version:

$$R(t) = \tilde{R}-[\tilde{R}-R(0)] \quad \exp(-\rho t) \tag{11}$$

and

$$S(t) = \tilde{S}-[\tilde{S}-S(0)] \quad \exp(-\sigma t) \tag{12}$$

where

$\tilde{R}$ = maximum resistance and $\tilde{S}$ = maximum stress.

Then

$$p(t) = k\{(\tilde{S}-\tilde{R})-[\tilde{S}-S(0)] \exp(-\sigma t)+[\tilde{R}-R(0)] \exp(-\rho t) . \tag{13}$$

In the translation considered here it is assumed that searching occurs only when p(t) is greater than some threshold value such as expected search costs. Thus,

$$ps(t) = \begin{cases} kS(t)-R(t) \text{ if } S(t)-R(t)\geq CS(t) \\ 0 \text{ if } S(t)\leq R(t) \end{cases} \tag{14}$$

where

ps(t) is the probability of search at time t,

and

CS(t) are the expected search costs at time t, and

k is a constant of proportionality.

A simple diagram illustrates the relationships which are posited in these expressions. As in the probability of moving, we imagine a saw-tooth function which describes the changing probability of search over time. But, only when the probability of moving reaches some sufficiently high level to stimulate active search behavior does the searching process become imbedded in the probability of moving. The same process as applies to mobility is assumed to apply to the searching process (in miniature). Individuals go through recognizable periods of search in which the likelihood of search is quite high while at other times the likelihood of search is much lower.

Verbally, resistance is couched in terms similar to those of the stress inertia model where resistance to moving increases with increasing duration. In the search version of the stress inertia model the resistance in this instance is generated by the increasing costs of search, both in money and time. The function is assumed to increase at a decreasing rate approximated by the exponential decay function. The cumulative resistance elements can include the effects of changes in the finance market, changes in health, changes in perceptions (high prices), and a lack of acceptable vacancies.

In the original model stress is generated from stage in the life cycle. Dissatisfaction with the residence and an increase in stress are related to the effects of household changes which increase the probability of moving. In the present adaptation, ps(t) is generated by the same life events but their cumulative impact has reached a threshold level such that p(t) is translated into ps(t).

From the original model it was possible to generate 4 distinct interpretations of p(t) with changes in S(t) and R(t). Assuming that we can translate ps(t) to LS(t) and treat LS(t) as either a likelihood or an intensity of search, these four cases can yield useful interpretations in the process of search. Graphically the cases represent varying levels and changes in those levels of stress and resistance:

(1) $\rho > \sigma$ and $S(0) < \tilde{S} - \frac{\rho}{\sigma}[\tilde{R} - R(0)]$,

(2) $\sigma > \rho$ and $S(0) > \tilde{S} - \frac{\rho}{\sigma}[\tilde{R} - R(0)]$,

(3) $\sigma > \rho$ and $S(0) < \tilde{S} - \frac{\rho}{\sigma}[\tilde{R} - R(0)]$,

(4) $\rho > \sigma$ and $S(0) > \tilde{S} - \frac{\rho}{\sigma}[\tilde{R} - R(0)]$.

Although the functions were applied to disaggregated populations, there is no reason to limit their application to such groups. We

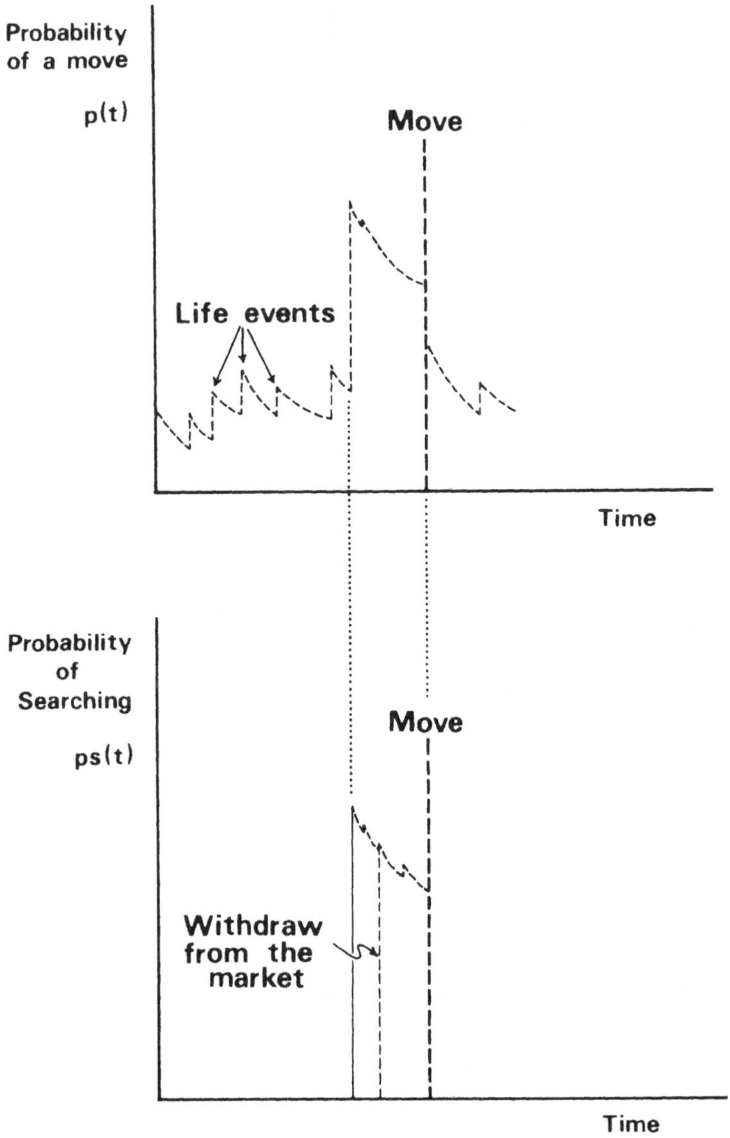

Fig. 1. A schematic illustration of hypothetical changes in the probability of moving and the likelihood of searching over time.

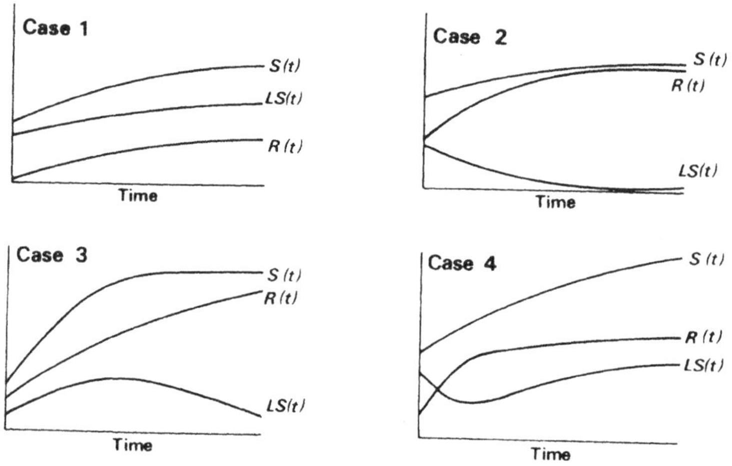

Fig. 2. The four general cases derived from Model (6) where $\frac{p(t)}{k}$ is translated to LS(t). (See Huff and Clark, 1978, p. 1110)

assume here that the functions apply equally well to individuals in the process of search.

In Case I search begins at moderate intensity and increases (moderately) at a decreasing rate. It could be interpreted as a relatively even level of search over time. Case 2, on the other hand, begins at a moderate intensity and decreases rapidly to a low intensity of search. In Case 3 search begins at a low level, increases rapidly, peaks and declines. Finally, in Case 4, search begins at a moderate to high intensity, decreases and then increased again. Certainly these patterns are consistent with qualitative descriptions of search behavior. There are obvious examples of individuals or households who begin searching enthusiastically and intensively and quickly lose interest or are forced to re-evaluate their chances of purchase. Other households search at varying levels of intensity with recurring peaks, while yet others maintain a relatively constant level of activity. Matching the behavior quantitatively will be more difficult.

The aim of this paper has been to review and interpret the search literature and models of search with the aim of providing a structure within which more dynamic modelling of search behavior will occur. It is quite clear that the speculations on a stress-inertia model of search will require considerable elaboration if they are to provide a dynamic model of search. In the meantime, they do emphasize the necessity of adding models of search which are representative of the processes of search as well as of the outcomes.

---

*This paper is based in part on a larger research project funded by National Science Foundation Grant SOC 77-27362. I would like to thank my colleagues in that endeavor, Jim Huff (University of Illinois), Terry Smith and Perry Shapiro (University of California, Santa Barbara) for their contributions to this paper and Jun Onaka (University of California, Los Angeles) for computer assistance and discussions of the joint interaction of residential mobility and search.

REFERENCES

Barrett, F., 1973. Residential Search Behavior. Geographical Monograph 1. Torondo, York University.

Bettman, J., 1979. An Information Processing Theory of Consumer Choice. Addison-Wesley, Reading, Mass.

Brown, L. and Holmes, J., 1971. Search Behavior in an Intra-Urban Migration Context: A Spatial Perspective. Environment and Planning 3: 307-326.

Clark, W. and Smith, T., 1979. Modelling Information Use in a Spatial Context. Annals of the Association of American Geographers 69: 575-588.

Courant, P., 1978. Racial Prejudice in a Search Model of the Urban Housing Market. Journal of Urban Economics 5: 329-345.

Cronin, F., 1979a. An Economic Analysis of Intra-Urban Search and Mobility Using Alternative Benefit Measures. Washington, D.C.: The Urban Institute.

Cronin, F., 1979b. Low-Income Households' Search for Housing: Preliminary Findings on Racial Differences. Washington, D.C.: The Urban Institute.

Enslow, P., 1966. A Bibliography of Search Theory and Reconnaissance Theory Literature. Naval Research Logistics Quarterly 13: 177-202.

Flowerdew, R., 1975. Search Strategies and Stopping Rules in Residential Mobility. Institute for British Geographers Transactions 1: 47-57.

Goodman, J., 1976. Housing Consumption Disequilibrium and Local Residential Mobility. Environment and Planning A 8: 855-874.

Goodman, J., 1980. Information, Uncertainty, and the Microeconomic Model of Migration Decision Making. The Urban Institute Working Paper 1384-2.

Hanushek, E. and Quigley, J., 1978a. An Explicit Model of Intra-Metropolitan Mobility. Land Economics 54(4): 411-429.
Hanushek, E. and Quigley, J., 1978b. Housing Market Disequilibrium and Residential Mobility. In W. Clark and E. Moore (eds.). Population Mobility and Residential Change. Evanston, Illinois: Northwestern University, Studies in Geography, 25.
Hempel, D., 1970. A Comparative Study of the Home Buying Process in Two Connecticut Markets. Storrs, Connecticut: Center for Real Estate and Urban Economic Studies, University of Conneticut.
Hempel, D. and Ayal, I., 1977. Transition Rates and Consumption Systems: A Conceptual Framework for Analyzing Buyer Behavior. In. A. Woodside, J. Sheth and P. Bennett (eds.). Consumer and Industrial Buying Behavior. Elsevier, North Holland.
Huff, J., 1980. Spatial Aspects of Residential Search. Unpublished Paper, University of Illinois.
Huff, J. and Clark, W., 1978. Cumulative Stress and Cumulative Inertia: A Behavioral Model of the Decision to Move. Environment and Planning A 10: 1101-1119.
Ioannides, Y., 1975. Market Allocation through Search: Equilibrium Adjustment and Price Dispersion. Journal of Economic Theory 11: 247-249.
Kohn, M. and Shavell, S., 1976. The Theory of Search. Journal of Economic Theory 9: 93-123.
McCall, J., 1970. Economics of Information and Job Search. Quarterly Journal of Economics 84: 113-126.
McCarthy, K., 1979. Housing Search and Residential Mobility. Draft Rand Report. Santa Monica, California: Rand.
MacLennan, D., 1977. Information Networks in a Local Housing Market. Scottish Journal of Political Economy 26: 73-88.
Meyer, R., 1980. Consumer Information Search and Choice Behavior: Two Models and Empirical Tests, University of Iowa, College of Business Administration, Working Paper 80-4.
Palm, R., 1976. Real Estate Agents and Geographical Information. The Geographical Review 66: 266-280.
Rothschild, M., 1973. Models of Market Organization with Imperfect Information: A Survey. Journal of Political Economy 81: 1283-1308.
Schneider, C., 1975. Models of Space Searching in Urban Areas. Geographical Analysis 7: 173-185.
Silk, J., 1971. Search Behavior: General Characterization and Review of the Literature in the Behavioral Sciences. University of Reading, England, Department of Geography, Discussion Paper.
Slovic, P., Fieschhoff, B. and Lichtenstein, S., 1977. Behavioral Decision Theory. Annual Review of Psychology 28: 1-39.
Slovic, P. and Lichtenstein, S., 1971. Comparison of Bayesian and Regression Approaches to the Study of Information Processing in Judgment. Organizational Behavior and Human Performance 6: 649-744.

Smith, T., Clark, W., Huff, J. and Shapiro, P., 1979. A Decision-Making and Search Model for Intraurban Migration. Geographical Analysis 11: 1-22.

Smith, T. and Mertz, F., 1980. An Analysis of the Effects of Information Revision on the Outcome of Housing Market Research, with Special Reference to the Influence of Realty Agents. Environment and Planning A 12: 155-174.

Smith, T. and Clark, W., 1980. Housing Market Search: Information Constraints and Efficiency. In W. Clark and E. Moore (eds.). Residential Mobility and Public Policy. Beverly Hills, California: Sage Publications.

Smith, T. and Clark, W., 1980. On the Measurement of Preferences for Housing. University of California, Los Angeles, Unpublished Manuscript.

Stigler, G., 1962. Information in the Labor Market. Journal of Political Economy 70: 94-105.

Weinberg, D., Atkinson, R., Vidal, A., Wallace, J. and Weisbrod, G., 1977. Housing Allowance Demand Experiment, Locational Choice, Part 1. Search and Mobility. Cambridge, Massachusetts: ABT Associates, Inc.

Weinberg, D., Friedman, J. and Mayo, S., 1979. A Disequilibrium Model of Housing Search and Residential Mobility. Cambridge, Massachusetts: ABT Associates, Inc.

Yinger, J., 1978. Economic Incentives, Institutions and Racial Discrimination: The Case of Real Estate Brokers. Discussion Paper D78-4. Cambridge, Massachusetts: Department of City and Regional Planning, Harvard University.

SPACE AND TIME IN THE GEOGRAPHY OF AGING

M. Sonis

Ruppin Institute

INTRODUCTION

The accelerated process of aging of world populations is a phenomenon unique to the twentieth century. The wave of aging, which arose in Europe, embraces North America and Japan and is approaching Latin America, Asia and Africa. It may be possible to predict the future of demographically young states by observing the changes in the age structure of older countries. In other words, the spatial diversity of age structure existing now reflects roughly the changes of age structure through time for each individual state. The hypothesis of interconnections between the spatial diversity and changes in time for age structures of population we will call the ergodic hypothesis (by analogy with the well-known principle of statistical mechanics).

In this paper we use a traditional division of population of country into three major age groups:

1) young: age 0-19;
2) adult: age 20-59; and,
3) elderly: age 60 and above.

For graphical description of spatial variations of age structures of population we use the triangular diagram (triangram) of Gibbs which was used in demographic analysis by Vincent (1947) [see also Dickinson (1973)]. The graphical representation of changes in age structure is given by the proposed hexagonal diagram (hexagram) presented here in Figure 1, which defines six main types of changes, and four of them are connected with the process of aging of population.

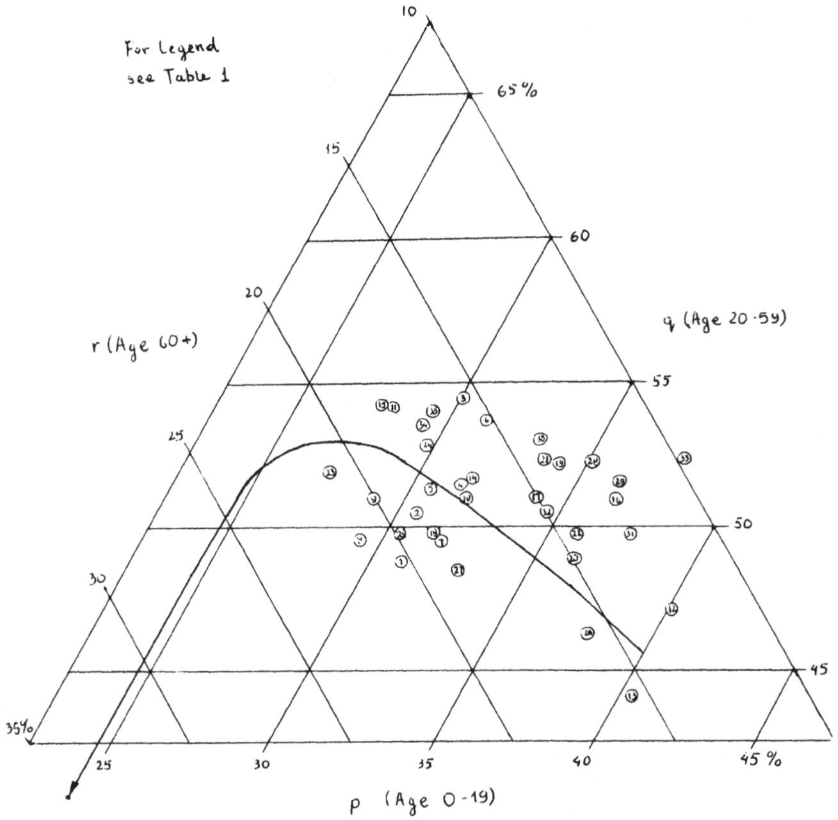

Fig. 1. The distribution of age structures and the aging curve for the European countries, 1975.

The analytical description of the spatial distribution of age structures for different countries is given by an econometric equation of aging curve which defines the connections between the portion (p) of young population and the portion (q) of elderly populations:

$$p = p_{min} + \exp[a-b/(r_{max}-r)] \quad .$$

The aging curve defines the upper bound r_{max} of an increase of portion of aged population and lower bound p_{min} of a decrease of portion of young population.

According to the ergodic hypothesis, the aging curve describes the secular trend of changes of age structure for an individual country. Therefore, a Markov process was constructed, depending on real time, for which the trajectory of behavior of age structure coincides with the above-mentioned aging curve.

A hodograph of aging curve on the hexagram makes it possible to classify types of aging by different combinations of "aging at the apex" and "aging from the base." A cartographic description of the wave of aging can be obtained by dividing the aging curve into parts and then assigning a shading to each part of the curve (and nearest domain) on the triangram. Choosing the sequence of shading according to density of lines, we introduce the unique parameter--the shading on the map--which determines the change of age structure in space and time. The transfer of shading from the triangram to the geographical map gives the spatial description of a wave of aging.

The statistical basis for this paper is the United Nations data for age structure of all European countries near the time-period 1970-1975 (see Table 1).

A CURVE OF AGING PROCESS AND THE ERGODIC HYPOTHESIS

The modern aging process involves two main tendencies: "aging at the apex"--continuous growth of the portion of elderly population, and "aging from the base"--decrease of the portion of young population. Sundbarg (1900) introduced the simplest type of relation between the portions of elderly and young population in the form of "the law of equilibrium of age structure." According to Sundbarg, the portion of parents remains stable (in space and time), while the portions of children and elders vary inversely: a population with a large share of children includes a small share of elders and vice versa. At present we are far away from the situation described by Sundbarg. The share of the adult population is not stable, and connections between the young and aged population are more complicated. Moreover, in contrast to Sundbarg's time, the changes of age structure have in the present period an essential economic content, measurable in terms of "social burden."

The statistical description of the connection between the portions of young and aged population can be done with the help of a triangular diagram (triangram). Let p, q and r be the percentages of young population (age 0-19), adult population (age 20-59) and elderly population (age 60+), so that $0 \leq p,q,r \leq 100\%$ and $p + q + r = 100\%$. The age pyramid of each country for the above mentioned division of population into three age groups can be represented by a point on the triangram whose

TABLE 1

Age Structure of Population of European Countries (in %)

Country	Years	Portions of population in the age groups 0-19 p	60+ r
1. Austria	1971	31.3	20.2
	1975	30.9	20.5
2. Belgium	1970	31.1	19.0
	1974	30.5	19.2
3. Bulgaria	1972	30.3	15.3
	1974	29.9	15.6
4. Czechoslovakia	1970	32.1	16.9
	1973	31.3	17.4
5. Denmark	1969	31.2	17.4
	1973	30.3	18.3
6. Finland	1970	33.5	14.4
	1974	31.0	15.3
7. France	1968	32.2	18.7
	1972	31.7	18.7
8. German Democratic Republic	1971	31.0	22.0
	1976	29.1	21.3
9. Federal Republic of Germany	1970	29.8	19.2
	1975	28.8	20.1
10. Greece	1971	32.4	16.3
	1975	31.7	17.4
11. Hungary	1970	30.0	17.0
	1975	27.9	18.3
12. Iceland	1970	42.4	12.2
	1975	40.1	12.8
13. Ireland	1971	40.3	15.6
	1975	40.3	15.7
14. Italy	1971	31.6	15.7
	1974	31.5	17.2
15. Luxembourg	1970	28.8	18.5
	1974	27.8	18.4
16. Malta	1972	37.0	13.5
	1974	36.4	12.7
17. Netherlands	1971	36.2	14.6
	1975	33.9	15.1
18. Norway	1970	32.2	18.2
	1975	31.5	19.1
19. Poland	1970	37.1	12.7
	1975	34.1	13.8
20. Portugal	1972	37.0	14.6
	1974	36.1	15.1

TABLE 1 (continued)

Country	Years	Portions of population in the age groups 0-19 p	60+ r
21. Romania	1972	34.4	13.8
	1975	33.6	14.3
22. Spain	1970	35.8	14.1
	1974	35.9	14.4
23. Sweden	1970	27.7	19.7
	1975	27.2	21.0
24. Switzerland	1971	30.9	16.6
	1976	29.6	17.6
25. United Kingdom:	1971	30.5	19.2
England & Wales	1975	30.3	19.9
26. UK: Northern	1971	38.5	15.4
Ireland	1975	37.9	15.8
27. UK: Scotland	1971	33.4	17.9
	1975	32.7	18.8
28. Yugoslavia	1971	36.5	12.2
	1974	35.1	12.7
29. USSR, Russian	1966	36.4	10.5
SFSR	1971	36.0	11.9
30. USSR, Ukrainian	1966	33.7	12.2
SSR	1971	33.1	13.9
31. USSR, Byelorus-	1966	37.9	11.9
sian, SSR	1971	37.5	13.1
32. USSR, Lithuanian	1966	33.4	13.5
SSR	1971	34.5	15.0
33. USSR, Moldavian	1966	40.0	8.7
SSR	1971	38.1	9.7
34. USSR, Latvian	1966	29.6	16.1
SSR	1971	29.2	17.3
35. USSR, Estonian	1966	29.8	16.0
SSR	1971	29.5	16.8

Source: UN Demographic Yearbook, 1973, 1979.

projections on the sides of the triangle give the shares p,q,r of three age groups.

Figure 1 describes the age structure of 35 European countries at about 1975 (see Table 1). A set of points of age structures for different countries forms the path on the triangram, which reflects the aging process of the European population. The "oldest" end of the path includes the oldest states of Europe: Sweden, France and England; between lies Bulgaria, the Estonian SSR and Switzerland. The "youngest" end of the path includes some European Republics of USSR, Ireland and Iceland.

The existence of the path on a triangram justifies the construction of a functional dependence $p = f(r)$ between the portions p and r of young and aged population of Europe. Also it is very interesting to evaluate the upper bound of aging at the apex--r_{max}--the maximal portion of aged population, and the lower bound of aging from the base--p_{min}--the minimal portion of young population. Therefore we require the following properties of f(r):

1) $f(r) > 0$;

2) $\frac{df}{dr} < 0$, i.e., the portion of young population decreases with an increase in the portion of elders; and,

3) There exists an upper bound of aging at the apex r_{max} and a lower bound of aging from the base p_{min} such that

$$p_{min} = \lim_{r \to r_{max}} f(r) \quad ,$$

and the portion of aged population increases monotonically $(r \to r_{max})$ and the portion of young population decreases monotonically $(p \to p_{min})$.

A function which satisfies the above mentioned properties is

$$p = p_{min} + \exp[a - \frac{b}{r_{max} - r}] \quad , \tag{1}$$

where the parameters p_{min}, r_{max}, a, b are positive and can be found with the help of the least-squares method by the minimization of some function F such that

$$\text{MIN:} \quad F(a,b,p_{min},r_{max}) = \sum_r [\ln(p-p_{min}) - a + \frac{b}{r_{max}-r}]^2 \quad .$$

The minimization can be achieved by the gradient method, choosing the step of approximation by the expansion of the function $F(a,b,p_{min},r_{max})$ into a Taylor series.

The aging curve representing the age structures of European countries in about 1975 with minimal error has the following econometric form:

$$p = 24.627 + \exp[7.6041 - \frac{100.299}{35.019-r}], \tag{2}$$

where all coefficients are significantly different from zero. Equation (2) gives the following estimation of the upper bound of aging at the apex and the lower bound of aging from the base: $r_{max} = 35.019\%$, $p_{min} = 24.627\%$.

An essential property of the aging process is that the portion of the aged population of European countries grows monotonically with time. The portion of the young population oscillates, but has the secular tendency to decrease [see, for example, Russet (1964)]. Figure 2 represents on the triangram

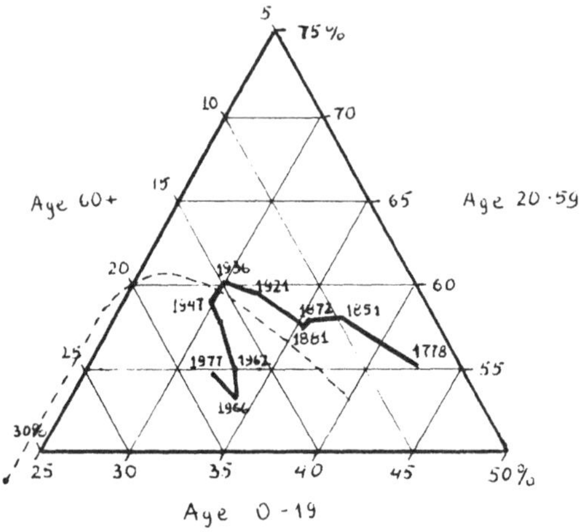

Fig. 2. Changes in the age structure for France, 1778-1977.

temporal changes during 200 years in the age structure of France--the "oldest" European country, in which the aging process was first shown in the 1850's [Reinhard (1949)]. We can see the monotone increase of the percentage of elders, but the portion of the young population oscillates. Figure 3 describes the changes in

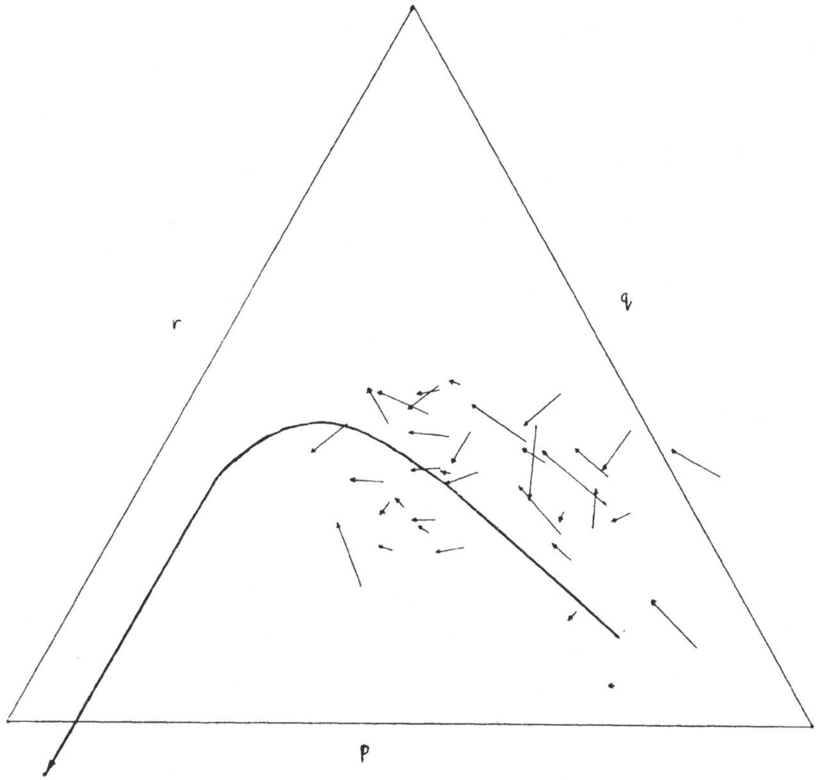

Fig. 3. Changes in the age structure for the European countries, 1970-1975.

age structure of 35 European countries during the period around 1970-1975. Note that for most European countries the age structure changes in the direction of aging at the apex and from the base, as on the aging curve.

It is possible to consider the aging curve as the presentation not only of the spatial diversity of age structures during a fixed time-period, but also as a reflection of the long-term changes--secular trend--for each individual country. Here we come across the idea of connection between temporal and spatial characteristics of changes in age structure: the aging curve, corresponding to the distribution of the countries on the triangram in a fixed time-period, is analogous to the secular trend of temporal behavior of age structure of an individual state. This hypothesis we will call the ergodic hypothesis by rough analogy with the ergodic theorem of Boltzman-Gibbs from statistical mechanics.

From the viewpoint of the ergodic hypothesis the parameters r_{max} and p_{min} represent states of the age structure to which the age structure of each country moves with time along an aging path. Points indicating the age structure of different countries move through time with oscillations near the curves of the analogical form. Individual countries differ each from the other by the initial state and by the velocity of the aging process; therefore, the choice of the time-period stops the points of age structures in different places on the same aging path.

THE MARKOV PROCESS OF AGING

We give the analytical presentation of the ergodic hypothesis by choice of the temporal parameter t, representing the real time. Let us write the equation of the aging curve (1) in the form

$$\begin{aligned} p = p(t) &= p_{min} + (p_o - p_{min}) \exp(-\alpha t) \\ &= p_{min} + (p_o - p_{min}) \delta^t \quad , \end{aligned} \tag{3}$$

where $p_o = p(0)$ is an initial portion of the young population,
t is the real time,
α is the scale coefficient, which will be defined below,
$\delta = \exp(-\alpha)$ is the rate of change, which defines the velocity of the aging process, and

$$\Delta t = \frac{1}{\alpha} [\ln(p_o - p_{min}) - a + \frac{b}{r_{max} - r}] \tag{4}$$

is the monotone function of r. We define the scale parameter α with the help of statistical data of the age structures of European countries to two different time points [approximately 1970 and 1975, since the UN data does not always correspond to the same time period (see Table 1)].

Equation (4) gives the following presentation for the time difference

$$\Delta t = \frac{b}{\alpha} [\frac{1}{r_{max} - (r + \Delta r)} - \frac{1}{r_{max} - r}] \quad .$$

Therefore

$$\alpha = \frac{b}{\Delta t} \left[\frac{1}{r_{max}-(r+\Delta r)} - \frac{1}{r_{max}-r}\right] ,$$

and we can find the scale coefficient α with the help of the least-squares method by minimization of the function

$$G(\alpha) = \sum_{\Delta r} \left[\frac{b}{\Delta t} \left(\frac{1}{r_{max}-(r+\Delta r)} - \frac{1}{r_{max}-r}\right) - \alpha\right]^2 .$$

The estimated value of the scale parameter is

$$\alpha = \frac{b}{m} \sum_{\Delta r} \frac{1}{\Delta t} \left(\frac{1}{r_{max}-(r+\Delta r)} - \frac{1}{r_{max}-r}\right)$$

where m is the number of countries (m=35). The result of the calculation for European countries is $\alpha = 0.045446$, and, therefore, the rate of change of the portion of the young population is $\delta = \exp(-\alpha) = 0.95557$.

Choosing the initial state $p_o = p(0) = 40\%$ we obtain the following form of the aging equation:

$$p = 24.627 + 15.373(0.95557)^t \quad (t \geq 0) .$$

Equation (3) describes a Markov chain with continuous time t. In order to construct this chain let us consider the portion $[1 - p(t)]$ of adults and elders, corresponding to the portion $p(t)$ of the young population. Then, a stochastic matrix S, which defines the Markov chain with the same equation (3) will have the form

$$S = \begin{pmatrix} 1 - p_{min} + \delta p_{min} & p_{min} - \delta p_{min} \\ 1 - p_{min} - \delta(1-p_{min}) & p_{min} + \delta(1-p_{min}) \end{pmatrix} .$$

More precisely, equation (3) has a vector form:

$$\begin{pmatrix} 1-p(t) \\ p(t) \end{pmatrix} = \begin{pmatrix} 1 - p_{min} + \delta p_{min} & 1 - p_{min} - \delta(1 - p_{min}) \\ p_{min} - \delta p_{min} & p_{min} + \delta(1 - p_{min}) \end{pmatrix}^t \begin{pmatrix} 1-p_o \\ p_o \end{pmatrix}$$

$$= \begin{pmatrix} 1 - p_{min} + \delta^t p_{min} & 1 - p_{min} - \delta^t(1 - p_{im}) \\ p_{min} - \delta^t p_{min} & p_{min} + \delta^t(1 - p_{min}) \end{pmatrix} \begin{pmatrix} 1-p_o \\ p_o \end{pmatrix} . \tag{5}$$

It is easy to obtain the equivalence of equations (3) and (5) by transforming equation (5) into coordinate form. In the numerical form for all European states

$$S = \begin{pmatrix} 0.98906 & 0.01094 \\ 0.03349 & 0.96651 \end{pmatrix} .$$

The coefficients of this matrix define the portions that lose and acquire annually the young population and population of adults and elders, respectively. The young population loses annually 3.349%, which passes to the adult and aged population; in turn, the adult and elderly population loses only 1.051%. This relative change of the positions of population reflects a more complicated picture of change in the absolute number of population, which includes the following processes: (1) passage of the young population into the category of adults; (2) renewal of the young population at the expense of birth; (3) non-proportional mortality of the young, adult, and elderly; and, (4) non-proportional migration of the young and adult and elderly. The above constructed Markov chain defines a final distribution of young and adult and aged population:

$$\begin{pmatrix} 1 - p_{min} \\ p_{min} \end{pmatrix} = \begin{pmatrix} 75.373 \\ 24.627 \end{pmatrix} ,$$

herewith, the "tempo" of movement to the stable final distribution is measured by the eigenvalue $\lambda \neq 1$ of the stochastic matrix: $\lambda = \delta = \exp(-\alpha) = 0.955557$ [Feller (1966), Ch. 10, 15].

From (4) we can find the dependence of the share of the aged population on time:

$$r = r_{max} - \frac{b}{\alpha t + a - \ln(p_o - p_{min})}$$

which defines also "the velocity" of the reaching of r_{max}. Table 2 presents the rates of changes of age structure of a country with time. For example, Ireland will spend 40 years to reach the age structure of Sweden in 1975. The velocity of approaching to the p_{min} and r_{max} is different; it is easy to see from Table 2 that during 50 years European countries will almost reach the lower bound of the aging at the apex, but the way to the upper bound of aging from the base will take longer (see also Figure 1).

HEXAGONAL DIAGRAM OF CHANGES IN AGE STRUCTURE

It is convenient to represent the changes in age structure in space or time with the help of a hexagonal diagram--hexagram. The hexagram contains three coordinate axes originating from a central point with a 120° angle between the axes (Figure 4). Each of these axes represents the changes Δp, Δq, Δr of shares of each of the three age groups. Each vector on the hexagram which starts from the origin is a difference-vector between two different states (in space or time) of age structure. For example, if the age structure of population in two different points of space or time is given by the vectors (p,q,r) = (30%,50%,20%) and (p^1,q^1,r^1) = (35%,60%,5%) then the difference-vector is equal to $(\Delta p,\Delta q,\Delta r)$ = (-5%,-10%,+15%), and a sum of relative changes $\Delta p+\Delta q+\Delta r = 0$. Moreover, the signs of differences (- - +) give the characterization of the aging process: in our example we have aging at the apex ($\Delta p<0$) and from the base ($\Delta r>0$) with a simultaneous decrease of the portion of adult population ($\Delta q<0$).

The axes of the hexagram divide it into six sectors, where each sector represents a different type of aging process according to six possible combinations of the signs of $\Delta p,\Delta q,\Delta r$ (see Table 3 and Figure 4). The changes of types I, II, III are connected with the aging at the apex; the changes of types I, II, VI with the aging from the base; and changes of types I, II with both aging of the apex and from the base.

The worst type of change is type II in which aging at the apex and from the base is burdened by the decrease of the portion of adult population, thus reducing the potential work force while at the same time increasing the burden of the elderly. For a demonstration of changes in age structure of European countries we will build an approximate hodograph of the aging curve. Let us divide the aging curve into parts with the same distance (for example, 2.5%) between the dividing points. The approximate hodograph is a set of difference-vectors, defined by the dividing points (see Figure 5). For the European countries the approximate hodograph describes the wave of aging at the apex and from the base, which is the continuous passage from type I of change (- + +) to type II (- - +), and the convergence to the "pure" aging at the apex (0 - +) with the stable share of young population.

TABLE 2

Rates of Changes in Age Structure of European Countries with Time

Time (in years)	Proportions of population in the age groups		Countries (approximately)
	0-19	60+	
t	p	r	
0	40.0	14.2	Ireland
5	36.9	15.1	Portugal
10	34.4	15.9	Lithuanian SSR
15	32.4	16.7	Greece
20	30.8	17.5	Czechoslovakia
25	29.6	18.2	Switzerland
30	28.6	18.8	F R G
35	27.8	19.4	Luxemburg
40	27.1	19.9	Sweden
45	26.6	20.4	
50	26.2	20.9	

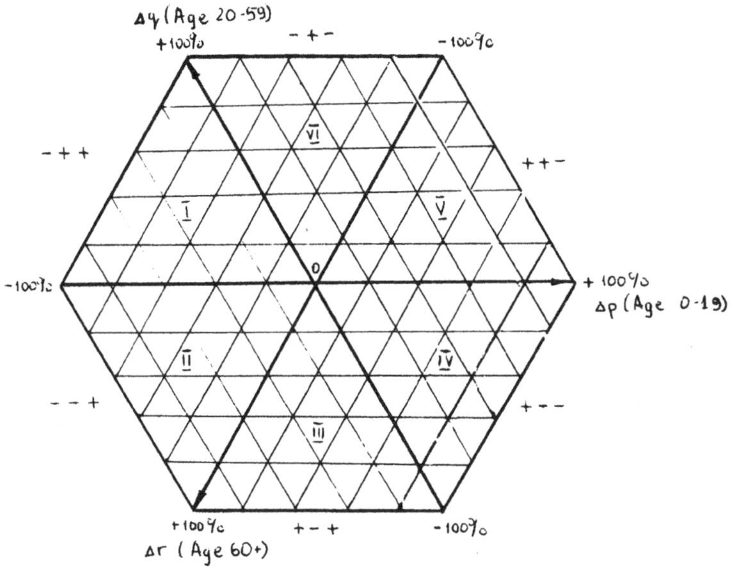

Fig. 4. Hexagonal diagram (hexagram) of changes in age structure.

TABLE 3

Six Possible Types of Changes in Age Structure

Type of Change	Direction of changes for the age groups 0-19 sign Δp	20-59 sign Δq	60+ sign Δr	Explanation
I	-	+	+	Increase in shares of the adult and elderly population at the expense of the young population.
II	-	-	+	Increase in share of the elderly population at the expense of the young and adult population.
III	+	-	+	Increase in shares of the young and elderly population at the expense of the adult population.
IV	+	-	-	Increase in share of the young population at the expense of the adult and elderly population.
V	+	+	-	Increase in shares of the young and adult population at the expense of the elderly population.
VI	-	+	-	Increase in share of the adult population at the expense of the young and elderly population.

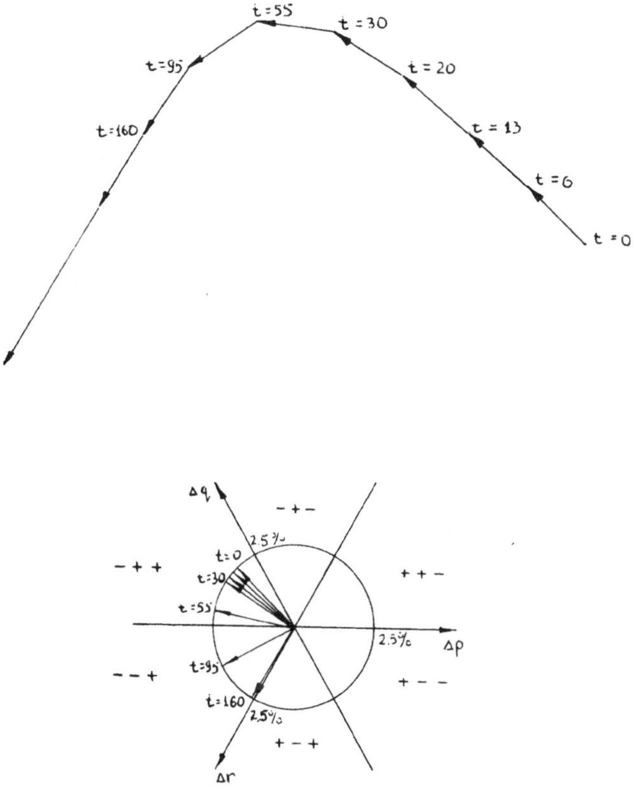

Fig. 5. Approximate hodograph of the aging curve.

The "velocity" of expansion of this wave depends on the position of a country with respect to the aging curve: the "youngest" countries grow old more rapidly than the "oldest". It is possible to evaluate the velocity of aging by the numerical value of a time-parameter of the corresponding Markov chain (see Figure 5).

The hexagram also gives the possibility to represent the direction and values of the real changes in age structure of each individual country. In Table 4 and Figure 6 there is a presentation of changes in age structure of the European countries. These real temporal changes are analogous to the changes on the approximate hodograph; in this way we obtain the additional evidence of the ergodic hypothesis.

CARTOGRAPHIC REPRESENTATION OF THE AGING PROCESS

The existence of the path of the aging process and the aging curve helps to present the wave of aging cartographically. For purposes of cartographic analysis, the aging path has been divided into progressive sections, in a somewhat arbitrary fashion and has been depicted in different shades, as shown in Figure 7. The transfer of shading from the triangram to the map gives a cartographic presentation of the wave of aging (see Figure 8). A concentric pattern around the central countries of Europe is clearly indicated. Moreover, the movement on the map from Central European countries to the periphery conforms to the movement from right to left on the aging curve, i.e., it conforms to the passage to progressively younger age structures.

The above chosen division of the aging path and curve presents the investigator's intuitive graphical perception of the aging process. The other choice of form of division can present cartographically, different parameters of aging connected, for example, with social and economic characteristics of aging.

Let us describe the division of the triangram based on the notion of the so-called demographic or social burden [Russet (1964)]. The demographic burden expresses the relationships between the portions of manpower--the adult population and the portions of the young and elderly. The numerical expression of the demographic burden is given, for example, by the following coefficients:

$k_1 = \frac{p}{r}$ - the demographical burden on the aged population by the young population;

$k_2 = \frac{p}{q}$ - the demographical burden on man-power by the young population;

TABLE 4

The Annual Average Changes in the Age Structure of the European Countries (in %)

Countries	Changes in the age groups 0-19 Δp	20-59 Δq	60+ Δr	Directions of change	Type of change
1. Austria	-0.1	+0.025	+0.075	- + +	I
2. Belgium	-0.15	+0.1	+0.05	- + +	I
3. Bulgaria	-0.2	+0.05	+0.15	- + +	I
4. Czechoslovakia	-0.267	+0.1	+0.167	- + +	I
5. Denmark	-0.225	0	+0.225	- 0 +	I,II
6. Finland	-0.625	+0.4	+0.225	- + +	I
7. France	-0.125	+0.125	0	- + 0	I,VI
8. GDR	-0.38	+0.52	-0.14	- + -	VI
9. FRG	-0.2	+0.02	+0.18	- + +	I
10. Greece	-0.175	-0.1	+0.275	- - +	II
11. Hungary	-0.42	+0.095	+0.325	- + +	I
12. Iceland	-0.46	+0.31	+0.15	- + +	I
13. Ireland	0	-0.025	+0.025	0 - +	II,III
14. Italy	-0.033	-0.342	+0.375	- - +	II
15. Luxembourg	-0.25	+0.275	-0.025	- + -	VI
16. Malta	-0.3	+0.7	-0.4	- + -	VI
17. Netherlands	-0.575	+0.45	+0.125	- + +	I
18. Norway	-0.14	-0.085	+0.225	- - +	II
19. Poland	-0.6	+0.38	+0.22	- + +	I
20. Portugal	-0.45	+0.2	+0.25	- + +	I
21. Romania	-0.267	+0.1	+0.167	- + +	I
22. Spain	+0.025	-0.1	+0.075	+ - +	III
23. Sweden	-0.1	-0.16	+0.26	- - +	II
24. Switzerland	-0.26	+0.06	+0.2	- + +	I
25. England & Wales	-0.05	-0.125	+0.175	- - +	II
26. Northern Ireland	-0.15	+0.05	+0.1	- + +	I
27. Scotland	-0.175	-0.05	+0.225	- - +	II
28. Yugoslavia	-0.467	+0.3	+0.167	- + +	I
29. Russian SFSR	-0.08	-0.2	+0.28	- - +	II
30. Ukranian SSR	-0.12	-0.22	+0.34	- - +	II
31. Byelorussian SSR	-0.08	-0.16	+0.24	- - +	II
32. Lithuanian SSR	+0.22	-0.52	+0.3	+ - +	III
33. Moldavian SSR	-0.38	+0.18	+0.2	- + +	I
34. Latvian SSR	-0.08	-0.16	+0.24	- - +	II
35. Estonian SSR	-0.06	-0.1	+0.16	- - +	II

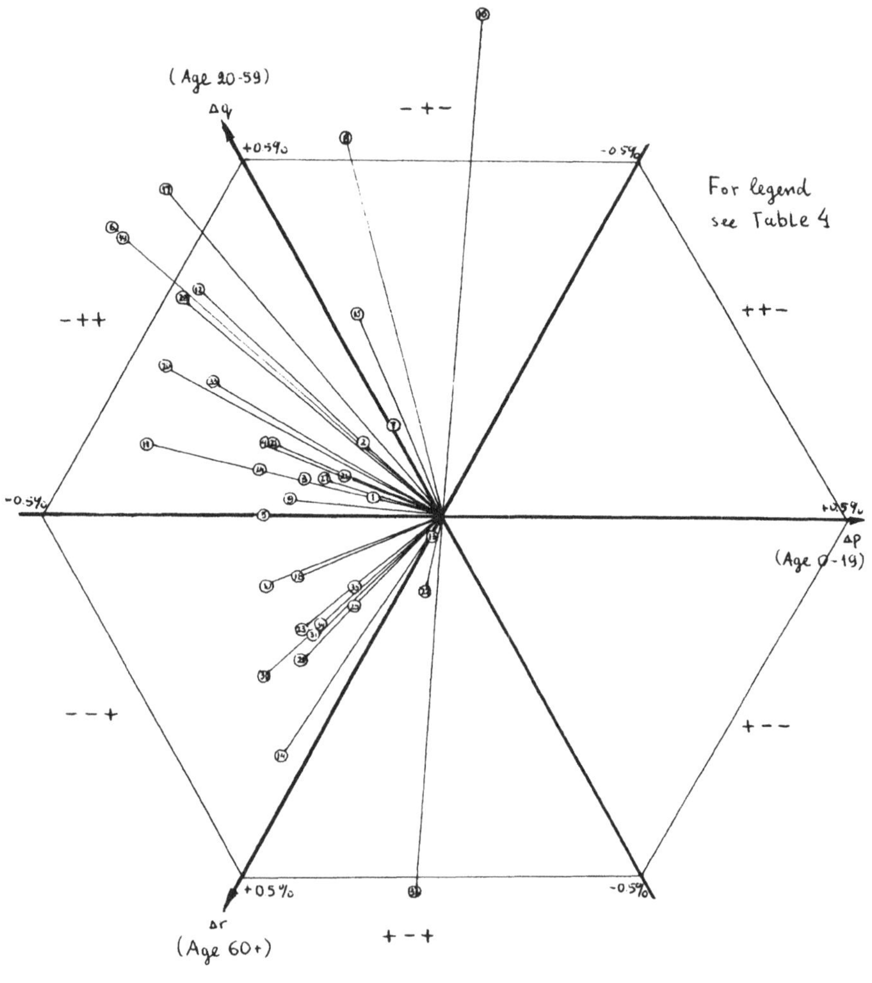

Fig. 6. Annual average changes in age structure for the European countries.

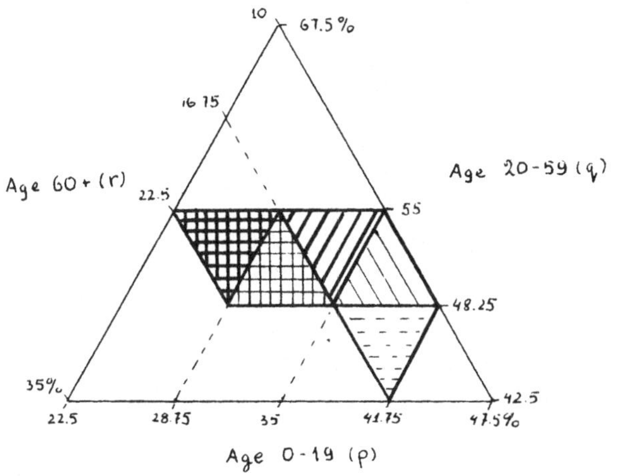

Fig. 7. Categories of age structure for the European countries.

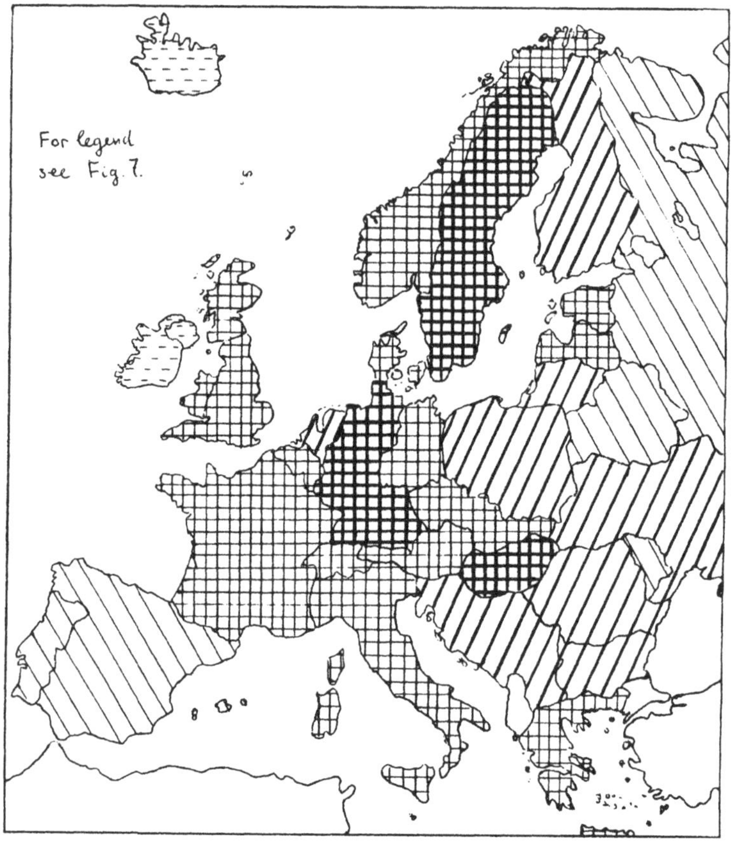

Fig. 8. Wave of aging of the European population.

$k_3 = \frac{r}{q}$ - the demographical burden on manpower by the elderly population, and

$k_4 = k_2 + k_3 = \frac{p+r}{q}$ - the common demographical burden on manpower by the non-productive population.

The lines of equal burden k_i = constant on the triangram are straight lines (see Figure 9). These straight lines form a lattice on the triangram, which defines the distribution of the demographic burden on the triangram. The non-equal cells of this lattice give the basis for the cartographic presentation of social burden.

For the European population, according to the secular trend of aging, the share of young population and the share of aged population monotonically converge to the bounds p_{min} and r_{max}. Therefore, the demographical burden on the aged population by children k_1 decreases monotonically from the value k_1 = 2.8 to the minimal value $k_{1\ min} = \frac{p_{min}}{r_{max}} = 0.7$. As we can see from Figure 9, the demographic burden on the adult population by the aged population k_2 increases monotonically from the value k_2 = 0.31 to the maximal value $k_{2\ max}$ = 0.87. The demographic burden on the adult population by the young population k_2 oscillates with the share of children; in the future the common demographic burden will increase to a maximal value $k_{4\ max}$ = 1.48, i.e., every 100 adults will support 148 children and elders.

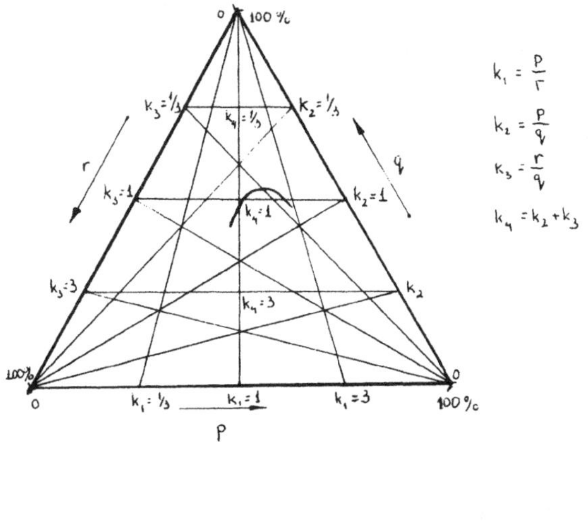

Fig. 9. Straight lines of equal demographical burden on the triangram.

A common basis for mapping all possible types of change in age structure is furnished by the hexagram shown in Figure 10. For purposes of refining this classification of changes it is possible to subdivide each of the six sectors of the hexagram into two (or more) subsectors, depending on the nature of the distribution of the difference-vectors of change and the need for mapping. We choose the division of sectors by three straight lines $\Delta p = \Delta q$, $\Delta r = \Delta p$ and $\Delta q = \Delta r$. Such lines are drawn from the origin and generate subsectors characterized by inequalities $\Delta q \gtrless \Delta p$, $\Delta r \gtrless \Delta p$, $\Delta q \lessgtr \Delta r$ (see Figure 10). For example, subsector Ia represents a relatively larger positive change in the proportion of those aged 20-59 than the change in the proportion of elders, while subsector Ib denotes a larger positive change in the proportion of those aged 60 and over, than the change in proportion of the adult population. Therefore, the subsector Ia could be regarded as presenting a lower level of aging than the subsector Ib. When changes in age structure in Europe are mapped according to the above categories, generated by the subsectors of hexagram (see Figure 10), a meaningful spatial pattern emerges, as shown in Figure 11.

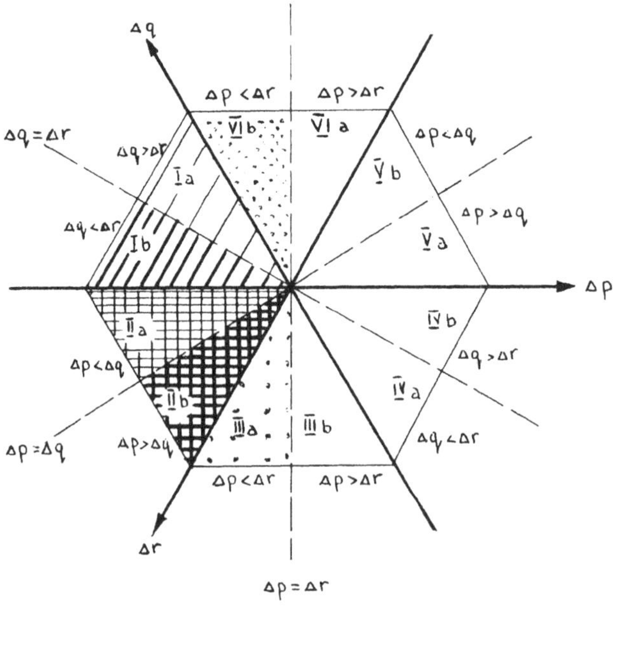

Fig. 10. Categories of change in age structure.

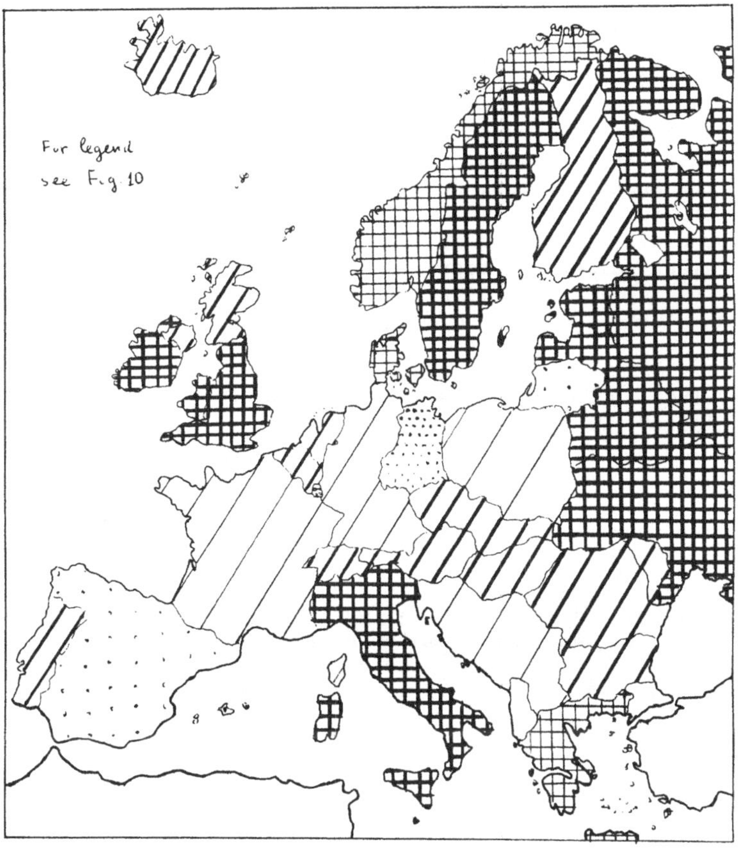

Fig. 11. Mapping of annual changes in age structure for the European countries.

In conclusion, we note that the ergodic hypothesis, which connects the spatial and temporal changes in age structure and implies the existence of the aging curve, demands additional statistical evidence and further theoretical development. But even now it can serve as a basis for the introduction of a universal uni-dimensional parameter--the shading of the map, which represents the direction, value and spatial and temporal behavior of the aging process, describes the social implications of aging in the form of coefficients of social burden, and gives the basis for a classification of types of aging processes in analytical, graphical, and cartographic form.

REFERENCES

Vincent, P., 1947. "Une interressante Application du Diagramme triangulaire", Population 2.

Dickinson, G, 1973. Statistical Mapping and Presentation of Statistics. London: Edward Arnold.

Sundbärg, G., 1980. "Sur la répartition de la population par âge et sur les taux de la mortalité." Bulletin de l'Institute International de Statistique.

Russet, E., 1964. Aging Process of Population, London: Macmillan.

Reinhard, M., 1949. Histories de la Population Mondiale de 1700 à 1948. Paris: Editions Domat-Montchrestien.

Feller, W., 1966. An Introduction to Probability Theory and Its Applications, Vol. I, New York: Wiley.

ECOLOGICAL PERSPECTIVES ON DYNAMIC SPATIAL MODELS

Much work concerned with spatial models has been undertaken in fields outside of geography and physics. In keeping with the multi-disciplinary nature of this volume, ecological contributions are presented in this section. In the mathematical theory of ecological communities two dominant themes have been pursued: (1) the temporal dynamics of interacting populations, and (2) the spatial pattern of a community. The first theme has been approached through the mathematical framework of dynamical systems, building upon the fundamental ideas of Lotka, Volterra and Kostitzin. It is earmarked by studies of predator-prey systems in, for instance, spatially discontinuous environments, spatially heterogeneous environments, or some continuous medium in which constant diffusion occurs. The key question here has to do with whether or not a predator-prey system describes selected space-time trajectories. Meanwhile, this second theme deals with the static and staistical description of communities as expressed through the corresponding spatial distributions of their populations. The primary question about dynamics asks whether inferences about space-time trajectories can be extracted from an observed map pattern. An example of each of these themes is presented in this section.

Curry employs the mathematical framework associated with temporal dynamics of interacting populations to study vacancy chains in spatially separated labor markets. His main concern is twofold: an analytic treatment of vacancy chains, and externalities. Occupations are equated with environmental niches. An analogy then is made between the labor market and the predator-prey model of ecology, followed by a description of interactions among occupations in a community matrix via the Lotka-Volterra equations. Next, the manner in which job search in space affects the stability of occupational structures is investigated. The basic spatial mechanism here is the spread of information, which is assumed to be a two-dimensional random walk on a square lattice in the plane. Although Curry demonstrates that the analytics of vacancy chains are formidable, subsequent research is needed to augment his findings. The very nature of these analytics has limited the findings to statements about trajectory stability.

Sokal and Wartenberg draw inferences about space-time trajectories from observed variations in geographic distributions of biological populations, such as pollen, plant seeds and selected animal species. Their basic tool of analysis is the correlogram, and they attempt to posit processes on the basis of similarities and dissimilarities of surfaces and correlograms that accompany these surfaces. This type of synoptic inference is used because population processes occur over a time horizon that is too long for practical observation or experimentation to take place. The reliability of this inferential mode stems from the kinds of geographic patterns ecological communities tend to generate. Although categories of patterns observed in the surfaces of biological variables vary endlessly, preventing the formulation of a complete taxonomy, some patterns are more likely than others. Patterns and biological processes are linked together in this paper.

These two papers show that general dynamic spatial models are forthcoming from studies dealing with both the dynamics of interacting ecological populations, and the spatial pattern of an ecological community. The fundamental ideas of Lotka and Volterra should help solve various non-ecological spatial problems, whereas evidence is furnished here in support of the viability of drawing inferences from map patterns. This latter topic is addressed later in this book in a paper by Haining.

MACRO-ECOLOGY OF VACANCY CHAINS

Leslie Curry

Department of Geography
University of Toronto

Markets are not the clickety-click automatically equilibrating price-regulated mechanisms to which our theory tends to habituate us. Difficulties in the transfer of information provide impediments to transactions which inhibit their formation. The micro-theory of the labor market is concerned with the search for jobs in terms of aspirations, uncertainty, information, and so on. There is considerable interest in it at present because of the social problem of unemployment and inflation, and because of its relation to economic development and migration. However, it is necessary to omit discussing the micro—theory as well as the macro-economic problems involved, and the diffusion processes and spatial patterns which result from the spatial search fields of individuals. Instead we shall concentrate on the structure of occupations via their recruiting interactions. To paraphrase White (1970), occupations are no more than categories until they are performed by workers while a worker in turn has his identity essentially defined by his simultaneous position in several networks and structures of positions provided by other persons. "To study mobility is to get at...the extent of the empirical duality between man and position." White specifically emphasizes his subject as the occupational mobility of persons that gives meaning to social structure, and criticizes the Holt-David representation of the job market because it does not use the idea of the job which needs to be filled when its occupant leaves. Normally one man is replaced by another in an existing job; random matching of vacancies and unemployed men is not plausible and unemployment is not a necessary intermediate stage between jobs. White introduces the vacancy chain where job vacancies move in a stationary absorbing Markov chain. Consider vacancies *sui generis* via the equation

$\underline{M} = \underline{MQ} + \underline{F}$. A vector $\underline{M}$ contains the arrivals of vacancies by stratum in a year. The matrix $\underline{MQ}$ counts by stratum the arrival of vacancies which left other strata except that some departures were to the outside. $\underline{F}$ counts arrivals of vacancies directly from outside into the system. This gives

$$\underline{M} = \underline{F}(\underline{I} + \underline{Q} + \underline{Q}^2 + \underline{Q}^3 + \ldots) \quad ,$$

which can be seen to have a multiplying effect.

Let M_0 be the total number of moves to the outside; it is a linear function of appearances, i.e., arrivals of vacancies in the various strata:

$$M_0 = \sum_{i=1}^{s} M_i q_{i0} \quad .$$

In vacancy terms,

$$\underline{P}' = (\underline{I} - \underline{Q})^{-1} \underline{q}_0' \quad .$$

Since the q_{ij}'s apportion a collection of vacancies among a set of s + 1 possible destinations (as probabilities or transition fractions),

$$q_{i0} + \sum_{j=1}^{s} q_{ij} = 1 \quad ,$$

i.e. $\underline{q}_0' = (\underline{I} - \underline{Q})\underline{1}'$,

where $\underline{1}'$ is a column vector whose s components are each unity, and

$$\underline{P}' = \underline{I1}' = \underline{1}' \quad ,$$

i.e., every price in the vacancy model is unity. This makes sense because the Markov structure is in equilibrium and, since the possibility of moves between any pair of occupations is allowed, all marginal evaluations should be equal.

Stewman (1976a,b) credits White with three results. The numbers of moves within and from each stratum are obtained as

$$\underline{M}(t) = \underline{F}(t)(\underline{I} - \underline{Q})^{-1} \quad ,$$

where $\underline{Q} = (q_{ij})$, the probability that a job vacancy in stratum i will move to stratum j, j = 1,...,k, and $\underline{F}(t)$ is the vector of vacancy arrivals per stratum. The probability distribution of chain length n, $\underline{P}_n = \underline{Q}^{n-1}\underline{Q}_{k+1}$ where $\underline{Q}_{k+1} = (q_{i.k+1})$, the probability that a job vacancy in stratum i will leave the job system. The mean length of a job vacancy chain by origin stratum equals $(\underline{I}-\underline{Q})^{-1}\underline{L}$, where $\underline{L}$ is a column vector of ones.

Although the reality of vacancy chains should be stressed, to insist that a new occupant of a job comes in while an old one leaves misses the importance of looking at economic changes through the operations of the labor market. Many jobs do end and others begin as the mortality of firms and changes in the numbers and character of their labor force adequately testify. More important for the present paper, however, is the fact that the vacancy chain is basically an empirical accounting scheme. Even supplying numbers for the empty boxes does not seem to add insights in spite of its promise. What we shall be doing here is to treat this topic analytically, apparently for the first time.

TWO SECTOR LOTKA-VOLTERRA EQUATIONS

The approach we adopt is ecological, a view espoused by Taschdjian (1975) who equates an occupation with an environmental niche. We shall be considering elsewhere the manner in which occupations or niches are defined. He insists that the niche is "organism-directed, organism-ordered and organism-timed." "Environmental factors not only select the organism but the organism selects and molds the environment." "The population density in an operational niche of a human ecosystem depends...on the recruitment of its members...which is due...to the process of acculturation." "In societies with social mobility, positive recruitment (occurs) from the indifferentiated pool of the next generation as well as from other already differentiated niches." As we have said, we shall be concerned with the structure of occupations through the vacancy chain concept.

Occupational Interactions

Our approach will be based on an analogy with the predator-prey model in ecology. The Lotka-Volterra equations describe the interactions between species in an ecosystem, while we shall discuss interactions between occupations in a community matrix. Hudson (1970, 1975) appears to have been the first to use these equations in geography for studying population growth and migration in urban and in two-region systems.

The interactions between individuals or groups can be described by a community matrix with negative, positive, or zero coefficients. In general, a system of differential equations,

$$dX_i/dt = f_i(X_1, X_2, X_3, \ldots, X_n)$$

gives rise to the matrix of first partials

$$M = \begin{pmatrix} a_{11} & a_{12} & \cdots & a_{1n} \\ a_{21} & a_{22} & \cdots & a_{2n} \\ \cdot & & & \\ \cdot & & & \\ \cdot & & & \\ a_{n1} & a_{n2} & \cdots & a_{nn} \end{pmatrix} ,$$

where each element is given by:

$$a_{ij} = \partial f_i / \partial X_j \quad .$$

The matrix elements are considered to be evaluated at an equilibrium point of the system. These coefficients may simply be regarded as phenomenological coefficients of the system macro-equations, but they could also be derived from micro-theories.

Let us first look at some of the possible 2×2 community matrices. The diagonal simply represents positive, negative or zero density dependence, whereas the off-diagonals ring the changes to give (+,+), (+,-), (+,0), (-,-), (-,0), (0,0). In fact these interrelations were detailed by Curry (1980b), quoting Bish and Nourse (1975), as the range of possible economic external interactions between agents. It turns out that ecologists analyze exactly the same types of interactions, of which competition (-,-) and predator-prey (-,+) are the two most important. Here we shall consider not only seeker and vacancy in the context of job search in analogy to the two species predator-prey case, but also the general occupational vacancy chain corresponding to the n species case. It should be noted that these are only mathematical equations describing interactions and have nothing inherently biological about them. There is no suggestion that men resemble animals or plants--except in the models of these phenomena. So as not to confuse the reader, we shall not mention the biological phrasing of the mathematical formalism and apologize to the population geneticists and ecologists for the injustice this does to their creations.

Periodic Case

In their simplest form the Lotka-Volterra equations are [Christiansen and Fenchel (1977)]:

$$dn_1/dt = n_1(r - a_1n_2) \quad , \text{ and}$$

$$dn_2/dt = n_2(-d + a_2n_1) \quad ,$$

where r is the rate of appearance of vacancies and d is the rate at which searchers find jobs. The rate of matching vacancies and searchers depends only on the product of their numbers. The assumption that vacancies are only limited by searchers and that, in the absence of the latter, vacancies would grow indefinitely is of course wrong. Density dependent regulation of both vacancies and searchers can be introduced, yielding

$$dn_1/dt = n_1(r - \alpha_1n_1 - a_1n_2)$$

$$dn_2/dt = n_2(-d + a_2n_1 - \alpha_2n_2) \quad .$$

Alternatively, searchers can be limited by the number of vacancies per searcher rather than the number of vacancies:

$$dn_2/dt = n_2[s - n_2(a_2n_1)] \quad .$$

This is like a logistic model with density capacity being a function of vacancy abundance. It is analyzed below.

The equilibrium population sizes may be written (Leigh, 1968) as

$$N_1 = e_2/a_{12} \equiv q_1 \quad \text{and} \quad N_2 = e_1/a_{12} \equiv q_2 \quad .$$

Let $x_1 = N_1 - e_2/a_{12}$ and $x_2 = N_2 - e_1/a_{12}$.

By neglecting products x_1x_2 of second order, linear approximations are:

$$dx_1/dt = -e_2x_2 = -a_{12}q_2x_2 \quad ,$$

$$dx_2/dt = e_1x_1 = a_{12}q_1x_1 \quad ,$$

$$d^2x_1/dt^2 = -e_1e_2x_1 = -a_{12}^2q_1q_2x_1 \quad .$$

Solutions are of the form $x_1(t) = (1/e_1^{0.5})\ [A \cos (e_1e_2)^{0.5}t - B \sin (e_1e_2)^{0.5}t]$ where $A = e_1^{0.5}x_1(0)$ and $B = e_2^{0.5}x_2(0)$. The populations oscillate with frequency $(e_1e_2)^{0.5} = (a_{12}q_1q_2)^{0.5}$. The time average of $x_1^2 = (1/2e_1)(A^2 + B^2) = \sigma_1^2$, of $x_2 = (1/2e_2)(A^2 + B^2) = \sigma_2^2$, $x_1x_2 = 0$, and of $x_1(dx_2/dt) = 0.5(A^2 + B^2) = (e_1e_2)^{0.5}\sigma_1\sigma_2 = a_{12}(q_1q_2)^{0.5}\sigma_1\sigma_2$. This latter equation may be written as $(d/ds)[\lim_{T\to\infty} (2T)^{-1} \int_{-T}^{T} x_1(t)x_2(t+s)dt]_{s=0}$ which is the covariance $r_{ij}(s)$ and which allows evaluation of the coefficient of interaction a_{12}. Hence

$$r_{12}(s) = (e_1e_2)^{-0.5}[e_1x_1^2(0) + e_2x_2^2(0)] \sin (e_1e_2)^{0.5}s \quad .$$

For $s = 0$ covariance is nil, while for $(e_1e_2)^{0.5}s \simeq 3$ it is quite high.

Periodicity and the Phillips Curve

The interactive behavior of a generalized two sector labor market composed of non-specific vacancies and searchers can be represented by the Kolmogorov scheme [Rescigno and Richardson (1973)]

$$dN_1/dt = N_1K_1(N_1,N_2) \quad , \text{ and}$$

$$dN_2/dt = N_2K_2(N_1,N_2) \quad .$$

The conditions underlying this representation for our purposes here are listed below while the results are shown in Figure 1:

a. an increase of vacancies is slowed down by an increase of job seekers, $\partial K_1/\partial N_2 < 0$;

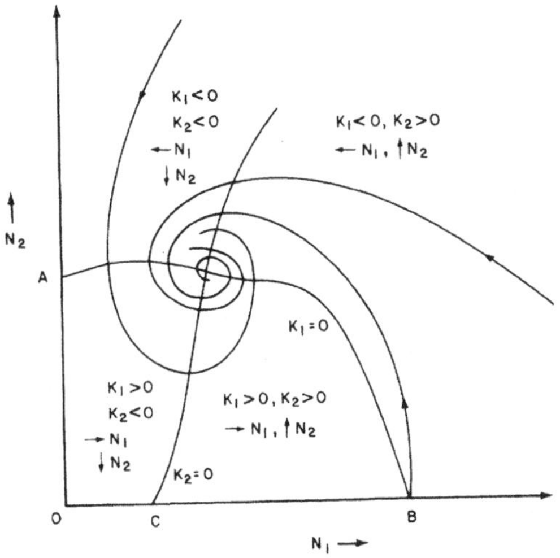

Fig. 1. Zones of interaction of vacancy N_1 and searcher N_2. [Resigno and Richardson (1973)].

b. even for constant N_1/N_2, multiplication of vacancies is slowed down by increasing seekers because there is frequent mating of jobs and seekers with higher population densities, $dK_1/dS < 0$;

c. if the populations of searchers and vacancies are small, the vacancies will multiply because they are harder to find, $K_1(0,0) > 0$;

d. if there are too many searchers, vacancies cannot multiply, $A > 0$ exists such that $K_1(0,A) = 0$;

e. if there are too many vacancies, they cannot multiply even in the absence of searchers: as vacancies cannot be filled, more are not created, $B > 0$ exists such that $K_1(B,0) = 0$;

f. multiplication of searchers decreases with their number because lack of jobs forces people to stay on the job, $\partial K_2/\partial N_2 < 0$;

g. even for a constant ratio of N_1/N_2, the multiplication of searchers is increased by an increase in the number of vacancies as dissatisfied workers now seek jobs, $dK_2/dS > 0$; and,

h. if there are not enough vacancies, seekers cannot multiply; i.e., a critical number of vacancies is required before seekers can increase, $C > 0$ exists such that $K_2(C,0) = 0$.

This provides a plausible analogy between a labor market cycle and the Lotka-Volterra equations. The Phillips curve is an empirical relation that can be expressed in terms of vacancies and unemployment and thus by the vacancy-search equations above. In fact, the Phillips curve has a cyclical component with about a nine year period which in its idealized form appears as Figure 2. It is realized that the existence of this relationship is questioned and that other hypotheses are entertained; however, we merely wish to point to there being some factual justification for an equation which can have cyclic solutions.

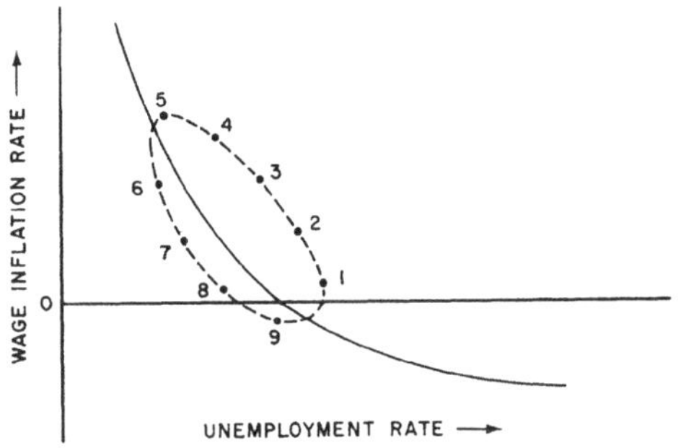

Fig. 2. Idealized Phillips curve and nine year cyclical component.

Micro-Behavior

The value of the coefficient in the Lotka-Volterra system of equations may be regarded as given phenomenologically. However, since these interactions are due to a highly complex web of, on the one hand, education, training, social flexibility and occupational complementarity, and, on the other, information dissemination, job search and the geographical distribution of occupations, the coefficients could presumably be derived from a micro-behavioral theory. However, such a theory requires a separate study, and here we shall follow Lewis (1977) in deriving the Lotka-Volterra coefficients from an elementary search model.

Consider a situation of equally effective searchers examining a set of equally likely vacancies: some fraction of searchers do not find a vacancy. Suppose $N_s(\tau)$ is the number of vacancies in interval τ and q_{sm} is the probability of never finding a particular vacancy (<1). Let

$q_{sm}^{N_s(\tau)} = \exp[(\log q_{sm})N_s(\tau)]$, the probability of never finding any vacancy. If $P_m(\tau) = (1 - q_{sm}^{N_s(\tau)})$, the probability of finding some vacancy, then it equals $1 - \exp[-\alpha N_s(\tau)]$ where $\log q_{sm} = -\alpha$.

The probability that all searchers miss a particular vacancy is q^{N_m} so that the probability that a particular vacancy is filled is $1 - q^{N_m}$. The probability that a particular searcher misses all vacancies is q^{N_s} so that the probability that he obtains a job is $1 - q^{N_s}$. As only a tiny fraction of either population is involved in encounters in one interval, $1 - q^{N_m}$ and $1 - q^{N_s}$ are very small so that they can be expressed approximately as $(\log q)N_m$ and $(\log q)N_s$. The rate of vacancy filling is $N_m(\log q)N_s$ just as the rate of searchers finding jobs is $N_s(\log q)N_m$, which is written $\alpha N_s N_m$. This is the Lotka-Volterra relationship and emerges only for small intervals, identical searchers or vacancies and independent individuals within each population. We can introduce non-uniform search, such as differences in efficiency or opportunities, without changing the results. By specifying the state of effectiveness (i) of a searcher we obtain

$$p_{mi}(\tau) = 1 - \exp|-\alpha_i N_s(\tau)|, \text{ where } \alpha_i = |\log q_{sm}(i)|, \text{ and}$$

$$p_m(\tau) = \sum_{i=1}^{k} p_i\{1 - \exp[-\alpha_i N_s(\tau)]\} \quad .$$

This function is shown in Figure 3: its shape varies widely depending on α_i and the corresponding p_i, but it always starts at the origin, has a monotonically decreasing positive slope and asymptotically approaches unity.

Stability of Vacancy Chains

What is meant by stability of occupational structures? These economic interactions are externalities and thus prices do not provide immediate equilibrating forces. Consequently we surmise that instability means the rules on which the structures are based are impossible and thus such structures cannot exist. If inter-

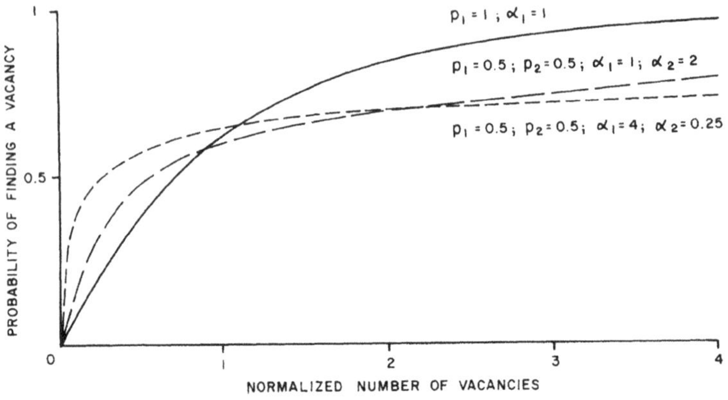

Fig. 3. Success rate as a function of target population in non-uniform search [Lewis (1977)].

actions are stable then the configurations can exist and the equations are meaningful. We have a system characterized by some differential equation $dx_i/dt = f_i(x_1,x_2,...,x_n)$. Its local stability in the neighborhood of equilibrium depends on the matrix of a_{ij}'s $= \partial f_i/\partial x_i$. When all the characteristic roots of the matrix are negative, the system is stable and asymptotically returns to equilibrium after displacement. When any roots are positive there is local instability, and for complex roots the system will oscillate, with damping if the real part is negative and increasing if the real part is positive.

A community of n interacting occupations may be modelled [Rescigno and Richardson (1973)] by allowing the per capita growth rate to be a linear function of the population sizes n_1:

$$n_i^{-1}(dn_i/dt) = w_i + \sum_{j=1}^{n} w_{ij}n_j \quad .$$

Consider a small perturbation from the equilibrium of such a system so that

$$n_i = \hat{n}_i + e_i \; ; \; e_i \text{ small.}$$

Then

$$de_i/dt = (\hat{n}_i + e_i) \sum_{j=1}^{n} w_{ij}e_j \quad .$$

Neglecting terms of order e^2, a linear system results:

$$de_i/dt = \sum_{j=1}^{n} (w_{ij}n_i)e_j \quad .$$

Behavior around equilibrium is determined by the community matrix $\underline{A}$ with elements $a_{ij} = w_{ij}\hat{n}_i$. The solution of the preceding equation gives

$$e_i = \sum_{j=1}^{n} k_{ij} \exp(\lambda_j t) \quad , \tag{1}$$

where λ_i, $i = 1,2,\ldots,n$, are the eigenvalues of $\underline{A}$ determined as the real or complex roots of the equation

$$\det(\underline{A} - \lambda\underline{I}) = 0 \quad , \tag{2}$$

and k_{ij} are constants determined by the initial conditions. In (1) if $\lambda_i > 0$ or the real root of $\underline{A}$, $Re(\lambda_i) > 0$, then the term $\exp(\lambda_i t)$ will increase without limit. However, if $Re(\lambda_i) < 0$ then the terms will vanish toward zero with time, and populations will return to their equilibrium values. True local stability occurs when the eigenvalues of the matrix $\underline{A}$, the roots of (2), all have negative real parts. Consider the matrix [Levins (1973)]

$$\underline{M} = \begin{pmatrix} a_{11} & a_{12} \\ a_{21} & a_{22} \end{pmatrix} \sim \begin{pmatrix} - & - \\ + & - \end{pmatrix} \quad .$$

Its characteristic equation is

$$(a_{11} - \lambda)(a_{22} - \lambda) - a_{12}a_{21} = 0 \quad , \text{ or}$$

$$\lambda^2 - (a_{11} + a_{22})\lambda + a_{11}a_{22} - a_{12}a_{21} = 0 \quad .$$

The roots are

$$\lambda = 0.5\{a_{11} + a_{22} \pm [(a_{11}+a_{22})^2 - 4(a_{11}a_{22}-a_{12}a_{21})]^{\frac{1}{2}}\}$$

$$= 0.5\{a_{11} + a_{22} \pm (a_{11}-a_{22})^2[1 + 4a_{12}a_{21}/(a_{11}-a_{22})^2]^{\frac{1}{2}}\}.$$

Expanding the term in square brackets

$$\lambda = a_{11}+2a_{12}a_{21}/(a_{11}-a_{22}) + \text{terms of order } (a_{12}a_{21})^2/(a_{11}-a_{22})^3$$

and greater, and

$$\lambda = a_{22}-2a_{12}a_{21}/(a_{11}-a_{22}) + \text{terms of order } (a_{12}a_{21})^2/(a_{11}-a_{22})^3$$

and more. When $|a_{11}-a_{22}|$ is large compared to the coupling term $a_{12}a_{21}$, the characteristic roots are roughly the diagonal terms a_{11}, which they would be exactly if the variables were independent. The criterion for negative real parts of both roots is that the coefficient to λ and the constant term in the characteristic equation are both positive. This will always be the case for the simple vacancy-search matrix shown above. However, if a_{11} is positive, then the values of the parameters will be needed to decide the issue.

VACANCY CHAINS

General Equations

We now turn to the heart of this inquiry which involves a large transition matrix, the n species case [Rescigno and Richardson (1973)]. The number of encounters between occupations r and s is $\alpha_{rs}N_rN_s\Delta t$; $\alpha_{rs} \geq 0$, $\alpha_{rs} = \alpha_{sr}$. Let b_r be the fraction of encounters resulting in vacancies in r being filled. Then

$$\Delta N_r = \varepsilon_r N_r \Delta t - b_r \alpha_{rs} N_r N_s \Delta t \quad .$$

But vacancies in r may be filled from several occupations so that

$$\Delta N_r = (\varepsilon_r - b_r \sum_s \alpha_{rs} N_s) N_r \Delta t \quad .$$

Taking account of the fact that r provides both vacancies and searchers, then

$$\Delta N_r = (\varepsilon_r - b_r \sum_s \alpha_{rs} N_s + d_r \sum_v \alpha_{rv} N_v) N_r \Delta t \quad ,$$

where the first summation refers to all occupations which supply r and the summation over v includes all occupations supplied by r. Dividing by Δt and taking the limit gives the system equations.

When a search results in an increase in occupation r, let $a_{rs} = \alpha_{rs}$. Occupation r increases by $b_r a_{rs} N_s N_r \Delta t$ and occupation s descreases by $d_s a_{rs} N_r N_s \Delta t$ because of this interview. Assume for simplicity that the probability of being supplied by a searcher during one operation is the same as the probability of filling a vacancy during a search, i.e.,

$$b_r = d_r = 1/\beta_r ; (\beta_r > 0) \text{ and thus } |a_{rs}| = \alpha_{rs} \text{ and } a_{rs} = -a_{sr} .$$

At an interview, the increase in occupation r is $B_r^{-1} a_{rs} N_r N_s \Delta t$, and the decrease in s vacancies is $B_s^{-1} a_{sr} N_s N_r \Delta t = -\beta_r^{-1} a_{rs} N_r N_s \Delta t$. At an interview, B_s^{-1} individuals of occupation s are transformed into B_r^{-1} individuals of occupation r. The sign of a_{rs} determines the direction of the change. The system equations are

$$dN_r/dt = N_r(\varepsilon_r + \beta_r^{-1} \sum_{s=1}^{n} a_{rs} N_s) \quad , \tag{3}$$

where $a_{sr} = -a_{rs}$, $a_{rr} = 0$, $\beta_1 > 0, \beta_2 > 0, \ldots, \beta_n > 0$.

β_r^{-1} is the mean change in population size per interview.

This model fails for two reasons: there turns out to be an artificial distinction between communities with an even or an odd number of occupations; and, certain of the populations can become infinite. A limiting term is required to decrease the coefficient of increase as the population increases. We cannot have a coefficient of self-increase which is constant so that the rate of increase is made proportional to population size: a term can be added which varies linearly with size, altering the coefficient to $(\varepsilon_r - \lambda_r N_r)$; $(\lambda_r > 0)$. The system equations (3) become

$$\beta_r dN_r/dt = (\varepsilon_r \beta_r - \lambda_r \beta_r N_r + \sum_{s=1}^{n} a_{rs} N_s) N_r \quad ,$$

where at least one $\lambda_r \neq 0$. A stationary state exists if positive roots $N_r = g_r$ occur for the equation of the bracketed term set equal to zero. It may be shown that the numbers in each occupation where the limiting coefficient $\lambda_r \neq 0$ will tend to a stationary state and that those where $\lambda_r = 0$ have upper and lower bounds.

Random Structure

Next consider the case of a random matrix of m occupations [May (1971), Maynard Smith (1974)]. Let each occupation have a stable equilibrium density which it approaches at a constant rate; i.e., $a_{jj} = -1$ for all j. All other $a_{jk} (j \neq k)$ are randomly distributed with mean of zero and variance σ^2 so that σ is the mean intensity of interaction between occupations. (-,-) and (+,+) interactions are equally common, and together these are as common as (+,-). If $m >> 1$ and if $\sigma < (2m)^{-0.5}$ then the system is stable, while the system is unstable if $\sigma > (2m)^{-0.5}$. If the a_{jk} are selected with probability c and put equal to zero with probability (1-c) so that c can be taken to represent the probability of direct interaction between any two occupations, the proportion of non-zero elements in the matrix or "connectance," then May believes the stability criterion to be $\sigma < (2mc)^{-0.5}$. Thus, in these circumstances, as m or c increases the average intensity of interactions, σ, compatible with stability, decreases: the transition between the two modes of behavior is sharp for $m >> 1$. A complex community is stable only when connections are non-random. Since σ and c are inversely proportional, May suggests that in existing systems where the web of connections, c, is large, the intensity of interactions is small and vice versa. He also asserts that for a system with given σ and c, stability is increased by forming smaller subsystems with most of the connectance concentrated within rather than between them.

Competitive Vacancies

Consider first the vacancies generated by a set of competitive occupations and filled equally by a single homogeneous group of searchers [Leigh (1975)]. The commodity matrix is

$$\underline{A} = \begin{pmatrix} -M'ab_{11} & -M'ab_{12} & \cdots & -M'ab_{1n} & -Pf \\ -M'ab_{1n} & -M'ab_{11} & \cdots & -M'ab_{1,n-1} & -Pf \\ \cdot & \cdot & & \cdot & \cdot \\ M'cf & M'cf & \cdots & M'cf & -r \end{pmatrix},$$

where b_{ij} is from the overlap matrix $\underline{B}$, i.e., the degree of competition between pairs of occupations, M' the average population of each competitive occupation, P the average population of searchers r the strength of the searchers' "self-damping" or territoriality, c the proportion of vacancies filled/searchers, f the proportion of successful searchers/vacancies. It can be shown that the eigenvalues are

$$-M'a\lambda_n\ , -M'a\lambda_{n-1}\ , \ldots, -M'a\lambda_2\ , \text{ and}$$

$$-0.5\{r + M'a\lambda_1 \pm [(r - M'a\lambda_1)^2 - 4ncPM'f^2]^{\frac{1}{2}}\} \quad ,$$

where $\lambda_i = \sum_j b_{ij}$ is the largest eigenvalue of the overlap matrix $\underline{B}$ and the λ_i, $i > 1$, are the others. The minimum eigenvalue of $\underline{A}$ is likely to be lower than one of those involving λ, so that searchers will affect stability only through their effect on M'.

Turn now to a vacancy chain ordered by levels. The log-linear approximation for this community is

$$dx_i/dt = k_i(t) - M'a \sum_{j=1}^{n} b_{ij}x_j(t) - P \sum_{j=1}^{n} f_{ij}y_j(t) \text{ , and}$$

$$dy_i/dt = k_{n+i}(t) - ry_i(t) + cM' \sum_{j=1}^{n} f_{ji}x_j(t) \quad .$$

M' is the average number of vacancies so that $x_i = \log(N_i/M')$ is the relative excess of i's population over its average. P is the average number of searchers so that $y_j = \log(N_j/P)$ is the relative excess of j's population above mean, c and r are as above. If the searchers are not territorial, the community matrix will be

$$\begin{pmatrix} -M'aB & -PF \\ -M'cF & 0 \end{pmatrix} .$$

$\underline{B}$ is the matrix b_{ij} of competitive overlap among vacancies, $\underline{F}$ is the matrix f_{ij} of searchers' success rates. Both of these are assumed to be symmetric and cyclic. The minimum eigenvalue will be one of k eigenvalues

$$\lambda_k = 0.5\{M'a\lambda_{1k} - [M'^2a^2\lambda_{1k}^2 - 4cQ^2\lambda_{2k}(M'/PB)]^{\frac{1}{2}}\} \quad ,$$

with $\lambda_{1k} = \sum_j b_{1j}\exp[-2\pi i(k-1)(j-1)/n]$ is the k-th eigenvalue of the vacancies' competitive overlap matrix $\underline{B}$.
$\lambda_{2k} = \sum_j B_{1j}\ \exp[-2\pi i(k-1)(j-1)/n]$ is the k-th eigenvalue of the searchers' degree of competition matrix, $B_{ij} = \sum_k f_{ik}f_{jk}/\sum_n f_{ik}^2$,

B is the searchers' breadth $(\sum_j f_{1j})^2/\sum_j f_{1j}^2$, and Q is the rate of filling of vacancies. The community stability can be no greater than 0.5 $M'a\lambda_1$, where λ_1 is the minimum eigenvalue of the vacancies overlap matrix. If there is a lot of overlap among searchers some λ_{2k} will be small, or if the overlap matrices differ so that λ_{1k} and λ_{2k} are minimum for different k then stability will be much less. Assuming equal stability of existing communities, then the complexity of a vacancy chain is a variable which can afford to be large only in a stable environment.

Preferred Occupations

For m competing occupations with vacancies $N_i(t)$ and m occupations supplying seekers $P_i(t)$, the quadratically non-linear equations are [May (1974)]

$$dN_i/dt = N_i(k_i - \sum_j \alpha_{ij}N_j - \sum_j \beta_{ij}P_j) \quad , \text{ and}$$

$$dP_i/dt = P_i(-g_i + \sum_j \gamma_{ij}N_j) \quad .$$

Assume: (1) all $k_1 = k_o$; (2) all $g_i = g_o$; (3) the competition coefficients, $\alpha_{ij} = 1$ if $i = j$, and $\alpha_{ij} = \alpha$ if $i \neq j$; and (4) the supply intensities, $\beta_{ij} = \beta_o$ and $\gamma_{ij} = \gamma_o$ if $i = j$, and $\beta_{ij} = \beta$ and $\gamma_{ij} = \gamma$ if $i \neq j$. This latter condition states that seekers from a given occupation will have a preferred occupation to go to. Otherwise the other occupations are equally acceptable. The 2m-by-2m matrix is

$$\underline{A} = \begin{pmatrix} -N^*\alpha & -N^*\beta \\ P^*\gamma & 0 \end{pmatrix},$$

where $N_i^* = N^*$ and $P_i^* = P^*$ are equilibrium populations, and the m-by-m matrices $\underline{\alpha}$, $\underline{\beta}$, $\underline{\gamma}$, are as defined above. The eigenvalues relevant to the stability of individual populations are given by

$$\lambda^2 + \lambda N^*(1 - \alpha) + N^*P^*(\beta_o - \beta)(\gamma_o - \gamma) = 0 \quad .$$

In general, the solution tends to dampen oscillations with damping rates $0.5N^*(1-\alpha)$, especially if each occupation with vacancies has its own specific search occupation ($\beta=\gamma=0$). Where there are no preferences ($\beta = \beta_o$; $\gamma = \gamma_o$) the situation is the same as for the many vacancy occupations-single seeker occupation case with oscillations. No model is needed to demonstrate that seekers who switch preferences from occupations with vacancies becoming relatively scarce to those becoming relatively plentiful will tend to stabilize the communities they dwell in.

Vacancy Chain Fluctuations

Communities of r species are described by the equation

$$dN_i/dt = e_i N + \sum_{i=1}^{r} a_{is} N_i N_s \qquad (4)$$

the a_i's being interaction coefficients ($a_{is} = -a_{si}$). Writing N_i in terms of the other N_s occupations yields

$$d(\log N_i)/dt = e_i + \sum_{s=1}^{r} a_{is} q_s (N_s/q_s)(N_s/q_s) \quad ,$$

where q_i are the equilibrium values of the N_i, i.e., for which $dN_i/dt = 0$. Setting $x_i = \log (N_i/q_i)$, it can be shown that

$$dx_i/dt = \sum_{s=1}^{r} a_{is} q_s [\exp(x_s) - 1] \quad .$$

These equations may be integrated to give

$$\sum_{i=1}^{r} q_i [\exp(x_i) - x_i] \equiv G = \text{constant} \quad .$$

Thus it is possible to write

$$dx_i/dt = \sum_{s=1}^{r} a_{is} (\delta G/\delta x_s) \qquad (5)$$

which are now the equations of motion of the system [Leigh (1968)]. From (5) it can be shown that the time average of N_i is q_i, and the time average of $(N_i - q_i)^2/q_i$ is the same for all occupations so that the plot of population size against its variance for each occupation in the community is a straight line. When there are a large number of equations, x_i can be shown to be distributed as

$$C \exp\{-(q_i/\theta)[\exp(x_i) - x_i]\}$$

so that N_i is distributed as

$$P(N_i) = [\Gamma(q_i/\theta)]^{-1} \theta^{-q_i/\theta} N_i^{q_i/\theta-1} \exp(-N_i/\theta) \quad .$$

The average of $x_i(dx_i/dt)$ is $a_{ri}\theta$, which equals the covariance

$$r_{is}(s) = (d/ds)[\lim_{T\to\infty} (2T)^{-1} \int_{-T}^{T} x_i(t)x_r(t + s)dt]_{s=0} \quad .$$

Substituting $N_i = q_i + v_i$ and using a linear approximation in (4) results in

$$(dv_i/dt) = \sum_s q_i a_{is} v_s \quad .$$

Setting $G = v_i^2/2q_i\theta$ yields

$$(dv_i/dt) = \sum_s a_{is} q_i q_s (\delta G/\delta v_s) \quad ,$$

where the coefficients are the rates of flow between the occupations involved.

$$P(v_i) = (2\pi q_i\theta)^{-\frac{1}{2}} \exp(-v_i^2/2q_i\theta) \quad ,$$

with variance in population size $q_i\theta$. It may be shown that the frequency with which the population curve $v_i(t)$ crosses the arbitrary threshold $v_i = a$, i.e., frequency of zeros of the function $v_i(t) - a$, is obtained as a function of the variance and covariances

$$\pi^{-1} \exp(-a^2/2q_i\theta)(\sum_r q_i q_r a_{ir}^2)^{\frac{1}{2}} \quad .$$

Alternatively, consider population fluctuations relative to the equilibrium size. Letting $v_i/q_i = x$, the frequency of zeros of $x_i(t) - a$ is

$$\pi^{-1} \exp(-q_i a^2/2\theta)(\sum_r q_i q_r a_{ir}^2)^{\frac{1}{2}} \quad .$$

Minimizing the average of $\sum_{r=1}^{k} a_{ir}^2 q_i q_r$ for k occupations provides a rough approximation of the minimum of the frequency of gross fluctuations. The a_{is}'s which do this will specify in some sense the community structure of optimal stability. It is preferable to add the constraint that the productivity of the community, i.e., the total flow through the system, is equal to P where $P = 0.5 \sum_{i,s} |a_{is}| q_i q_s$. To minimize $\sum_{i,s} a_{is}^2 q_i q_s - \lambda(\sum_{i,s} |a_{is}| q_i q_s - 2P)$ is thus the aim. Using Lagrangian multipliers,

$$(\Sigma q_i)^2 - \Sigma q_i^2 = (\Sigma q_i)^2 = B^2 \quad ,$$

where B is the total number of vacancies in the community,

$$|a_{is}| = 2P/B^2 \quad ,$$

and the minimum of the average of $\sum_{r=1}^{k} a_{is} q_i q_s$ is found to be

$$k^{-1} \sum_{i,s=1}^{k} a_{is}^2 q_i q_s = 4P^2/kB^2 \quad .$$

The average frequency of population decline per occupation per unit time is proportional to

$$(2P/Bk^{1/2})\exp(-q_i a^2/2\Theta) \quad .$$

Stability thus increases with k, the number of occupations tapped for recruits, it decreases with P/B, the rate at which occupations are renewed and it decreases with Θ which is the variance.

SPATIAL MODELS

Spatial Operator

So far we have not been concerned with the spatial aspects of vacancy chains to any extent. It should be recalled that this approach, by definition, is investigating a steady state and so cannot follow the spatial manifestations of change. The diffusion of fortuitous change brought about by accidents of sampling has been analyzed elsewhere [Curry (1980a)] and it is hoped that purposeful selection in space will also be treated. We have the very much more limited role of investigating how job search in space will affect the stability of occupational structures. We conceive of information about vacancies passing from a workplace through acquaintanceship networks to applicants and eventually a recruit. Equally, the desire for a job can be signalled by a worker and disseminated via the same network until either interviews are arranged and a new post obtained or the search is abandoned.

The basic spatial mechanism is the spread of information that is thought of as a two-dimensional random walk on a square lattice in the plane. The probability of the message being at location (m,n) at time s+1, given that it could have been in a neighboring location at time s is given as

$$P(m,n,s+1) = .25[P(m+1,n,s)+P(m-1,n,s)+P(m,n+1,s)+P(m,n-1,s)].$$

Going to the limit the partial differential equation is

$$\partial P/dt = K[(\partial^2 P/\partial x^2) + (\partial^2 P/\partial y^2)]$$

$$= K\nabla^2 P \quad .$$

∇P is the gradient of $P = (\partial P/\partial x)i + (\partial P/\partial y)j$, where i and j are the unit component orothogonal vectors. Its solution is

$$P(m,n,s|m_o,n_o) = (2\pi s)^{-1} \exp\{[-(m-m_o)^2 + (n-n_o)^2]/2t\} \quad .$$

Diffusion and Stability

We may examine the conditions of linear stability for diffusion equations. Let the change in density of a population in a bounded one-dimensional space [Percus (1977)] be

$$\partial u/\partial t = 0.5\, \sigma^2(\partial^2 u/\partial x^2) + \alpha u \quad ,$$

where α is the intrinsic growth rate. Solutions are required of the form

$$u = \exp(ikx + pt) \tag{6}$$

which amounts to taking a Fourier transform in position x and a Laplace transform in time t; k is the wave number and p is the amplification rate. The dispersion relation

$$p = -0.5\, \sigma^2 k^2 = \alpha \quad (p > 0)$$

must be satisfied. The closed space requires boundary condition

$$u(-L,t) = u(L,t) = 0 \quad ,$$

and this allows only certain k values to appear in the solution.

$$u(x,t) = \sum_{n=1}^{\infty} A_n \exp(p_n t) \sin(n\pi x/L) ,$$

so that $k_n = \underline{+}n\pi/L$ and $P_n = \alpha - 0.5k_n^2\sigma^2$. A wavelength is unstable, satisfying the condition $(\pi/L)^2 0.5\sigma^2 < \alpha$. There is a critical length below which $L < L^*$ and the population dies out, whereas above it $L < L^*$ and the population grows. That there should be a minimum size of area for a population to exist, given that σ and α are favorable, is somewhat unexpected. Dubois (1975) emphasizes the stabilizing effects of the mobility of the populations. To introduce spatial effects he brings in the diffusion of a vacancy and of a searcher by modifying the Lotka-Volterra equations to:

$$dN_1/dt = k_1N_1 - k_2N_1N_2 + D_1\partial^2N_1/\partial r^2 \quad , \text{ and} \qquad (7)$$

$$dN_2/dt = -k_3N_2 + k_4N_1N_2 + D_2\partial^2N_2/r^2 \quad , \qquad (8)$$

where D_i is the appropriate coefficient of diffusion. For a stationary situation in uniform space,

$$N_{1,0} = k_3/k_4 \quad \text{and} \quad N_{2,0} = k_1/k_2 \quad .$$

In order to examine their stability, solutions of the form

$$N_j(r,t) = N_{j,0} + n_j \exp(wt + ir/\lambda) \qquad (j = 1,2)$$

are required. n_j, w and λ are constants. Population j will be stable only if the real part of w is negative or xero. (8) can be approximated by

$$n_1w \simeq -(k_2k_3/k_4)n_2 - D_1n_1/\lambda^2 \quad , \text{ and}$$

$$n_2w \simeq \underline{+}(k_1k_4/k_2)n_1 - D_2n_2/\lambda^2 \quad .$$

Values of w are obtained as

$$w = -(D_1 + D_2)/2\lambda^2 - [(D_1 + D_2)^2/4\lambda^4 - D_1D_2/\lambda^4 + k_1k_3]^{\frac{1}{2}} \ . \quad (9)$$

Equation (9) shows that w has a negative real part so that the addition of diffusive terms has forced asymptotic stability on the equations.

Percus (1977) applies this approach to vacancy-search interactions with different diffusion rates in one dimension, such that

$$\partial u/\partial t = \mu(\partial^2 u/\partial x^2) + au + bv \quad , \text{ and}$$

$$\partial v/\partial t = \nu(\partial^2 v/\partial x^2) + cu + dv \quad .$$

Using the transforms of (7) he obtains

$$p\underline{I} = \begin{pmatrix} a-\mu k^2 & b \\ c & d-\nu k^2 \end{pmatrix} = \underline{B}$$

It is thus possible that while an interaction matrix

$$\underline{A} = \begin{pmatrix} a & b \\ c & d \end{pmatrix}$$

is stable, $\underline{B}$ could be unstable for some k. Consequently, diffusion can excite instability. In general terms, when

$$\partial u/\partial t = \underline{\nu}(\partial^2 u/\partial x^2) + \underline{A}u$$

where ν is a diagonal matrix having non-negative entries and $\underline{A}$ is an n-by-n matrix, then if $0.5(\underline{A}^{Tr} + \underline{A})$ is stable there will be no diffusive instability.

Migration Waves

Kerner (1972) is interested in migration, not only that connected with the job vacancy-search process but also that separate from it, say house vacancy-search. In both cases the movement of information can be regarded as performing a random walk so that in sum there will be diffusion down the gradient of acquaintanceship. This will provide a net flow of house applicants unconnected with occupation so that from within an occupation it will appear as self diffusion:

$$j_i = -D\nabla\rho_i \quad .$$

There will also be a flow of job acceptances from various occupations

$$j_i = -\sum_j D_{ij}\nabla\rho_i \quad ,$$

where D_{ij} are density ρ dependent. For the simple vacancy (1)-search(2) case,

$$j_1 = -D_1\nabla\rho_1 - d_1\nabla\rho_2 \quad \text{, and}$$

$$j_2 = -D_2\nabla\rho_2 + d_2\nabla\rho_1 \quad .$$

Thus there develops a set of diffusional rates composed of the self diffusion of each occupation and the drifts between jobs experienced via the local vacancy-search gradients for each pair of occupations. It is these complementary occupational relationships operating via spatial diffusion which prevent spatial differences from being eliminated, as they would for simple diffusion. Under Volterra mechanics, this will lead to

$$\partial\rho_1/dt = \varepsilon_1\rho_1 - \alpha_1\rho_1\rho_2 + D_1\nabla^2\rho_1 + d_1\nabla^2\rho_2 \text{ , and} \tag{10}$$

$$\partial\rho_2/dt = -\varepsilon_2\rho_2 + \alpha_2\rho_1\rho_2 + D_2\nabla^2\rho_1 - d_2\nabla^2\rho_1 \quad . \tag{11}$$

In equilibrium, $\rho_1 = \varepsilon_2/\alpha_2$ and $\rho_2 = \varepsilon_1/\alpha_1$, so that the first two terms on the right hand side of both (10) and (11) sum to zero. Consider small deviations from equilibrium Q_1, Q_2 by writing

$$\rho_1 = (\varepsilon_2/\alpha_2) + Q_1 \quad \text{and} \quad \rho_2 = \varepsilon_1/\alpha_1 + Q_2 \quad .$$

Then

$$\partial Q_1/dt - D_1\nabla^2 Q_1 - d_1\nabla^2 Q_2 + \lambda_1 Q_2 = 0 \quad \text{, and}$$

$$\partial Q_2/dt - D_2\nabla^2 Q_2 + d_2\nabla^2 Q_1 - \lambda_2 Q_1 = 0 \quad ,$$

where

$$\lambda_1 = \alpha_1\varepsilon_2/\alpha_2 \quad \text{and} \quad \lambda_2 = \alpha_2\varepsilon_2/\alpha_1 \quad .$$

Because of the imposed linearity Q_1Q_2 is not present, and Q_1 and Q_2 can be decoupled, satisfying

$$(\partial^2 Q/\partial t^2) - (D_1+D_2)\nabla^2(\partial Q/\partial t) + (D_1D_2+d_1d_2)\nabla^4 Q - (\lambda_1 d_2+\lambda_2 d_2)Q = 0.$$

Where self-diffusion involving the mixed term $\nabla^2(\partial Q/\partial t)$ is negligible ($D_1, D_2 \simeq 0$) so that dissemination and search are the sole driving forces for movement, plane waves propagate $Q = \exp[i(kr-wt)]$, where k is the wave number and w is frequency. Let $(d_1d_2)^{\frac{1}{2}} = c$, $[k^2 + (\lambda_1/d_1)]^{\frac{1}{2}} = a$, and $[k^2 + (\lambda_2/d_2)]^{\frac{1}{2}} = b$. Then $w = cab$. Let $a/b = A$. Then

$(\partial w/\partial k) = kc(A^{-1} + A)$ with all k present. Then self-diffusion propagation of the waves is damped according to $w = \alpha - iB$, where $B = 0.5(D_1+D_2)k^2$. Since $-k^2$ is involved it is the larger wave numbers which are most strongly damped. Hence,

$$\alpha^2 = k^4[d_1d_2 - 0.25(D_1-D_2)^2] + k^2(\lambda_1 d_2+\lambda_2 d_1) + \lambda_1\lambda_2$$

so that if $d_1d_2 < (D_1-D_2)^2/4$ the coefficient of k^4 is negative and the spectrum is truncated at larger wave numbers.

Uniformity and Patchiness

Whereas so far we have pointed to spatial behavior as the factor which stabilizes a vacancy-search occupational matrix, we now examine it as a destabilizing activity on a prior uniform spatial distribution [Levin (1976)]. Denoting vacancies by V, searchers by E and the spatial gradient operated by ∇:

$$\partial V/\partial t = V(K - \alpha V - \beta E) + \nabla\cdot(\mu\nabla V) \quad , \text{ and}$$

$$\partial E/\partial t = E(-L + \gamma V - \partial E) + \nabla\cdot(\nu\nabla E) \quad .$$

Only conditions near equilibrium are analyzed so that linearity may be assumed. Let $K > 0$, $\beta > 0$, $\gamma > 0$ and $L = 0$. In one-dimensional form,

$$\partial\bar{v}/\partial\bar{t} = (1 + \overline{kv}) - \overline{aev} + \nabla^2\bar{v} \qquad (a > \bar{k}) \text{ , and}$$

$$\partial\bar{e}/\partial\bar{t} = \overline{ev} - \bar{e}^2 + \theta^2\nabla^2\bar{e} \quad .$$

θ^2 is the ratio between the diffusion ability of the searcher compared to vacancies. There is a steady state spatially uniform condition

$$\bar{e} = \bar{v} = p^{-2} \ ,$$

where $p^2 = a - \bar{k} > 0$. Taking the deviations from the steady state

$$-\partial \underset{\sim}{V}/\partial\bar{t} + \underset{\sim}{M}_\theta \underset{\sim}{V} = \underset{\sim}{N}(\underset{\sim}{V}) \quad ,$$

where $\underset{\sim}{V} = (v,e)^\tau$, the deviations may be written as

$$\underset{\sim}{N}(\underset{\sim}{V}) = (aev - \bar{k}v^2, e^2 - ev)^\tau \quad , \text{ and}$$

$$\underset{\sim}{M}\theta = \begin{pmatrix} \bar{k}p^{-2}+\nabla^2 & -ap^2 \\ p^2 & -p^2+\theta^2\nabla^2 \end{pmatrix}.$$

Pertrubations of the form

$$\underset{\sim}{V} = \underset{\sim}{C} \exp(\sigma t) \exp(i\underset{\sim}{w}\cdot\underset{\sim}{x})$$

are used to investigate the stability of the uniform state.

If $\bar{k} \leq 0$, the equilibrium is stable;

$\bar{k} > 1$, the uniform distribution is unstable even when the perturbation is of infinite wavelength, i.e., $\underset{\sim}{w} = 0$ in which case no diffusion occurs; and,

$0 < \bar{k} < 1$, other than for zero frequency perturbations when stability occurs, stability depends on θ, the ratio of the diffusivities.

If $\theta = 1$ or, indeed, if $\theta < 1/[a^{\frac{1}{2}} - (a-\bar{k})^{\frac{1}{2}}]$ the equilibrium is stable [note that $a > \bar{k}$ and $1 > \bar{k} > 0$, so that $a^{\frac{1}{2}} - (a-\bar{k})^{\frac{1}{2}} < 1$]; and,

$\theta > 1/[a^{\frac{1}{2}} - (a-\bar{k})^{\frac{1}{2}}] = \theta_c$, perturbations of wave number $q = [a^{\frac{1}{2}} - (a-\bar{k})^{\frac{1}{2}}]/(a-\bar{k})^{\frac{1}{2}}$ will destabilize the uniform distribution.

Thus, diffusion can break up uniform vacancy-search distributions provided that searcher mobility is sufficiently larger than that of vacancies. However, as perturbations grow, non-linear effects lead to patchy distributions that provide a new steady state. According to Levin this also may happen for $\bar{k} > 0$.

While not intuitively obvious, it seems not unreasonable that patchiness could develop from uniformity. In the acquaintance network, if job information moves slowly and locally while search is much wider ranging, polarization could well occur as spatial differentiation. There seems to be an analogy here with the land-use differentiation that can occur from trade, depending on the relative mobilities of labor and traded goods [Curry (1970)].

CONCLUSION

We began our approach to labor markets looking for a valid approach to externalities in a positivistic framework. When interactions between economic agents are not regulated by prices it is common practice to adopt a normative stance by inventing fictional side payments to optimize some welfare function. As a criterion for allocation which eschews price we have sought positivistic validity by seeking temporal viability. It turns out that the very information flows that cannot be fitted into the classical matching of given preferences with given resources and technology are in fact an important ordering principle of the geography of these markets.

According to Boulding (1962) social systems are not dynamically stable in the small so that a set of simultaneous differential equations has only a very limited applicability to them. He waxes lyrical about "distant goals that are contemplated in the imagination" as providing stability, but one suspects that it may be romanticism rather than logic which is guiding him. It is unfortunate that the analytics of vacancy chains have proved so formidable that we are limited to statements about their stability. However, it is fair to say that we have learned more about the operation of this type of market than is possible from calculation of a numerical model.

Some of the general points made are: structures cannot be random; density dependence is required to limit increase; and, intricate connectivity requires weak interactions and/or the break-up into sub-systems. Stability is reduced when there is competition among searchers. It is increased when the number of occupations tapped for recruits is large. Stability declines with the rate at which occupations are renewed. Complexity does not increase stability and may in fact decrease it. However, it is clear that considering the community in its spatial context implies greater stability than when it is neglected. That waves can occur, that there must be minimum areas for existence or that patchiness can develop from uniformity seems remarkable. Clearly when attention is focused on the evolution of occupational structures rather than their existence properties their geography will prove equally basic.

REFERENCES

Bish, R. and Nourse, H., 1975. Urban Economics and Policy Analysis, McGraw Hill, New York.

Boulding, K., 1962. Conflict and Defense, Harper & Bros., New York.

Christiansen, F. and Fenchel, T., 1977. Theories of Populations in Biological Communities, Springer-Verlag, Berlin.

Curry, L, 1970. Geographical Specialisation and Trade, in: Studies in Regional Science, 2: 85-95.

Curry, L., 1979. Demand in the Spatial Economy: I Homo Deterministicus, Geographica Polonica, 42: 184-212.

Curry, L., 1980. Diffusion, The Spatial Structure of Occupations and Inflation, unpublished manuscript, Department of Geography, University of Toronto.

Dubois, D., 1975. Learning, Adaptation and Evolution of the Environment--Ecosystem Couple, in: Progress in Cybernetics and Systems Research, Vol. 1, R. Trappl and F.R. Pichler (eds.) John Wiley, New York.

Hudson, J., 1970. Elementary Models for Population Growth and Distribution Analysis, Demography, 7:3.

Hudson, J., 1975. Some Observations on Migration Theory for an Urban System, in: People on the Move, L.A. Kosinski and R.M. Prothero (eds.), Methuen, London.

Kerner, E., 1972. Gibbs Ensemble: Biological Ensemble, Gordon & Breach, New York.

Leigh, E., 1975. Population Fluctuations, Community Stability and Environmental Variability, in: Ecology and Evolution of Communities, M.L. Cody and J.M. Diamond (eds.), Harvard University Press, Cambridge, Mass.

Leigh, E., 1968. The Ecological Role of Volterra's Equations, in: Some Mathematical Problems in Biology, Vol. 1.

Levin, S., 1976. Spatial Patterning and the Structure of Ecological Communities, in: Some Mathematical Questions in Biology, Vol. 8.

Levins, R., 1973. The Limits of Complexity, in: Hierarchy Theory, H.H. Patee (ed.), George Braziller, New York.

Lewis, E., 1977. Network Models in Population Biology, Springer-Verlag, Berlin.

May, R., 1971. Stability in Multispecies Community Models, Bulletin of Mathematical Biophysics. 12: 59-79.

May, R., 1974. How Many Species: Some Mathematical Aspects of the Dynamics of Populations, in Some Mathematical Questions in Biology, Vol. 6.

Maynard Smith, J., 1974. Models in Ecology, Cambridge University Press, Cambridge, Mass.

Percus, J., 1977. Combinatorial Methods in Developmental Biology, Courant Inst. Math. Sci., New York University.

Resigno, A. and Richardson, I.W., 1973. The Deterministic Theory of Population Dynamics, in: Foundations of Mathematical Biology, Vol. III, R. Rosen (ed.).

Stewman, S., 1976a. Markov Models of Occupational Mobility: Theoretical Development and Empirical Support. Part 1: Careers, Journal of Mathematical Sociology. 4: 201-245.

Stewman, S., 1976b. Markov Models of Occupational Mobility: Theoretical Development and Empirical Support: Part 2: Continuously Operative Job Systems, Journal of Mathematical Sociology. 4: 247-278.

Taschdjian, E., 1975. Operational Niches in Human Ecosystems, in: Progress in Cybernetics and Systems Research, Vol. II, R. Trappl and F. de P. Hanika (eds.), John Wiley, New York.

White, H., 1970. Chains of Opportunity, System Models of Mobility in Organizations, Harvard University Press, Cambridge, Mass.

SPACE AND POPULATION STRUCTURE*

Robert R. Sokal and Daniel E. Wartenberg

Ecology & Evolution Department
State University of New York at Stony Brook

INTERPRETATION OF GEOGRAPHIC VARIATION PATTERNS

Biological processes at the population level occur in a spatial context. Pollen and seeds from plants are transported over various distances and in different directions. Animals disperse from their places of birth and some migrate over great distances. Populations of organisms of the same species and those of different species compete for resources which are distributed over space, and one of which may indeed be space itself. Frequently individuals or populations demarcate home territories which they define and whose resources they exploit.

It is quite rare that biologists can witness spatial events at the time of their occurrence. Barring spectacular population phenomena such as the periodic migrations of lemmings or monarch butterflies, or the population explosions in some of the pest species, spatial population processes such as dispersal, diffusion or competition are difficult to document observationally and to quantify. Moreover, many such processes occur over a time span large enough to tax an observer's patience and longevity. For these reasons, much of our knowledge of population processes in nature must come from inferences based on observed patterns of variation over space (synoptic inference).

The problems and limitations encountered in these studies are, of course, thoroughly familiar to analytic geographers concerned with human geography and demography. Although many of the parameters of contemporary dynamic spatial processes can be estimated by censuses and questionnaires, various aspects of the movement of humans, ideas, goods and artifacts are difficult to ascertain, especially in primitive areas and in an historical context. The methods employed by statistical geographers to explore problems of this nature have obvious relevance and application to ecology and evolution, but until very recently there has been almost no intellectual contact between these disciplines. Similarly, recent work by population biologists that takes advantage of the unique properties of biological systems to aid in the interpretation of the dynamic spatial processes may help geographers in interpreting such processes in humans in cases where biological properties are examined, and by analogy may suggest new ways of approaching traditional spatial problems even in a non-biological context. Thus this paper is presented with the aim of increasing this reciprocal exchange of ideas between population biology and analytical geography.

Although there are numerous ways of approaching biological processes in space, the discussion presented here is restricted to the examination and analysis of geographic variation patterns. The geographic variation relates to values of a variable mapped in a plane representing geographic space. By analogy with topographic surfaces in geography or response surfaces in biostatistics, we can speak of the biological variable as representing a surface. Although a continuous surface may exist for any one biological variable for a given geographic area, we are unlikely to know its value at any given geographical location, and we must therefore rely on estimates of the value of the variable at suitably chosen stations over the area. Biologists have contributed to solving this problem of the reliability of inferences about the entire surface from sampling stations, and it will not be further pursued in this paper. For our subsequent discussion we shall consider the sampling stations as representative of the surface.

Given a set of such samples, the student of geographic variation first asks whether there are indeed differences in surface level among the sampling stations. Expressed more precisely, there must be an initial test of the null hypothesis that the surface is a horizontal plane. If this null hypothesis cannot be rejected, there is little point in further analysis of the data, although it may be of considerable interest biologically why a variable that is potentially subject to variation should be equivalent at all sampling stations.

Given that there are differences in surface level, one examines if the values Z_{ij} at each of the coordinate points i, j are independent of the values of the variable at points $i + \delta$, $j + \varepsilon$ where δ and ε are suitably chosen increments in latitude and longitude. If values at a given sampling station lack independence from those at contiguous sampling stations, then we can speak of pattern in the surface. The nature of the pattern will depend on the type of observed departure from independence.

Some inferences about population structure can be made from the depiction and description of patterns in a single surface. More information is usually provided by the analysis of several surfaces representing either different variables or parameters of the same biological population, or the same variable in a given population observed at different times. Yet another profitable approach is to study the same variable in two different but comparable and contemporaneous populations.

KINDS OF SURFACES

Study areas may be continuous, or discontinuous and fragmented. Examples of the former range from regularly laid out field plots to continental land masses, whereas the latter are most clearly exemplified by islands, whether these are real islands as in an archipelago, or ecological islands such as isolated copses in a grassland area, individual creosote bushes in the desert or an assemblage of unconnected pools or lakes. The distinction between continuous and discontinuous areas is far from absolute. In fact, most real cases would fall between these extremes because habitats within areas are not equally favorable for all organisms, and dynamic spatial processes will consequently be anisotropic.

An added consideration is the peculiarity of the shape of an area and the relation this has to the connectivity graph frequently employed in its analysis. Points affecting one another must be connected in some objective manner, the connectivity graph describing the nature of the interaction. The method for finding a connectivity graph is bound to be affected by the constellation of sampling stations, by the configurations of the boundaries of the land area and by the placement of ecological barriers. In turn, the one locality dominates others in terms of the degree of its connections with these localities will affect the evaluation of spatial relationships.

The conventional classification of variables employed in statistics--continuous (e.g., measurements, weights, percentages), ordinal (rank orders), and nominal or categorical (qualitative) variables--applies to geographic variation analysis as well. Computational procedures for continuous and ordinal variables are the same, but differ from those for nominal or categorical variables.

The continuous and ordinal variables may be divided into those of primary ecological importance and those of evolutionary or genetic importance. A measure of abundance of a species, most directly estimated as population density at the sampling location, is a common ecological variable, as are various indirect measures of density such as biomass, optical density (for microorganisms and planktonic populations), pest damage, and the like. Other variables of ecological importance may be weight, fecundity, fertility, reproductive rate, mortality, and similar biotic parameters.

Evolutionary or genetic studies are based on morphometric variables (measurements of various structures or, on occasion, ratios between such measurements) and gene frequencies in population samples. Such gene frequencies may be based on gene loci identified by traditional, morphological variants. Examples of gene frequencies calculated from such visible markers are studies of the banding pattern in the banded snail *Cepaea nemoralis* [Jones, Selander and Schnell (1980)], various human gene loci identified by morphological means [Cavalli-Sforza and Bodmer (1971)], or flower color in *Phlox* [Levin and Sokal]. Most current studies employ one of the modern biochemical or serological techniques for identifying alleles at various gene loci. Electrophoretic techniques distinguish variants among enzymes by placing them in an electric field resulting in the enzymes' migrating toward the pole of opposite charge. Different enzymic variants, called allelomorphs, will have different molecular weights and will migrate to the attracting pole at different rates. This method, pioneered in population genetics by Lewontin and Hubby (1966), has become the single most powerful tool of the population geneticist, and studies on organisms ranging from man to microbes have flooded the evolutionary literature.

Two kinds of categorical variables are encountered frequently. Species identifications of individuals whose spatial location is determined are of interest in ecological studies investigating the random versus patterned distribution of various species associations [Pielou (1977); Sokal and Oden (1978b)]. In evolutionary studies (as in ecological genetics) the genotypes of individuals may be used to study departures from randomness in the spatial location of genetically different individuals [Selander (1970); Sokal and Oden (1978b)].

PATTERNS

The types of patterns to be observed in the surfaces of biological variables vary endlessly, defying any attempt at an exhaustive classification or even catalog. Nevertheless there are some patterns that are observed more frequently than others. A brief discussion of these follows.

Statistically Homogeneous Surfaces

Spatially random surfaces. The statement that a surface is homogeneous for a measurement variable over the sampling stations implies acceptance of the null hypothesis that the means at the sampling stations are sampled from the same parametric population. Such an hypothesis is usually tested through analysis of variance. If at the same time the means are spatially independent of each other, their variation pattern is random. Conventionally expressed there is "no pattern" to the observed values. The surface is a horizontal plane.

An alternative model relevant to population density is a Poisson process, homogeneity of the surface being construed as fit to a Poisson distribution by the observed population sizes at the localities sampled. Gene frequencies can be tested for homogeneity by a *G*-test [Sokal and Rohlf (1981)] or by the family of *F*-statistics developed by Wright (1978).

Spatially patterned surfaces. The seemingly paradoxical case of a homogeneous surface exhibiting a statistically significant spatial structure has been discussed by Sokal and Oden (1978b) and by Sokal (1979a). Plausible biological models can be constructed for such a case, but in fact no example has been observed in nature.

Statistically Heterogeneous Surfaces

Spatially random patterns. In a random surface the observations are spatially independent over all distances. Such a surface lacks a pattern. Tests of spatial randomness depend on the spacing of the locality samples. If the environment is very fine grained, as in a fine mosaic, and the biological population tracks the environment closely, spacing the locality sites at distances appreciably greater than the average diameter of each environmental patch will result in observations that lack spatial pattern. However, finer spacing of locality samples would demonstrate the environmental heterogeneity.

Gradients. These are the most regular of patterns. They are commonly observed in biological populations because organisms respond to gradients in climate or in substrate, or because they are subject to directional pressures as in many instances of migration. The simplest such gradient is a cline, which in its most elementary manifestation is a linear slope over distance. Observations of a cline, or at least variates changing monotonically with distance, are often made for populations studied along a single spatial dimension, such as along a coast line, along rivers or along a transect across an area. In the more usual case of a two-dimensional space, a clinal surface is represented by an inclined plane. Other gradients are found in surfaces representing depressions or hills (bowls or inverted bowls). In all these cases, the implication of a gradient is that values of the observed variable near any given location are going to be similar to the value of the variable at that location, whereas variables farther away will be different. Depending on the nature of the gradient, the farthest values may be dissimilar or again similar.

Patches. A patch in population biology is a homogeneous region for an environmental or biological variable. Patches can only be detected if the spacing of the locality samples is on the average less than the average diameter of the patch. The closer the spacing of the samples, the better the definition of the patches. The clearest definition of patches is obtained in an island model where the patches are imbedded in interstices lacking observations because populations do not occur there. Thus when the gaps between patches are greater than patch diameters (true for many island models), the definition of patches is simple and clear.

Arrangement of patches. By definition, patches are areas homogeneous within their boundaries relative to the entire surface. If the patches do not differ among themselves for the variable studied they can be recognized only in the island model, that is, with neutral interstices. Patches that differ in observational values may be arranged randomly over an area or they may again have patterns of the nature discussed above. Thus there may be homogeneous sets of patches, which might well be called regions; the taxonomy of such patterns is limited only by their complexity and their detection by the number of observation stations available for a verification of the pattern.

PROCESSES

The following population-biological processes often create spatial patterns.

Stasis

Populations and the individuals composing them remain in place. In an absolute sense that is, of course, unrealistic. All organisms or their gametes or propagules disperse to some degree. Nevertheless, if the amount of dispersal is minor with respect to the spatial and temporal scale of the observations, differences among the locality samples are likely due to a static spatial process. The most obvious explanation of such patterns is that they reflect the tracking environmental patterns. Such tracking may simply be a physiological response of individuals to a climatic gradient, such as the increased stunting of plants up the side of a mountain, or the decrease of growth rate of bivalve moluscs in response to an increase in salinity. Alternatively, such tracking may be a biological response of populations to environmental diversity, as in the response of population size to variation in quality or quantity of a resource. Finally, such tracking may represent genetic variation in response to an ongoing selection process that fits local populations to their environments. These possibilities are not mutually exclusive, and it is a continuing and complex problem of evolutionists to partition environmental from hereditary components in the responses of organisms.

Dispersal

When the dispersal of individuals is sufficiently great, given the space and time frame of the observations, various diffusion processes take place. Two major cases may be distinguished. In the first case, the diffusing individuals come from populations that differ in the variable being observed. Examples would be diffusion between two areas differing initially in population density, in gene frequency, in the means of a morphometric variable, or in species composition. In genetic and morphometric work, diffusion when observed over a relatively narrow zone of contact between the two differing populations is known as a zone of secondary intergradation. If time has been sufficient for the process to go to completion, a homogeneous surface should result.

In the second case of no initial differences in the diffusing populations, one might suppose no observable effects, yet various biological properties of individuals and populations result in spatial dependence of the observed variables. Wright (1943, 1969) has shown that populations that do not differ initially in gene frequencies will begin to differentiate on a stochastic basis, and moreover form homogeneous areas that are differentiated from other such regions, as long as they are finite in size and their ability to disperse is limited with respect to the extent of the entire species. This well known isolation-by-distance model has been verified by computer simulation [Rohlf and Schnell (1971); Sokal and Wartenberg].

Another model that should be quite frequent in population biology is extinction of local populations and their replacement from nearby sources. If these nearby sources differ in the biological property under study, spatially contagious distributions may result from a purely stochastic process. Patches of homogeneous populations with respect to densities or gene frequencies could result in an area that at the start of the process might have had a random distribution of the different source populations.

Migration

When movements of individuals are so massive as to involve the majority of the members of a population we speak of migration rather than dispersal [Endler (1977)]. Migration may result only in the speeding up of the diffusion process considered above, so that the end result of the process may be more rapidly attained. However, migration may in some cases result in the displacement or replacement of previously existing populations. Replacement by a genetically different population along a wave of forward migration (with little or no interbreeding between the immigrants and the residents) will result in the successive relocation of a steep zone of intergradation, a so-called step cline. The eventual result would be a homogeneous surface in which the residents had been replaced by immigrants.

METHODOLOGY OF ANALYSIS

The simplest methods of testing for geographic variation patterns are tests of specific alternative hypotheses, such as whether or not there are clines in the data. This problem is dealt with more easily when the clines are one-dimensional, such as along coast lines or rivers, or in line transects across areas. In such cases the computationally simplest approach is the fitting of a linear regression of the variable against distance, but such a procedure loses considerable information on variable rates of change over distance through the extent of the cline. In special cases, such as when a specific diffusion hypothesis is postulated, various nonlinear equations can be fitted to the data with some profit. Piecewise fitting of regression lines such as by the method of splines is also used to overcome the problem of differential rates of change in the separate regions of a cline. Another approach to this problem is through nonparametric methods, such as rank order techniques that test the monotonicity of the change. Tests for such null hypotheses can be constructed from simple probabilistic considerations based on the permutation of n ranked variates [Quenouille (1952)], or based on more elaborate assumptions such as Page's tests for ordered alternatives [Sokal and Rohlf (1981)].

There are serious reservations to parametric as well as nonparametric approaches to testing for clines using the methods outlined above. These relate to the likelihood of spatial autocorrelation among the observations whose trends are being observed. In spatial models, such as those considered here, predictor (regressor) as well as criterion (regressed) variables are subject to an autoregressive process. In the latter case one would expect the regression residuals to be autocorrelated. In the former case the predictor itself is autocorrelated and one would expect the residuals to show no further autocorrelation. There will, of course, be many instances in which both predictor and criterion variables will be separately and interlinkedly autocorrelated, and it may be very difficult in such cases to disentangle the complex nexus of cause and effect in determining the lack of independence. For instance, let us imagine a cline of body size of a species of animal in response to mean annual temperatures. These temperatures, which are likely to exhibit a north-south trend, will be autocorrelated to some degree because of weather patterns above and beyond the variation in latitude. The size of the animals may be genetic adaptations to these climatic conditions. Yet, the size variable may be independently autocorrelated because of diffusion (gene flow) among adjacent populations. The statistical problems are treated in some detail by Cliff and Ord (1973).

To test for clines in areal or two-dimensional data one can use multiple regression analysis to fit the data to an inclined plane. However, this model is usually too simple. The general tests for patterns discussed below do result in classes of outcomes which permit the recognition of two-dimensional clines. Yet another approach to the analysis of areal clines could be based on concepts from differential geometry. An early attempt in this direction was the work of Womble (1951), who devised a method for piecewise estimation of rates of change in gene frequencies of human populations.

A more general approach testing departures from spatial randomness of variables was developed independently by Mantel (1967) and Royaltey, Astrachan and Sokal (1975). Patterns are tested only over those localities that are considered neighbors in some sense [see also Sokal (1979b)]. An alternative approach to the testing of patterns, especially for categorical data, is provided by various random line and transect methods discussed extensively in Pielou (1977) and Cormack (1979).

Although not formally identical, the method of spatial autocorrelation [Cliff and Ord (1973)] subsumes many of the earlier mentioned approaches to testing for patterns in surfaces. Spatial autocorrelation is the dependence of the values of a variable on values of the same variable at geographically adjoining locations. Early biological work by Whittle (1954) and Matern (1960) recognized the importance of these concepts to biology, but the recent interest is probably largely due to the important monograph by Cliff and Ord (1973) that furnishes measures of estimation of the spatial dependence and tests of its significance. Biologists quickly recognized the relevance of these methods to their work. Jumars, Thistle and Jones (1977) and Jumars (1978) introduced spatial autocorrelation to marine ecology by an analysis of abundance data of marine benthos. Sokal and Oden (1978a,b) independently presented the technique to population biologists, extended it to include the analysis of spatial correlograms, and analyzed and interpreted a series of data sets from population genetics, population ecology and community ecology.

The computation of spatial autocorrelation requires a set of localities represented as points in the plane. Mutually exclusive and collectively exhaustive areal units into which a plane has been partitioned can be used as well. For purposes of analysis each unit is also considered as a point. One or more variables are mapped onto these points with one value per variable for each point. The autocorrelation for each variable is computed for those pairs of points that are connected, i.e., considered neighbors in some sense. Rules for connecting localities are given by various authors. Tobler (1975) lists eight different techniques. One of these frequently employed by us is Gabriel graphs, developed by Gabriel and Sokal (1969). Two localities A and B are connected if and only if the square of the distance between A and B is less than the sum of the squares of their distances to any other locality C. The properties of Gabriel graphs have recently been examined by Matula and Sokal (1980).

By any technique, localities considered connected are indicated graphically by a line (edge) joining the pair of localities concerned, whereas localities not considered neighbors will not be so connected. The resulting graph is represented in matrix form by a so-called connectivity or adjacency matrix $\underline{W}$. The simplest such matrix is binary, indicating absence of connections between pairs of localities by zeros and the presence of connections by ones. In many instances the adjacency matrix becomes a weight matrix, with the ones in the matrix replaced by weights $\underline{w}_{ij}$ which are some function of the geographical or ecological distance between locations $\underline{i}$ and $\underline{j}$ (as for example, inverse distance squared weighting).

The details of the computation of spatial autocorrelation are furnished by Cliff and Ord (1973), and in simplified form by Sokal and Oden (1978a) and by Sokal (1979a). The two common autocorrelation coefficients for continuous or meristic variables are Moran's *I*, which is a product-moment coefficient, and Geary's ratio *c*, a pair-comparisons coefficient. For nominal or categorical data, such as individuals representing different species or different genotypes, one calculates join counts, a join being a synonym for an edge connecting two localities considered neighbors. Joins can connect localities with like or unlike values of the nominal variable. Thus, if there are two color morphs, black and white, a count of the number of edges connecting like individuals (BB and WW), and those connecting unlike morphs (BW) can be compared against expected counts on the assumption of spatial independence of the observations. Departures from expectation can be tested for both the interval and the nominal case by dividing the deviations by their appropriate standard errors.

Autocorrelation can be computed for all pairs of points in the study as long as the set of points and edges constitutes a connected graph. If the weight matrix is binary then the edge length between a pair of directly connected localities is one, and the minimum distance between any pair of localities *i* and *j* is the least number of edges one needs to traverse to move from locality *i* to locality *j*. Alternatively one can assign real values to edge lengths, usually the distance between the localities or some transformation of this distance. One can then compute the minimum distance between all pairs of localities measured along the connecting graph. A matrix of such minimum distances is then used as the weight matrix for the computation of the autocorrelation coefficients.

For correlogram analysis a frequency distribution of interlocality distances is set up for a given study. For each class of such distances a new binary weight matrix is constructed such that any two localities whose minimal distance falls within the range of the distance class are shown as connected, all others as not connected. Autocorrelation coefficients are then calculated separately for each distance class using the new weight matrix. The graph of these spatial autocorrelation coefficients against the distance classes is known as a *spatial correlogram*. Such correlograms summarize the pattern of isotropic geographic variation exhibited by the surface of a given variable, and are thus simple analogs of spectral analysis of surfaces. Correlograms describe the underlying spatial relationship for a surface rather than its appearance, and for this reason they are probably closer guides to some of the processes that have generated the surfaces than are the surfaces themselves.

Spatial autocorrelation analysis is not dependent on any one method of connecting localities. The connectivity matrix postulates the null hypothesis against which the surface is tested for pattern. It is possible and legitimate to test the same pattern against two or more null hypotheses representing different dynamic spatial models. For instance, consider the study by Sokal and Menozzi (1981) in which 21 allele frequencies for the HLA blood groups were analyzed in 58 human populations from Europe and the Middle East. These 21 frequency surfaces had been analyzed earlier by Menozzi, Piazza and Cavalli-Sforza (1978) who believed that observed trends in these data, specifically a northwest-southeast cline, were related to the spread of agriculture into prehistoric Europe. The origin of European agriculture in the Middle East has been well documented, and its main routes into Europe have been ascertained by fossils of grains and domestic animals, archaeological remains (Bandkeramik) and radiocarbon dating. Ammerman and Cavalli-Sforza (1971, 1973) postulated that this advance took place by means of demic rather than cultural diffusion, that is, agriculturists rather than the knowledge of agriculture were migrating. If intermingling with the native population took place, as is likely, a gradient of allele frequences should be established along the path of advance. Menozzi, et al. (1978) believed that those allele frequencies exhibiting a northwest-southeast cline reflect this process of demic diffusion. Sokal and Menozzi (1981) tested these hypotheses by connecting the 58 localities so as to test for a northwest-southeast trend. They were unable to confirm the existence of a special northwest-southeast gradient. Only five allele frequencies in the overall correlation shown in Figure 1a exhibit clines. These same frequencies also show north-south clines. However, the hypothesis relating trends to HLA frequencies to the origin and spread of agriculture is not necessarily dependent on the demonstration of a northwest-southeast cline. In an alternative formulation of the demic diffusion hypothesis, Sokal and Menozzi (1981) connected the 58 sample localities by a directed network representing the supposed spread of agriculture into these regions. This network was based on sources in Ammerman and Cavalli-Sforza (1971, 1973) and Barraclough (1979). The overall correlogram based on these hypothesized connections is shown in Figure 1b and indicates gradients along the hypothesized routes for the majority of the allele frequencies. The results support the hypothesis that demic rather than cultural diffusion took place. Clearly the results of this analysis are dependent on the connection network against which the data were tested.

A second example of testing the same pattern against two alternative networks is found in the work of Sokal (1979a) analyzing the results of decennial censuses in the Republic of Ireland starting in 1871. By connecting putative "source"

Fig. 1. Spatial correlograms of the frequencies of 21 HLA alleles in Europe and the Middle East. Data from Sokal and Menozzi (1981).

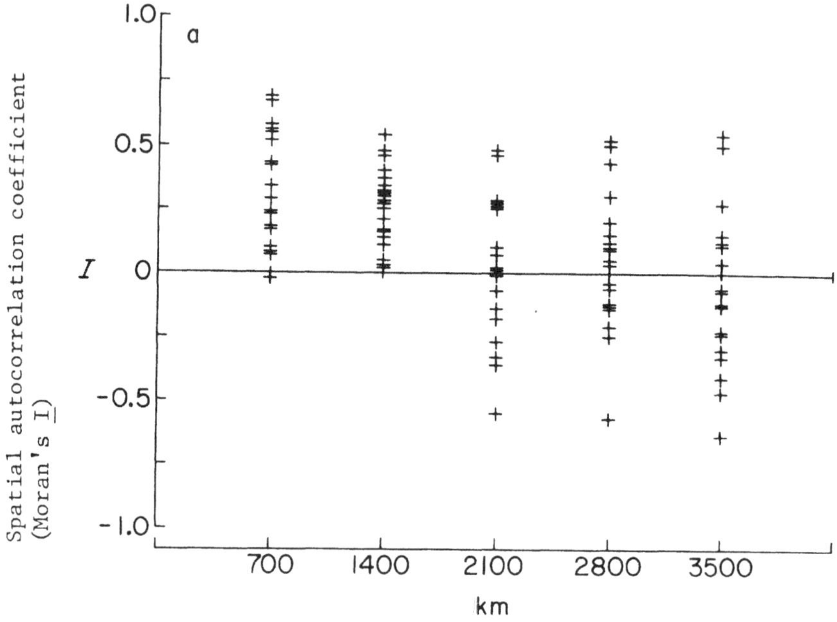

Distances indicated are upper limits of the distance class (minus 20 km) rather than class marks.

a. Correlograms computed using a connectivity graph testing for a northwest-southeast cline.

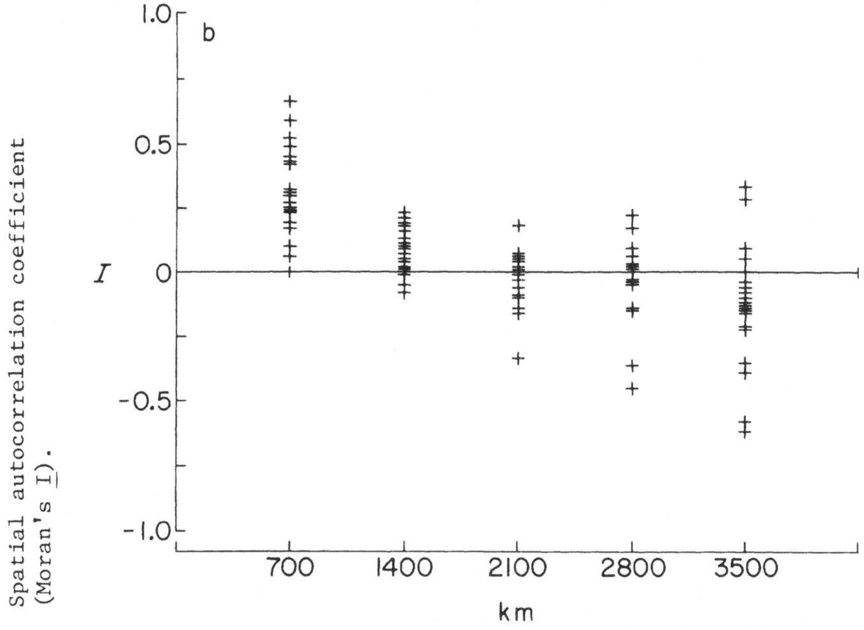

Distances indicated are upper limits of the distance class (minus 20 km) rather than class marks.

b. Correlograms computed using a directed connectivity graph representing the spread of agriculture from the Middle East into Europe.

counties (mainly agricultural) with putative "sink" counties (urbanized) he was able to show that characteristic patterns of relative population increase in the urban east and southeast and decrease in the largely rural west, are due to migration from source to sink counties.

When the structure of the surface is patchy, the actual value of the autocorrelation coefficient is a function of the similarity of patches at given distances, and of the percentage of within-patch connections for a given distance along the correlogram. As patch size increases, the proportion of within-patch connections of a given length will necessarily increase. Based on these self evident relations Sokal (1979a) carried out a simulation experiment with artificial surfaces differing in patch size. He showed that the correlogram was an accurate indicator of patch size. In square patches the distance class with the first negative autocorrelation defines the size of the patch as measured by one side. In rectangular patches the length of the narrower side approximates the distance at which the first negative autocorrelation is observed, whereas for variable sized patches the first negative autocorrelation coefficient corresponds in distance to the average diameter of the patch. Similar relations hold for patches along a one-dimensional cline (Sokal).

We can report several applications of this notion of patch size as the maximum area of homogeneity observable by autocorrelation analysis. Sokal and Friedlaender (1981), studying human variation on Bougainville Island, noted that for 20 variables with significant spatial structure (out of 121 blood group, anthropometric, dermatoglyphic, and dental variables studied) the modal patch size was 30 km. In this largely linear pattern of geographic variation this means that the maximum distance one can travel up and down the island is 30 km and still remain in an area similar to the point of origin for the variable studied. This figure can be compared to an earlier finding [Friedlaender (1975)] that the majority of adult individuals in these Melanesian villages had migrated no more than 10 km from their birthplace. Sokal and Menozzi (1981) found that in the HLA allele frequencies of European populations, and at a structural level considerably above that of the Bougainville study by Sokal and Friedlaender, positive autocorrelation occurs always up to 1400 km, frequently up to 2100 km, for all variables observed. Thus migrational processes in all directions in these human populations seem to ensure homogeneity at a higher level of variation at these distances. Studies on the banded snail _Cepaea nemoralis_ in north Wales and in the Pyrenees yielded patch sizes for gene frequencies of these rather diversified animals ranging from 6 to 50 km [Jones, Selander and Schnell (1980); Caugant, Jones and Selander (1981)]. In studies of three related forms of gall-forming aphids, Sokal, Bird and Riska (1980) found patch sizes in _Pemphigus populicaulis_ to range between 200 and 400 km, whereas Sokal and Riska (1981) noted patch sizes ranging from 800 km to 1200 km for the elongate morph of _P. populitransversus_, and of 400 km for the globular morph of the same species. These differences may be related to the size of the environmental patches that determine the nature and composition of the secondary hosts.

NATURE OF INFERENCES

Useful inferences concerning dynamic spatial processes in biology are complicated by one's inability to map the relations between process and outcome on a one-to-one basis. As we have seen, there are various population biological processes that will result in a gradient. We shall have to attempt to overcome this problem of equifinality by distinguishing among the several processes all of which could give rise to the same outcome. The converse problem is less serious in this type of work. Multifinality, different surfaces resulting from identical underlying processes, is generally relevant only in stochastic models in population biology.

Similar Patterns

Similarities between variation patterns can be studied as correlations between the variables concerned over the set of localities under study. We generally have employed product-moment correlations in our studies. The usual statistical caveats need to be pronounced. Not only may the variables have to be transformed to correct for departures from normality, but since the inherent variation over geographic space may depart markedly from that of a normal or even symmetrical distribution, other means of calculating correlations, such as nonparametric rank correlations, may prove of value.

To make inferences about the process involved we also need measures of similarity between correlograms. Since magnitude as well as slope of the coefficients against time is involved, we have employed the average Manhattan distance [Sneath and Sokal (1973)] between pairs of correlograms in some of our studies. Note that both Geary's ratio $\underline{c}$ and Moran's $\underline{I}$ coefficient are standardized in the denominator by the sum of squares of the variable over the geographic locations. Gradients over the same geographic area whose absolute slopes would be very different will thus yield the same spatial correlograms.

Various biological processes may give rise to similar patterns between pairs of variables. Barring spurious correlations, putative causes can be grouped into three major categories. The first is necessary association between the variables concerned. This will be true whenever the variables respond to a common developmental factor, such as upper arm length and lower arm length which share a response to a general size factor as well as to an appendage factor in many organisms. Similarity of geographic variation patterns in this case implies that there is a single response pattern of the underlying morphological factors which is being assayed by two or more variables. An illustration of such a necessary association between morphological variables is found in the study of 18 villages on Bougainville by Sokal and Friedlaender (1981). These villages are situated in a chain along the central and southern portions of the island. Figure 2a shows a profile of mean values of finger ridge counts of males on the third right and fourth left digit along this chain. Their patterns from village to village correspond very closely (correlation $\underline{r} = 0.80$). The correlograms of these patterns shown in Figure 2b are equally similar (average Manhattan distance $\underline{d} = 0.20$).

Fig. 2. Geographic variation of finger ridge counts in human populations sampled from eighteen villages on Bougainville Island. Data from Sokal and Friedlaender (1981).

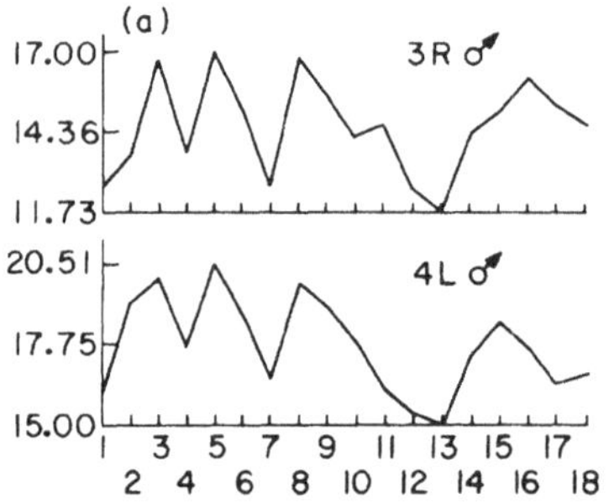

Number of village from north to south.

a. Profile of means of the two variables for the eighteen villages representing an approximate north-south transect of the central portion of Bougainville.

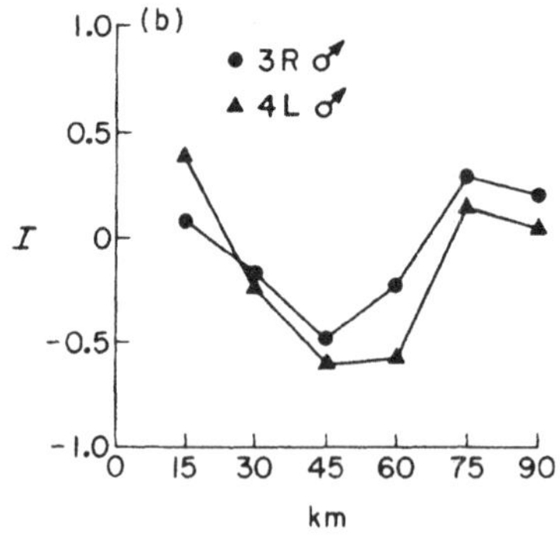

Distances in kilometers.

b. Spatial correlograms for these two variables.

A second class of similar patterns may be due to structural properties of the underlying populations. By this we mean ways in which the units determining the variable information are packaged. Two models might make this clear. If two genes coding for different enzymes are so close to each other on the same chromosome that their free random assortment is inhibited (geneticists describe this as tight linkage), then the dispersion of this particular chromosome into neighboring populations through diffusion (gene flow) will necessarily associate the two genes. And if the initial proportions of the various combinations of different states (alleles) of each gene depart from random expectations (this situation is known as linkage disequilibrium in population genetics), then in the short run, until the disequilibrium is removed by crossing-over among chromosomes, one would expect the frequencies of these genes to have correlated variation patterns over the area. As a second model we postulate properties associated with a local population even though not on the same chromosome. If populations differ initially in many of their characteristics, diffusion patterns based on dispersal and migration of individuals from these populations will necessarily affect not just one but all of the properties of these populations. Thus a dispersing individual carries with it all its genetic complement: it does not selectively transfer some genes and leave behind others. This property will also result in correlation between variables in which the initial populations differ until such time as there is random reassortment of the various properties.

Examples of the first model are hard to come by because there are few known gene frequency distributions involving linked genes. The second model is more easily demonstrated although any examples we can cite are inferential and not proven beyond reasonable doubt. Allele frequencies at the HLA loci in human populations in Europe, that load heavily on the first principal axis of surfaces [Menozzi, et al. (1978); Sokal and Menozzi (1981)] are believed to exhibit their parallel clinal pattern shown in Figure 3 because of the demic diffusion of early farmers into Europe from the Middle East with the spread of agriculture into Europe. The striking parallelism of frequencies at several allozyme loci reported by Sokal and Oden (1978a) for the snail Helix aspersa (see Figure 4) is believed by these authors to be due to the gene diffusion following original colonization by founder colonies from differing populations of this introduced species.

A third category of pattern similarity is due to externally imposed conformity. An environmental pattern imposes a response pattern on two or more variables of the organisms in the area. The response may be due to physiological adaptive responses or to genetic responses, as already noted. Although numerous such

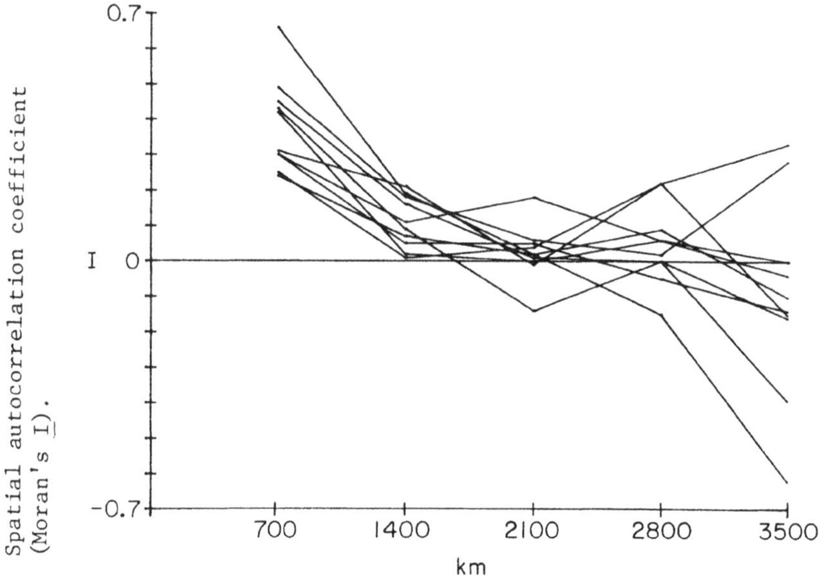

Distance in kilometers.

Fig. 3. Spatial correlograms based upon Gabriel graphs for ten HLA allele frequencies in Europe and the Middle East.

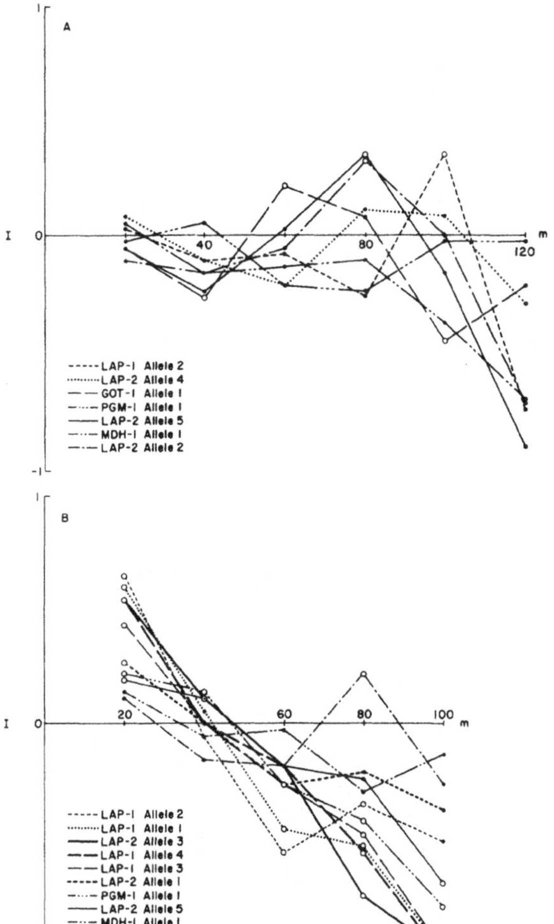

Spatial autocorrelation coefficient (Moran's I).

Distances in meters.

Fig. 4. Spatial correlograms for geographic variation patterns of allozyme frequencies of the snail Hexlis aspersa in two blocks in Byran, Texas. From Sokal and Oden (1978a).

instances must exist in geographic variation patterns, it is difficult to furnish a putative example of a similarity due to this cause. For most morphometric or gene frequency variables analyzed in geographic variation analysis, we are ignorant of the immediate environmental factors to which they are responding either physiologically or genetically. An exception is the micro-geographic variation study in the slender wild oat Avena barbata by Hamrick and Holden (1979). A spatial autocorrelation analysis of electromorph frequencies on a hillside in California demonstrated moderate clines in 4 of 5 frequencies examined. Habitat conditions were rated xeric to mesic on a 5-point scale and showed a marked cline. In view of the absence of pattern in one electromorph, and the fact that these populations are largely selfing, diffusion as an explanation is unlikely and the four electromorph frequencies seem to be tracking the habitat conditions.

Closely correlated patterns yield similar correlograms. We illustrate in Figure 5 the scattergram of the relations between the average Manhattan distances of pairs of I-correlograms and

Similarities between correlograms computed as average Manhattan distances between I-correlograms.

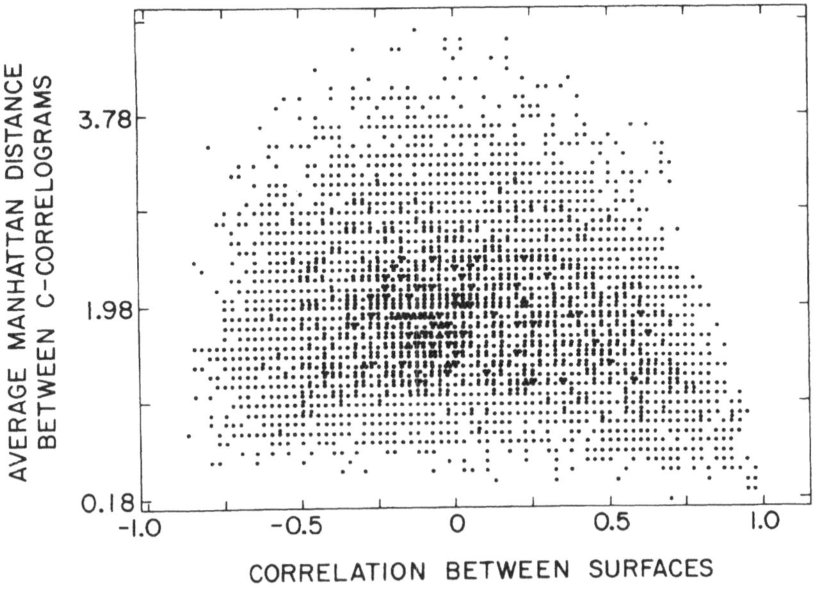

Similarities between character surfaces expressed as correlation coefficients

Fig. 5. Correspondence between correlograms and surfaces of variables based on all pairs of 121 variables in samples of human populations from eighteen villages in Bougainville

correlations between variable surfaces in 121 pairs of variables in the study of 18 villages on Bougainville [Sokal and Friedlaender (1980)]. Those variables with similar surfaces can be shown to have similar correlograms as well, whereas those surfaces that are dissimilar may or may not have similar correlograms.

Dissimilar Patterns

Three major groups of phenomena will bring about dissimilarities in patterns. The first are random processes affecting independent variables. Thus if the genetic composition of populations is subjected only to chance sampling phenomena and dispersal among neighboring populations, a random surface will result for a given gene frequency. This will be uncorrelated with the random surface for a second gene frequency subjected to the same stochastic forces. A second situation giving rise to uncorrelated geographic variation patterns is migration of differing source populations at different rates due to varying environmental resistances in diverse directions of the compass. A third and perhaps the most common cause for dissimilarities among geographic variation patterns would be the tracking of different environments by different variables. If one variable is responsive to temperature whereas another one is responsive to organic content of soil, then clearly the pattern of variation of these two characters will differ greatly, and any remaining correlation between the patterns will reflect correlation between the environmental variables.

Most dissimilar patterns will also result in dissimilar correlograms, reflecting their different origins. However, one class of dissimilar patterns that will result in similar correlograms is random processes with the same generating functions. These should yield surfaces independent of each other yet with the same distribution. Thus the correlation between such surfaces will be low or zero while their correlograms will be similar or identical. This can be visualized most easily by considering an inclined plane that when rotated by 90° would yield two surfaces which would be independent of each other. Yet the correlograms to be obtained from these surfaces--clines in this case--would be the same. Recent experimental results by Sokal and Wartenberg have verified these conclusions for the isolation-by-distance model [Wright (1943, 1969)].

In these studies we adapted the computer simulation model of Rohlf and Schnell (1971) that consists of a population of 10,000 individuals arranged as a 100 x 100 lattice. This population was settled at random by individuals sampled from a population of gene frequency 0.5 at Hardy-Weinberg equilibrium. In each generation individuals were replaced by offspring from parents chosen at random from within defined neighborhoods around the location of the offspring. Four sets of experiments were carried out. In sets 1 and 2, one sex (the males) was taken at random from a neighborhood of size $\underline{n} = 9$, whereas the other sex (the females) was considered to be sessile; that is, the individual at the same location in the previous generation was designated as the female parent. In set 3, both sexes were vagile and were taken at random from a neighborhood of size $\underline{n} = 9$, a three-by-three sublattice centering on the individual to be replaced. In set 4 the same model was used except that the neighborhood size has been increased to $\underline{n} = 25$ (a five-by-five sublattice). The results of each "mating" were computed as stochastic realizations of the Mendelian expectations for the given cross.

Five independent runs, each for 200 generations, were carried out for each of sets 1, 2, 3 and 4, with set 2 being treated as a variant of set 1. The results of the experiment were summarized every fifth generation by dividing the 100 x 100 lattice into 40 five-by-five sublattices and computing the gene frequency for each of these quadrats containing 25 individuals. Additionally, variances of gene frequencies among quadrats and Wright's $\underline{F}$-statistics, as well as correlograms for the surfaces, were computed. There is a general increase in variance among quadrats over the course of the experiment, with some evidence that the surfaces have reached quasi-stationarity after generation 150.

As anticipated, the surfaces resulting from the initial settlement and those at generation 200 are uncorrelated with each other for pairs of runs within any set of experiments. The correlograms for the runs within one set are quite similar, however, as is illustrated for set 4 in Figure 6. As time passes the patchiness of the surface increases. Correlograms appear diagnostic of the parameters of the generating process of the surface. As the parameters of the generating process change, the correlograms change. Figure 7 shows the mean autocorrelations for each distance class for the five replicates of each set. Whereas sets 1 and 2 do not differ consistently, sets 3 and 4 are differentiated from set 1 and from each other. Average Manhattan distances for pairs of $\underline{I}$-correlograms for the four sets range from 0.030 for sets 1 and 2 to 0.120 for set 1 and 4.

Dissimilarities arising from different rates and directions of migration will result in differing correlograms. This situation is illustrated by the HLA frequencies in Europe whose surfaces can be decomposed into several classes [Menozzi, et al. (1978); Sokal and Menozzi (1981)]. Although they all result in positive autocorrelation up to 1400 km, the correlograms of each class differ at greater distances. This presumably reflects various migratory events that took place in the prehistory and early history of Europe.

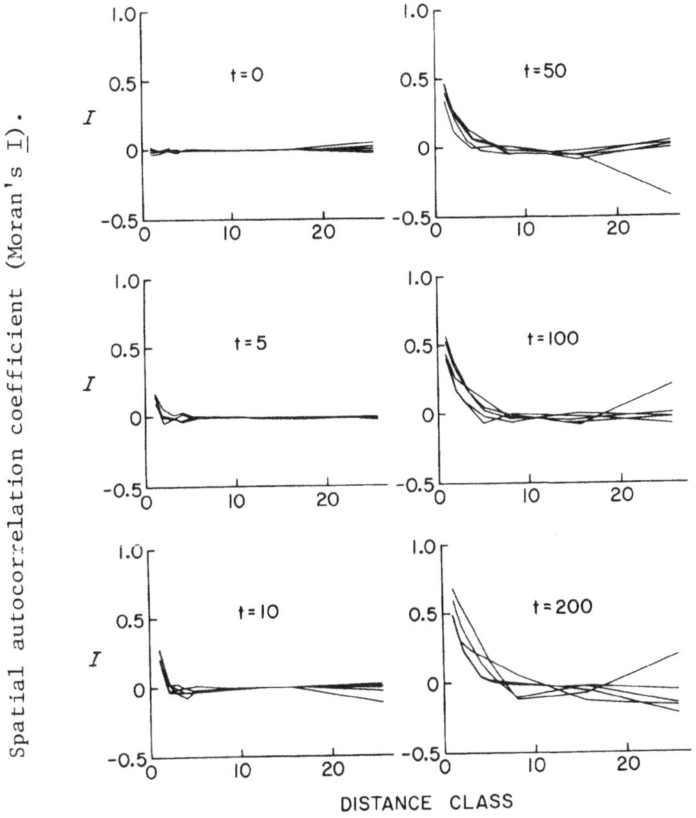

Distances in lattice units.

Fig. 6. Spatial correlograms representing results of a simulation study on isolation-by-distance. Parameter set 4: both sexes vagile, taken at random from a neighborhood of size n = 25.

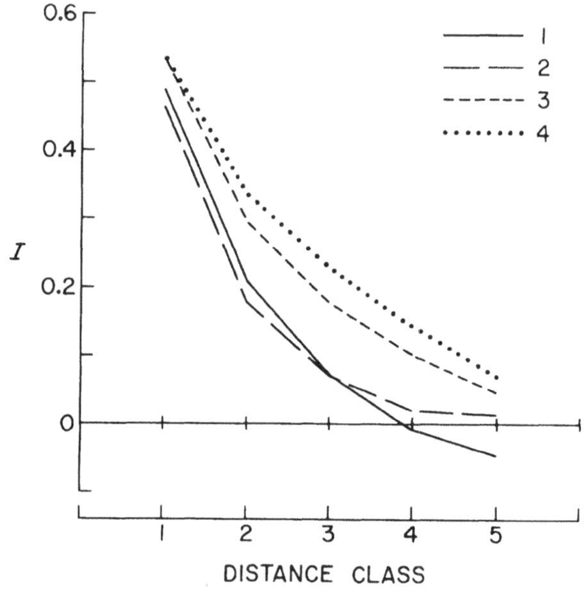

Spatial autocorrelation coefficient (Moran's I).

Distances in lattice units.

Fig. 7. Spatial correlograms summarizing results of the isolation-by-distance simulation study for four different sets of parameters. Sets 1 and 2, males taken from a neighborhood of size $n = 9$, females sessile. Set 3, both sexes vagile, taken at random from a neighborhood of size $n = 9$. Set 4, both sexes vagile, taken at random from a neighborhood of size $n = 25$.

Similarly, nine separate classes of surface patterns on Bougainville [Sokal and Friedlaender (1981)] are unlikely to have been selected by nine separate environmental agents. A repreated series of migrations that have given rise to the particular pattern observed is a more likely explanation, given our knowledge of the biology and history of these populations.

It will be difficult to prove in any given case that the combination of statistically heterogeneous samples with different geographic variation patterns is caused by the tracking of different environmental factors by different variables. Such proofs require a detailed analytical demonstration that these environmental variables do indeed affect the measured response variable. Experimental proof of this relationship obviously is preferred. Yet this particular situation is likely to occur commonly in nature, and should at least be postulated as a working hypothesis in instances where the genetic or physiological tracking of environmental factors seems plausible. Such cases are reported by Sokal, Bird and Riska (1980) and Sokal and Riska (1981) in geographic variation studies of species of the aphid genus Pemphigus. In three forms analyzed the significant spatial structure in many variables is associated with significant statistical differentiation for virtually all the morphometric variables studied. There tend to be classes of patterns with different autocorrelation structure with good correspondence between patterns and correlogram classes. Patch size differences may be related to the size of the environmental patches that determine the nature and composition of the secondary hosts. Pattern differences on a continental scale presumably are determined by climatic, edaphic and primary and secondary host plant differences over this large area.

CONCLUSIONS

The underlying philosophy of our approach has been to make inferences about dynamic spatial models from observed geographic variation patterns, rather than from a predictive approach based on mathematical models of the processes. In some cases, as in the isolation-by-distance model, we have employed computer simulation to aid in the interpretation of our findings. The approach has been a multivariable rather than a multivariate one. The surfaces are analyzed singly, and inferences are made based on the similarities of surfaces and their correlograms. The type of reasoning employed by us for single variables can, with minor modifications, be applied as well in a more strictly multivariate way to linear combinations of variables [Sokal and Menozzi (1981)]. To evaluate the similarities of correlograms methods must be developed for testing the statistical significance of differences between correlograms. We are currently engaged in such an endeavor.

The comparative study of surfaces and correlograms of variables may be of value in non-biological areas of research. Study of the transport of merchandise, the spread of linguistic innovations, rituals and myths in anthropology, or of artifacts in archaeology may lead to fruitful hypotheses about the nature and origins of the underlying processes.

REFERENCES

Ammerman, A. and Cavalli-Sforza, L., 1971. Measuring the Rate of Spread of Early Farming in Europe, Man. 6: 674-688.
Ammerman, A. and Cavalli-Sforza, L., 1973. A Population Model for the Diffusion of Early Farming in Europe, in: The Explanation of Culture Change, C. Renfrew (ed.), Duckworth, London, 343-357.
Barraclough, G. (ed.), 1979. The Times Atlas of World History, Hammond, Maplewood, New Jersey.
Cain, A. and Currey, J., 1963. Area Effects in Cepaea. Philosophical Transactions of the Royal Society of London B, 246: 1-181.
Caugant, D., Jones, J. and Selander, R., 1981. Morphological and Molecular Polymorphism in Cepaea nemoralis in the Spanish Pyrenees. MS submitted to Biological Journal of the Linnean Society.
Cavalli-Sforza, L, and Bodmer, W., 1971. The Genetics of Human Populations, W.H. Freeman, San Francisco.
Cliff, A. and Ord, J., 1973. Spatial Autocorrelation, Pion, London.
Cormack, R., 1979. Spatial Aspects of Competition Between Individuals, in: Spatial and Temporal Analysis in Ecology, R. Cormack and J. Ord (eds.), International Co-operative Publ. House, Fairland, Maryland.
Endler, J., 1977. Geographic Variation, Speciation and Clines. Princeton Univ. Press, Princeton, New Jersey.
Friedlaender, J., 1975. Patterns of Human Variation, Harvard University Press, Cambridge.
Gabriel, K. and Sokal, R., 1969. A New Statistical Approach to Geographic Variation Analysis, Systematic Zoology, 18: 259-278.
Hamrick, J. and Holden, L., 1979. Influence of Microhabitat Heterogeneity on Gene Frequency Distribution and Gametic Phase Disequilibrium in Avena barbata. Evolution, 33: 521-533.
Jones, J., Selander, R., and Schnell, G., 1980. Patterns of Morphological and Molecular Polymorphism in the Land Snail Cepaea nemoralisis. Biological Journal of the Linnean Society (in press).
Jumars, P., Thistle, D. and Jones, M., 1977. Detecting Two Dimensional Spatial Structure in Biological Data. Oecologia, 28: 109-123.
Jumars, P., 1978. Spatial Autocorrelation with RUM (Remote Underwater Manipulator): Vertical and Horizontal Structure of a Bathal Benthic Community, Deep-Sea Research, 25: 589-604.
Lewontin, R. and Hubby, J., 1966. A Molecular Approach to the Study of Genic Heterozygosity in Natural Populations. II. Amount of Variation and Degree of Heterozygosity in Natural Populations of Drosophila pseudoobscura. Genetics, 54: 595-609.

Mantel, N., 1967. The Detection of Disease Clustering and a Generalized Regression Approach, Cancer Research 27: 209-220.
Matern, P., 1960. Spatial Variations: Stochastic Models and Their Application to Some Problems in Forest Surveys and Other Sampling Investigations, Matter Meddelanden fran Statens Skogsforskinginstitut, 49: 1-144.
Matula, D. and Sokal, R., 1980. Properties of Gabriel Graphs Relevant to Geographic Variation Research and the Clustering of Points in the Plane, Geographical Analysis, 12: 205-222.
Menozzi, P., Piazza, A., and Cavalli-Sforza, L., 1978. Synthetic Maps of Human Gene Frequencies in Europeans, Science 201: 786-792.
Pielou, E., 1977. Mathematical Ecology, Wiley, New York.
Quenouille, M., 1952. Associated Measurements, Butterworths Scientific Publ., London.
Rohlf, F. and Schnell, G., 1971. An Investigation of the Isolation by Distance Model, American Naturalist, 105: 295-324.
Royaltey, H., Astrachan, E. and Sokal, R., 1975. Tests for Patterns in Geographic Variation, Geographical Analysis, 7: 369-395.
Selander, R., 1970. Behavior and Genetic Variation in Natural Populations, American Zoologist, 10: 53-66.
Sneath, P. and Sokal, R., 1973. Numerical Taxonomy, W.H. Freeman, San Francisco.
Sokal, R., 1979a. Ecological Parameters Inferred from Spatial Correlograms, in: Contemporary Quantitative Ecology and Related Econometrics, G.P. Patril and M.L. Rosenzweig (eds.)International Co-Operative Publ. House, Fairland, Maryland, 167-96.
Sokal, R., 1979b. Testing Statistical Significance of Geographical Variation Patterns, Systematic Zoologist, 28: 227-232.
Sokal, R., Bird, J. and Riska, B., 1980. Geographic Variation in Pemphigus populicaulis (Insecta: Aphididae) in Eastern North America, Biological Journal of the Linnean Society, 14: (in press).

---

*Contribution No. 212 from the Program in Ecology and Evolution at the State University of New York at Stony Brook. This research has been supported by grants No. DEB 77-0824 from the National Science Foundation and No. GM 2826201 from the National Institutes of Health.

We are indebted to Joyce Schirmer for the art work and to Barbara McKay for typing several drafts of this manuscript.

THEORIES OF STRUCTURAL CHANGE

In recent years, catastrophe theory has swept through many of the physical and social sciences, being adopted to study a wide variety of problems, many of them previously resistant to any mathematical analysis. The degree to which the impact of catastrophe theory will be long-lived and fundamental is still a matter of debate. There is, however, little doubt that its ability to deal with and predict sudden, discontinuous and irreversible change is among its most attractive features.

Casetti's paper addresses one of the most important issues in modern society--the economic development process. How is it that economies can lie dormant for centuries and then suddenly experience explosive economic growth? Are there ways in which such growth can be stimulated? Are there ways in which growth, if achieved, can be dampened to proceed in a more orderly fashion? Bringing together relationships from economics and demography and presenting them graphically as well as mathematically, Casetti shows how relatively small but qualitative shifts in levels of technology and exploitation can result in rapid economic development rather than a stagnant low equilibrium level of income.

Papageorgiou similarly is concerned with the sudden change that often is associated with economic development--rapid, sometimes pathological, urbanization. Extending an earlier paper by Casetti, Papageorgiou provides a rationale for sudden urban growth based on micro-economic principles. He analyzes this phenomenon in the case of a single city that is homogeneous with respect to population. Some specifications concerning population heterogeneity, a system of cities and policy implications are also included. In particular the author is convinced that the dynamics of social and economic systems are such that we must be prepared to accept, predict and plan for sudden urban growth rather than attempt to prevent it from occurring.

Puu focuses upon the obverse of the discontinuity issue--structural stability in spatial systems, systems of two dimensional flows in particular. Like Beckmann in the first section of this volume, Puu analyzes geographical systems in continuous two dimensional space. He uses the notion of structural stability as a major criterion by which a model can be accepted or discarded. Thus a flow pattern that could be radically changed by any small disturbance is not of interest to us. By using theorems from catastrophe theory, Puu studies how spatial systems can move from one structurally stable flow pattern to another.

TECHNOLOGICAL PROGRESS, EXPLOITATION AND SPATIAL ECONOMIC GROWTH: A CATASTROPHE MODEL

E. Casetti

Department of Geography
The Ohio State University

INTRODUCTION

Modern societies are characterized by rates of economic growth much higher than those that prevailed in the premodern world. In fact, the explosive self-sustaining growth that brought the modern world into existence has come to be regarded as qualitatively different from its premodern counterpart. One of the most significant findings produced by Kuznets' analyses of historical data concerns the acceleration of economic progress in modern times. Kuznets [(1971), p. 22] convincingly argued that the rates of growth of product per capita experienced by the contemporary developed countries during their industrialization were larger by at least a factor of ten than those prevailing during premodern times. Many contemporary less developed countries entered modern economic growth after the end of World War II. During this period their rates of growth of product, as well as of product per capita, exceeded those of the developed countries at a comparative stage in their economic evolution [Pearson (1969), p. 27].

The transition from the premodern to the modern world can be characterized in broad terms as follows. Prior to the industrial revolution incomes per capita and levels of living tended to increase at rates very low by modern standards. In fact these increases are only apparent if very long time periods are considered. In the short- to medium-run the premodern societies appeared essentially static to the historians and to the people living in them. For instance, the secular improvements in productive capacity and in standard of living in Western Europe between the time of the Romans and the eighteenth century would be impossible to detect within a time horizon spanning, say, one generation.

Over one such time interval the secular trend would be entirely obliterated by fluctuations in crops, by the consequences of wars and civil convulsions, or by the temporary establishment of peaceful and orderly circumstances within the context of momentarily viable political units.

Modern economic growth was produced by a switch to a completely different dynamics. The barely discernible growth of premodern societies was replaced by explosive increases in productive capacity and, to some extent, in standards of living that changed dramatically the world during the two odd centuries since its inception.

The switch from premodern to modern growth needs to be considered in its spatial as well as in its temporal dimensions. At the macro-level, modern economic growth originated in England during the second half of the eighteeneth century, and from there spread throughout Europe. Then it leapfrogged to Japan and North America. In the contemporary world we are witnessing the spread of modern economic growth to essentially all the peoples of the earth.

At the micro-level the diffusion of modern growth is much more complex and involves an intricate leapfrogging to centers, social groups and economic activities that are more receptive to it, followed by a spread to other centers, social groups and economic activities via a quasi-filling out process. However, the key ingredient of modern economic growth, whatever the level of resolution at which we observe it, is the switch to a rate of growth much higher than the one prevailing in premodern societies, accompanied by increases in productive capacity and, to various degrees, in standards of living. The investigation of modern economic growth calls for analyzing the dynamics of a vast array of phenomena such as population and its spatial distribution; the mix, scope and extent of economic activities; transportation and communication networks; value systems, social institutions and policical structures. However, two themes consistently provided major points of focus from which modernization phenomena have been viewed. They are concerned respectively with technological progress and with exploitation.

The switch from premodern to modern growth can be usefully portrayed as the transition between two dynamics corresponding to a fold catastrophe [Poston and Stewart (1978), p. 173]. Certainly the awareness that the beginning of modern economic growth involves a switch between two dynamics predates catastrophe theory. On the other hand, the catastrophe theory paradigm, because of the very fact that it places the accent on the modelling and analysis of discontinuous change, has the potential for producing a deeper understanding of the essence of the "catastrophic" transition from the premodern to the contemporary world. The model that will be outlined in following sections interprets the onset and spatial spread of modern economic growth in terms of economic demographic interactions, incorporating explicitly technological progress and exploitation, within a catastrophe theory frame of reference.

THE MODEL

Denote a country's aggregate product by Y and its population by P. Indicate by y the "available" product per capita, The available product per capita y is defined as a function of P, Y, and of an "exploitation" factor p:

$$y = pY/P \quad . \qquad (1)$$

For p=1 the available product per capita is equal to the country's aggregate product Y divided by its population P. If the country is being exploited it will lose a portion pY of its aggregate product to one or more exploiting countries. In this case $0<p<1$. On the other hand, an exploiting country will dispose of a larger product pY than it produces, such that $p>1$. To sum up, the exploitation factor p is smaller than one in the case of exploitees, is larger than one in the case of exploitors, and equals one for a country that neither exploits nor is exploited.

Let t stand for time and assume that Y, P, p and y are smooth functions of t. Indicate by a quotation mark (") the logarithm derivative with respect to time of the variable to the left of the sign. For instance, for the variable p

$$p'' = (dp/dt)/p \quad , \qquad (2)$$

where p" denotes the percentage rate of change of the exploitation factor p. Since p" plays a key role in the mathematical model discussed in this paper it is desirable to dwell upon its significance. Suppose that a country is "exploited" in the sense given to p. A positive p" would indicate that at some subsequent point in time its p value will be larger. This in turn implies that the country is in the process of becoming exploited to a lesser degree, or that it is moving to an exploitor condition. Consider next an exploitor country, with $p>1$, which indicates that the product available to its inhabitants is larger than what they produce. A negative p" means that the country is in the process of becoming an exploitor to a lesser degree, or of turning into an exploitee, if its p value falls below unity.

The model discussed in this paper includes the identity (1) and the two differential equations

$$Y'' = Y''(P'',m) \tag{3}$$

and

$$P'' = P''(y) \quad . \tag{4}$$

These equations relate, respectively, the change in product to population change and technological progress, and the population change to the level of available product per capita. In the sections that follow equations (3) and (4) will be specified and discussed.

Let the relation between growth of output and growth of population be expressed as follows:

$$Y'' = nP'' + m \quad , \tag{5}$$

where n is an economies of scale parameter. The parameter m incorporates the effects of technological progress and capital accumulation on growth. P" stands for percentage rate of change of physical labor inputs under the simplifying assumption, valid for a first approximation, that these inputs are proportional to population size.

Both n and m reflect a society's ability to generate and apply "useful knowledge." This is obvious as regards m, which can be directly related to technological progress. According to equation (5) the parameter m induces growth of outputs in excess of what can be ascribed to the growth in labor inputs. The relation between the n parameter and a society's technological level is less obvious and requires some clarifications. Essentially, n determines which effects a given percentage rate of growth of labor inputs have on the percentage rate of growth of outputs. For $n=0$ such effect is nil. For $n=1$ a given P" enters in its entirety into Y". For $0<n<1$ P" contributes fractionally to Y".

Specific socio-economic contexts can be associated with distinct values or classes of values of n. A "hunting and gathering" society in equilibrium with its environment might be characterized by $n=0$, or by very small values of n. The aggregate product of hunting and gathering activities is a function of ecology and habitat, and in the long run would not change much if the labor inputs are increased. A primitive agriculture requiring very specific soils and micro-environments would be characterized by low positive values of n. On the other hand, manufacturing activities are often associated with increasing returns to scale, and therefore economic systems with a substantial manufacturing component may exhibit n coefficients with values close to or greater than one. Finally, whenever the depletion of natural resources dominates an economy negative n's would result. For instance, a primitive pastoral economy would exhibit a negative n coefficient at the point where overgrazing reduces outputs. Generally, economic progress ought to be associated with larger values of the n parameter. In the analysis that follows n will be assumed to be a constant satisfying the condition $0<n<1$.

Let us now discuss and specify equation (4). Assume that in the system considered there is no in-migration or out-migration, so that here P" stands for natural increase in population. Furthermore, assume that P" is a function of y characterized by the following properties: (a) at very low levels of product per capita deaths will exceed births so that P" is negative; and, (b) as the product per capita increases the excess of births over deaths will grow larger, up to a maximum, after which it will decline. The proposition that at very low levels of product per capita P" is negative is basically deductive, but it is supported by short run relations between income and vital rates for pre-modern societies [Goubert (1960)]. The proposition that P" declines at higher levels of product per capita corresponds to the experience of those countries that went through the demographic transition.

Let us specify P"(y) as a parabolic function in ln(y):

$$P'' = -a + b[\ln(y)] - c[\ln(y)]^2 \quad . \tag{6}$$

Equation (6) attains a maximum of PM" at y* and intersects the ln(y) axis at $\ln(y_1)$ and $\ln(y_2)$. Namely, $PM'' = P''(y^*) = \text{maximum}$ and $P''(y_1) = P''(y_2) = 0$. By definition $y_2>y_1$. Equation (6) also can be written in the following form:

$$P'' = v[\ln(y) - \ln(y_1)][(\ln(y_2) - \ln(y)] \quad , \tag{7}$$

where

$$a = v[\ln(y_1)][\ln(y_2)] \quad , \tag{8}$$

$$b = v[\ln(y_2) + \ln(y_1)] \quad , \tag{9}$$

$$c = v = 4(PM'')/[\ln(y_2) - \ln(y_1)]^2 \quad , \text{ and} \tag{10}$$

$$y^* = [\ln(y_2) + \ln(y_1)]/2 \quad . \tag{11}$$

Since

$$y'' = p'' + Y'' - P'' \quad , \tag{12}$$

from (5), (6) and (12) we obtain the following differential equation:

$$y'' = p'' + m + (n-1)\{-a+b[\ln(y)] - c[\ln(y)]^2\} \quad . \tag{13}$$

Equation (13) relates the percentage rate of change over time of the available product per capita to the available product per capita. This equation identifies a parabola in ln(y), with a minimum at the same value y* at which the population function P"(y) attains its maximum. For appropriate values of n, a, b and c, y"(y) will either intersect the horizontal axis, touch it, or lay above it, depending upon the values of p" and m. These patterns are qualitatively distinct, and respectively correspond to a dynamics characterized by the existence of a Malthusian trap, to a transitional situation, and to unbounded economic growth. Smooth changes in p" and/or m can bring about a catastrophic switch from a Malthusian dynamic to a modern dynamic. A discussion of the significance and implication of these patterns as regard the onset of modern economic growth is carried out in the ensuing paragraphs.

THE MALTHUSIAN TRAP

Differential equation (13) specifies y" as a parabola in ln(y), with a minimum at ln(y*). When the parabola intersects the ln(y) axis the equation possesses two equilibrium solutions, where one is stable and the other is unstable. When the parabola touches the ln(y) axis equation (13) has one semistable solution. It has no equilibrium solution when the parabola lays above the ln(y) axis. In this last case all solutions of (13) are monotonically increasing functions of time, irrespective of initial conditions. These three situations are "qualitatively" different and will be referred to as topologies A, B and C. Infinitely many combinations of parameter values can produce any one of them. On the other hand, all the combinations of the model's parameters associated with the existance of two equilibrium solutions, or of one semistable equilibrium solution, or with the absence of any equilibrium solution are "topologically" equivalent.

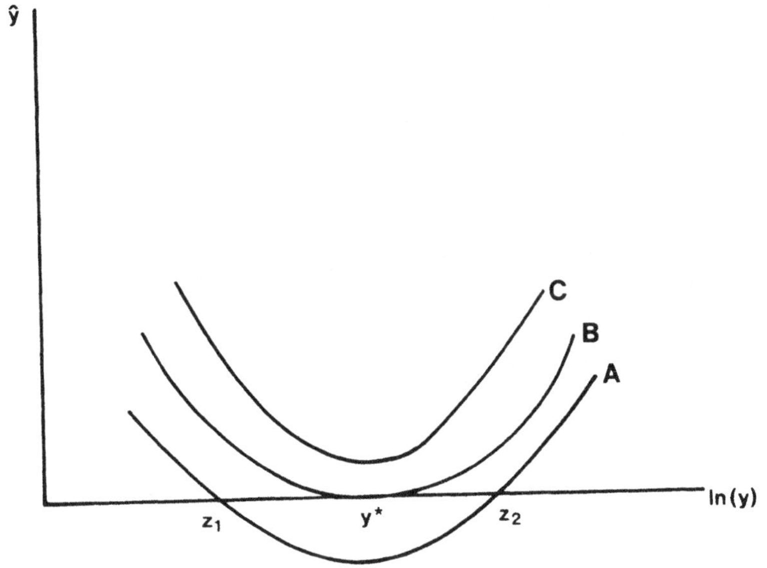

Fig. 1. Graphs of Equation (13) for Topologies A, B and C.

The three situations are illustrated in Figure 1 by curves A, B and C, that are labelled so as to match the topologies that they are designed to typify. Curve A in Figure 1 intersects the ln(y) axis at $\ln(z_1)$ and $\ln(z_2)$. z_1 and z_2 identify equilibrium levels of product per capita, since the rate of change of product per capita becomes zero whenever they are attained. Moreover, $y''(z_1) = y''(z_2) = 0$. z_1 is a stable equilibrium value because it is possible to specify a neighborhood around it such that any value of $y \neq z_1$ belonging to this neighborhood will tend to approach z_1 from above or from below. z_2 identifies an unstable equilibrium because it is possible to specify a neighborhood around it such that any value of $y \neq z_2$ belonging to the neighborhood will tend to move away from it. In particular, a product per capita greater than z_2 will grow beyond any bounds. Also, z_1 and z_2 are stable/unstable equilibrium values, respectively, because $dy/d[\ln(y)]$ is smaller/larger than zero when evaluated at z_1/z_2.

Curve B in Figure 1 touches the ln(y) axis at one point. The value y* of y at which $y''(y)$ is tangent to the ln(y) axis is an equilibrium point that is stable from the left and unstable from the right. y* is also the value of y at which y'' is a minimum.

Curve C is a parabola positioned above the ln(y) axis. This implies that whenever the C topology occurs, any solutions of equation (13) will have positive derivatives with respect to time throughout, and consequently will be characterized by a monotonic increase of product per capita over time.

The Malthusian trap or low level trap referred to in the literature corresponds to the lower stable equilibrium z_1 that exists when the y curve intersects the ln(y) axis. The upper unstable equilibrium z_2 will be referred to as the door of the trap. A lower stable equilibrium of product per capita accounts for the perverse dynamics that dominated the premodern world, whereby a country's product per capita would move from above or from below towards a low equilibrium level. Modern economic growth occurs because the product per capita y is larger than the door of the trap z_2, or because a type C curve specifies the dynamics of the system and the trap itself no longer exists.

To determine whether specific parameters produce a situation typified by the A, B or C curves in Figure 1, we set y" of equation (13) to zero and solve for the roots of the resulting quadratic in ln(y). The system is characterized by a type A topology if the quadratic has two real and distinct roots constituting the z_1 and z_2 equilibrium values. It is characterized by a type B topology if its roots are real and equal, which corresponds to the collapse of the low level trap into the door of the trap. And, it is characterized by a type C topology if the quadratic has imaginary roots. Imaginary roots imply that there are no equilibrium values of y", and that the dynamics of the system is characterized by unbounded growth of product per capita irrespective of initial conditions.

By a few manipulations, using equations (8), (9), (10) and (13), it can be shown that the roots $\ln(z_1)$, $\ln(z_2)$, of $y''(y) = 0$ are:

$$\ln(z_i) = \left(\ln(y_1)+\ln(y_2) \pm \sqrt{[\ln(y_2)-\ln(y_1)]^2(1-R)}\right)/2 \quad , \qquad (14)$$

where $i = 1$ uses the negative sign;
$i = 2$ uses the positive sign; and

$$R = (m+p'')/(n-1) \quad . \qquad (15)$$

Equations (14) and (15) relate the roots of $y''(y)=0$ to all of the system parameters, and allow an analysis of the effects of any parameters or combination of parameters on the dynamics of the economic demographic interactions considered. Here, however, for the sake of simplicity we will regard y_1, y_2, n and PM'' as constants and concentrate instead on the effects of the rate of technological progress m and of the rate of change of "exploitation" p''. If R<1, $\ln(z_1)$ and $\ln(z_2)$ are real and distinct, implying that the system is characterized by a type A dynamics. z_1 is the low level stable equilibrium corresponding to the Malthusian trap, and z_2 is the high level unstable equilibrium that constitutes the door of the trap. R=1, $\ln(z_1)$ and $\ln(z_2)$ real and equal produce a type B dynamics whereby the Malthusian trap and its door collapse into a single equilibrium level of product per capita. If R>1, y''=0 has imaginary roots, a type C dynamics prevails, and although the system has no equilibrium level of product per capita, it is characterized instead by unbounded growth.

Since n and PM'' are constants here, as the rate of technological progress m and the growth of exploitation p'' increase, so does R. Some combination of values of p'' and m will produce an R corresponding to a type A dynamics, characterized by the existence of a Malthusian trap, while larger values of p'' and m can produce an R>1 corresponding to a type C dynamics and to an unbounded growth of product per capita. It should be pointed out that the switch from a type A to a type C dynamics can be produced by larger values of the rate of technological progress when p''=0, or by a constant rate of technological progress associated with increasing rates of the exploitation factor. This, in turn, implies that either technological progress, exploitation, or some mix of the two can produce the disappearance of the Malthusian trap. It should also be noted that a premodern income dynamics requires a type A topology in which a trap exists, and that a country's product per capita be lower than the door of the trap. In contrast, modern economic growth can take place either within a type A topology because a country's product per capita is larger than the door of the trap, or because a type C topology exists.

The model discussed in this paper can be used to evaluate during a specified time interval whether a country was a prisoner of the Malthusian trap or a state of modern economic growth. One such investigation described in Casetti (1980) showed that England was prisoner of a Malthusian trap over the time interval 1695-1785, and was in a state of modern economic growth during the time interval 1801-1871. The model used in that study did not incorporate an exploitation factor.

The present model also can be used to conceptualize a country's transition to modern economic growth. At least two alternative formalizations of this transition are possible. The first one involves the following: (1) in premodern times the product per capita quickly approaches its low equilibrium level, so that only a small inaccuracy is involved if we regard it as identical to it; and, (2) the parameters determining the level and existence of equilibrium incomes change very slowly, and cause R in equations (14) and (15) also to increase very slowly. These assumptions imply that the country's income will increase slowly following the increase in z_1 induced by R, as long as R<1, and will increase rapidly instead when R>1 causes the low level trap to disappear. Within this frame of reference the trap's door does not play a significant role. The onset of modern economic growth results from the disappearance of the Malthusian trap rather than by the fact that y becomes positioned past the door of the trap, and the switch to modern economic growth coincides with the switch from an A to a C topology. The slow change of R is coupled first with the fast adjustment of y to its equilibrium level, and later with a fast rate of growth of y. This constitutes an increase of the concurrent fast dynamics of system variables, coupled with slow dynamics of system controls discussed by Zeeman (1977, p. 65).

A second conceptualization of the onset of modern economic growth is produced if we assume that in premodern times y oscillates "rapidly" around its long run low level equilibrium value, for instance as a result of random shocks, while the parameters determining the level of existence of the system's equilibria drift slowly. Here, again, these slow parameter changes would cause R to increase, and an upward shift in the parabolic y"(y) curve. This, in turn would cause the intersections of y"(y) with the ln(y) axis to grow closer to each other, and eventually to collapse into y* and disappear. If a country's available product per capita oscillates around the low level equilibrium z_1 these oscillations will bring y past the door of the trap z_2, thus switching on the modern economic growth when the dynamics of the system are still in a type A topology. Of course, within this framework the onset on modern economic growth is likely to extend over a sizeable time span, because of the likelihood of multiple switchings into and out of modern economic growth, before modern growth becomes solidly and irrevocably locked in. This second conceptualization of the onset of modern economic growth is perhaps more realistic. The first conceptualization can be regarded as a simplified approximation of the second.

SPATIAL DIMENSIONS OF ECONOMIC GROWTH

Let us now consider the onset of modern economic growth in a spatial context and in relation to the model discussed in the preceding sections of this paper. Modern economic growth originated in England, spread into continental Europe, and then leapfrogged to Japan and to the U.S. The countries that entered into modern growth experienced a surge of political and military power. The power imbalance between early modernizers and the rest of the world produced the partition of most of the rest of the world into colonies. Conflicts and wars broke out when later modernizers experienced their surge in power and set claims to a share of territories appropriated by earlier modernizers. In the period following World War II a swift process of decolonization took place, and it appears that by now modern economic growth has engulfed all countries on earth. Technological progress and exploitation are regarded as the key ingredients in these historical processes, although it is a subject for debate which ingredient had which role when. There is no question that higher rates of technological progress are a major determinant of the onset of modern economic growth. The model presented earlier leads itself very easily to "explain" the spatial spread of the onset of modern economic growth as the result of the spread of higher rates of technological progress [Casetti (1971)]. On the other hand, many indications suggest that transfers of product and services brought about by the direct or indirect use of force or by "favorable" political circumstances did play a significant role in the international transactions in modern times.

The salient aspects of the spatial temporal dynamics of modern economic growth, when viewed in the light of the model discussed in this paper, suggest the following hypotheses concerning the comparative roles of technological progress and exploitation:

(1) The first modernizing countries experienced an acceleration in technological progress (namely, an increase in m). This acceleration was associated with the surge in political and military power that brought into existence exploitative transfers of goods and services from less developed societies, that, in terms of the model considered, correspond to an increase in p". Both increases in m and p" contributed to, or accompanied the onset of modern economic growth in the first modernizing countries.

(2) The colonial dependencies did experience some acceleration in their rate of technological progress, m, as a result of their interactions with the respective "mother countries." On the other hand, they also experienced "unequal exchanges" corresponding to a negative p". Negative p" values counteracted the increasing m's, thus slowing down or hampering in these countries the onset of modern economic growth.

(3) The decolonization in the post World War II period produced an increase in p" in the former dependencies, and was associated with an acceleration in the rate of technological progress m. Both the increase in m and p" account for the higher rates of economic growth that these countries have experienced since then. In many instances the onset of modern economic growth may have started in former colonies during the past thirty years, as the result of these higher m's and p" 's.

(4) Finally, the increases in the price of oil by the OPEC countries can be viewed as exploitation in the sense used in this paper. It produced a large positive p" component to the OPEC countries, and negative p" components to the remaining countries, regardless of whether they were more developed or less developed. Over the past few years these positive and negative p" 's associated with the OPEC's policies probably constituted a significant determinant of rate of growth and, specifically, of the tempo of modern economic growth in the less developed countries.

CONCLUSIONS

Prior to the industrial revolution incomes per capita and standards of living all over the world tended to fluctuate about a trend characterized by a rate of increase so slight that it was practically nil in the short to medium run. After the industrial revolution the presently developed countries experienced an explosive growth of incomes per capita. In general, modern economic growth has come to be regarded as qualitatively different from its premodern counterpart, and its onset is characterized by an acceleration of economic progress.

In this paper technological progress and exploitation are discussed with respect to their impact on modern economic growth. More specifically, a model is presented that links the onset of modern economic growth to the smooth change of the rates of technological progress and of exploitation past a critical threshold. The mathematical structures used are of the type focussed upon by the catastrophe theory paradigm.

REFERENCES

Casetti, E., 1971. A Spatial Model of the Low Level Equilibrium Trap, paper presented at the Budapest Meeting of the Commission on Quantitative Methods, International Geographical Union, August 10-14, 1971.

Casetti, E., 1980. The Onset of Modern Economic Growth: A Catastrophe Model, paper presented at the eleventh Pittsburgh Modeling and Simulation Conference, May 1-2, 1980, and forthcoming in the Conference's Proceedings.

Goubert, P, 1960. Beauvaisis et le Beauvaisis de 1600 a 1730, Paris.

Kuznets, S., 1971. Economic Growth of Nations: Total Output and Production Structure, Harvard University Press, Cambridge, Mass.

Pearson, L., 1969. Partners in Development, Praeger, New York.

Poston, T. and I. Stewart, 1978. Catastrophe Theory and Its Applications, Pitman, London.

Zeeman, E., 1977. Catastrophe Theory: Selected Papers 1972-1977, Addison-Wesley, Reading, Mass.

Comments by Dr. D.S. Dendrinos are gratefully acknowledged. The errors are mine.

HISTORIES OF URBANIZATION

G.J. Papageorgiou

Department of Geography
McMaster University

INTRODUCTION

Sudden urban growth is a phenomenon that transcends political, cultural and regional barriers (see Figures 1-4). One may envisage silent urban "explosions" around the earth. These happen with increasing frequency and begin to accumulate precisely as an avalanche does, suggesting that this relatively recent pattern of urbanization will continue for some time to cause the greatest of urban problems on a global scale. In spite of its obvious significance, little if anything has been known about what drives the phenomenon: Why do some cities suddenly "explode?" Why do most cities fail to experience this "explosion?" Why can we observe population-stable cities amidst a rapidly growing urban and regional environment--and vice-versa? Is this "explosion" predictable? If yes, what are the associated difficulties? It is the aim of this paper to describe mechanisms that provide an answer to such questions. The phenomenon is analyzed in the case of a population-homogeneous single city. Speculations concerning population heterogeneity, a system of cities and policy implicators are also included.

THE MODEL

Sudden urban growth feeds on migration. This suggests a clear advantage of the city over the existing alternatives, an advantage that must be created suddenly and must persist with city growth. Hence in order to explain this phenomenon one is bound to begin with a study of the relationship between urban opportunities and urban size. The simplest construct that cap-

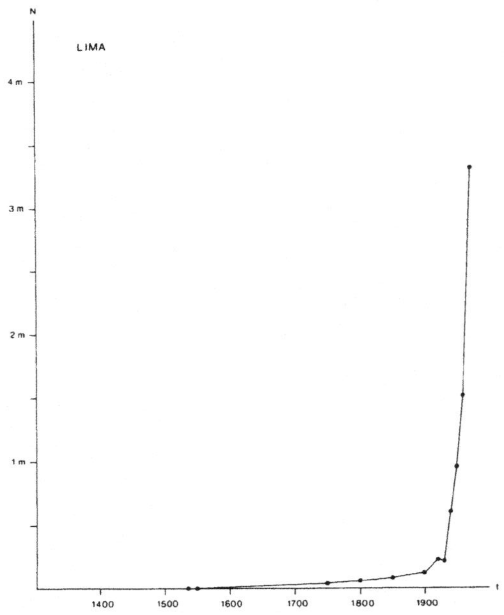

Fig. 2. Population Growth in Lima, Peru.

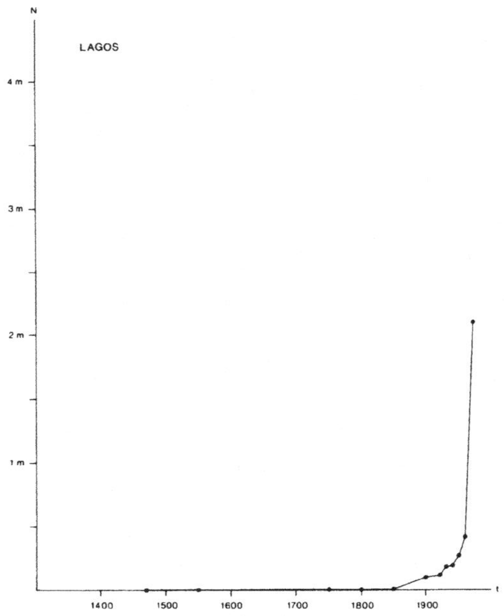

Fig. 1. Population Growth in Lagos, Nigeria.

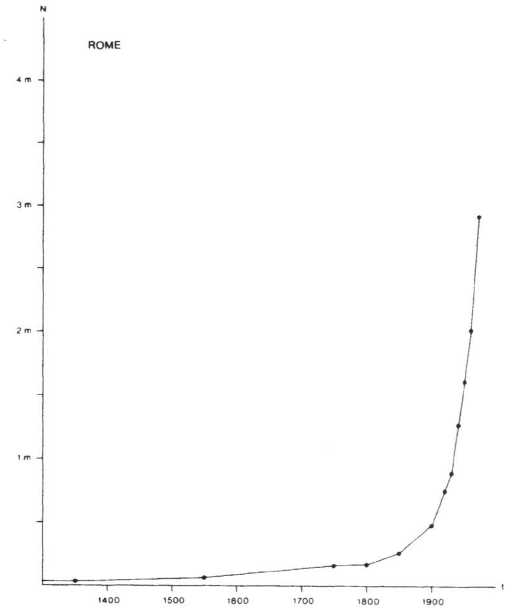

Fig. 3. Population Growth in Rome, Italy.

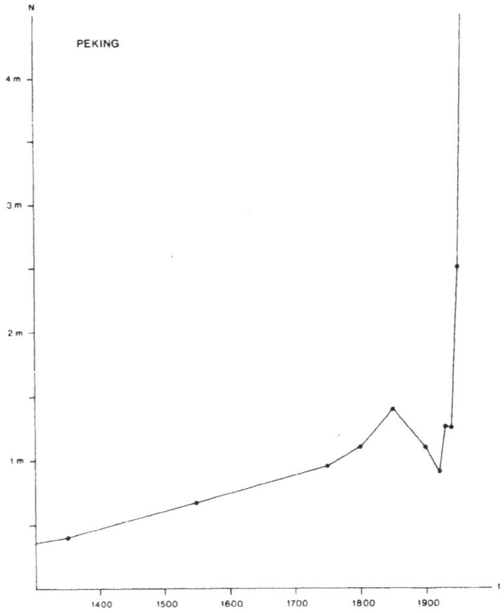

Fig. 4. Population Growth in Peking, China.

tures such a relationship is the following. There is a standard circular city partitioned into two concentric zones, [0,a] and [a,b], occupied by production and residences, respectively. Production refers to the private good and uses labor and land. No outside labor is used for production. The output of production is concentrated at the center where it is sold. If there is a surplus (deficit) in production, the necessary quantities are exported (imported) at the same, fixed price. Production is organized in accordance with the principle of maximizing the difference between the value of what is produced and the value of what is used for production:

$$\underset{N,a}{\text{maximize}}\ f[N,a] - \left(yN + \int_0^a R2\pi s ds + C[f]\right) \tag{2.1}$$

where f is the production function of the private good; N is the urban population; y and R are prices of labor per capita and land used for production; and, C is the total cost of transporting the output from the place of production to the center where it is sold.

Individuals are identical. They strive to optimize utility:

$$\begin{aligned} &\underset{x,q}{\text{maximize}}\quad u[x,q] \\ &\text{subject to}\quad x + rq + c[s] = y \end{aligned} \tag{2.2}$$

where u is utility; x,q are quantities of the private good and land consumed; and, 1,r,c[s] are prices of the private good, land and transportation at distance s from the center. Individuals regularly travel to work along a radial, dense, uncongested, spaceless transportation network and then continue toward the center in order to obtain the private good.

An equilibrium for this city is characterized by a common utility level $\bar{u}$; by the requirement that everyone who works has a place to live in the city:

$$\int_a^b \frac{1}{q}\ 2\pi s ds = N\ ; \tag{2.3}$$

and, by the requirement that locations at the border between two land-uses are bid on identically by those land-uses:

$$\begin{aligned} R|_a &= r|_a \\ r|_b &= \overset{v}{R} \end{aligned} \tag{2.4}$$

where $\overset{\vee}{R}$ is the opportunity cost of land.

To summarize, our initial objective is to obtain a relationship between urban opportunities and urban sizes. In terms of this model it is to determine $d\bar{u}/dN$, which represents how the equilibrium utility level changes with urban population. A comparative statics analysis of this model yields the following proposition concerning du/dN [Papageorgiou (1980)]:

If utility is strictly quasi-concave, if both goods have positive income effects and if

$$\frac{\partial r}{\partial s}/\frac{r}{s}\Big|_{\hat{a}} > -1 \tag{2.6}$$

then

$$\begin{aligned}&\left(k_1 - k_2 \frac{d}{dN}\frac{\partial X}{\partial \hat{s}} - k_3 \frac{\partial}{\partial \hat{s}}\frac{\partial X}{\partial \hat{s}}\right)\frac{d\bar{u}}{dN} = \\ &-k_4 - k_5 \frac{d}{dN}\frac{\partial X}{\partial \hat{s}} + k_6 \frac{d}{dN}\frac{\partial X}{\partial N} + k_7 \frac{\partial}{\partial \hat{s}}\frac{\partial X}{\partial \hat{s}} \\ &-k_8\left[\left(\frac{\partial}{\partial \hat{s}}\frac{\partial X}{\partial \hat{s}}\right)\left(\frac{d}{dN}\frac{\partial X}{\partial N}\right) - \left(\frac{d}{dN}\frac{\partial X}{\partial \hat{s}}\right)^2\right]\end{aligned} \tag{2.7}$$

with all $k_i > 0$

where $X(=f-C)$ is the value of what is produced net of the associated transportation costs.

Since this proposition provides the basis for everything that follows, it is important to understand what is intuitively implied by the related assumtpions and definitions. Strict quasi-concavity ensures the standard convex, downward sloping indifference curves. Positive income effect ensures higher consumption for higher income. Inequality (2.6) is a condition on the elasticity of residential rent at the interior border with respect to this border. It says that the residential rent at the interior border must decrease relatively slower than the movement of this border away from the center. The constants k_i in (2.7) are known functions of individual characteristics such as marginal utilities, and of observables such as the opportunity cost of land. Hence knowledge of these characteristics would permit a direct computation of the constants k_i. The quantity $d(\partial X/\partial s)/dN|_a$ describes how the marginal value of land is affected by the number of people working on it. The quantity $(\partial X/\partial s)/\partial s|_a$ describes returns to scale in land as a factor of production.

A value greater than (equal to, smaller than) zero indicates increasing (constant, decreasing) returns to scale. Finally, the quantity $d(\partial X/\partial N)/dN$ describes how the marginal value of labor changes with urban population.

IMPLICATIONS

Consider the case of constant returns to scale in land as a factor of production. Intuitively, the marginal value of land used in production must increase with city size. It is however unclear if it increases at an increasing, constant, or decreasing rate. In consequence let

$$\frac{d}{dN}\frac{\partial X}{\partial s}\Big|_{a} = m_1 \pm m_2 N \quad , \tag{3.1}$$

with all $m_i > 0$. Suppose that the structure of advantages in the organization of production causes the marginal value of labor to increase initially at an increasing rate with city-size:

$$\frac{d}{dN}\frac{\partial X}{\partial s} = m_3 + (m_4 - m_5 N)N \quad , \tag{3.2}$$

where all $m_i \geq 0$. Under these circumstances the qualitative behavior of $d\bar{u}/dN$ appears in Figures 5-7 (Papageorgiou, 1980). The particular behavior realized depends upon the particular combination of constants k_i and m_i. Hence, in principle, given a knowledge of individual characteristics, production technology, and the such, it would be possible to determine precisely how the equilibrium utility level changes with urban population.

In order to see how these results may imply sudden urban growth, consider a closed state of identical individuals with a unit total population partitioned between a single city (N) and a single alternative sector (1-N). Consider migration from the alternative sector to the city. Suppose that an increase in urban utility, with everything else remaining the same, causes further migration. That is, the better the urban opportunities are, the more population is attracted by the city. This assumption is expressed as $\bar{v}$ in Figure 8 which is based on the original contribution of Casetti (1980). Measured from left to right, $\bar{v}[N]$ represents urban utility as a function of urban population *at equilibrium* (where no further migration between sectors occurs). Measured from right to left, $\bar{v}[1-N]$ represents the utility of the alternative sector as a function of the population in that sector, otherwise, if utility in the alternative sector is lower (higher) than $\bar{v}$ then migration would continue toward (away from) the city--a contradiction. Therefore, $\bar{v}[1-N]$ in Figure 8 expresses the

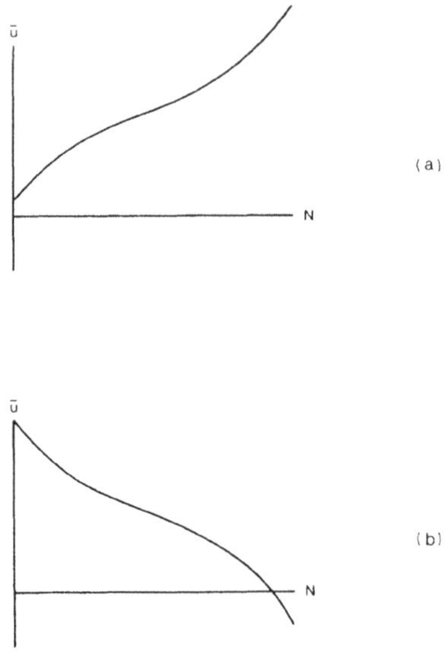

Fig. 5 Variation of Average Utility Over City Size: Near Linear

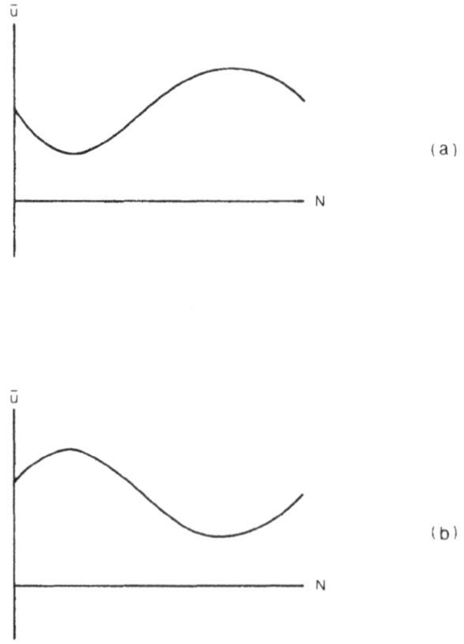

Fig. 6. Variation of Average Utility Over City Size: Cyclical

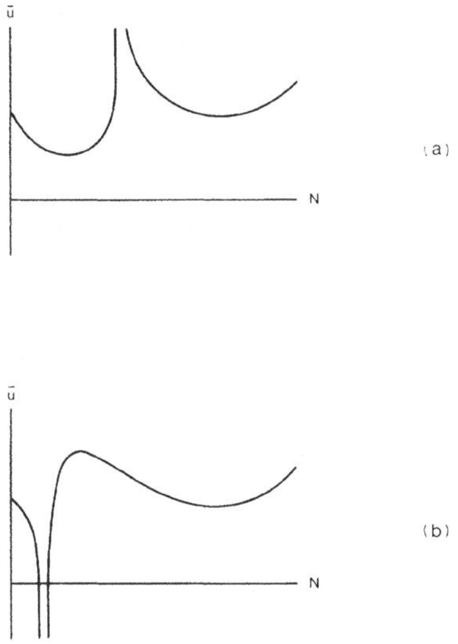

Fig. 7. Variation of Average Utility Over City Size: Discontinuous

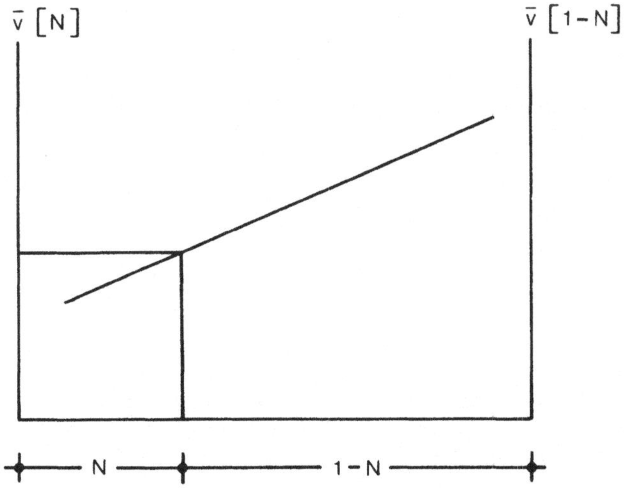

Fig. 8. Equilibrium Point Between Sectors.

assumption that agricultural occupations in economies with a high proportion of agricultural population enjoy a lower standard of living than in comparable highly-urbanized economies.

Functions $\bar{u}$ and $\bar{v}$ may be used simultaneously to determine why the relative advantages between the two sectors unfold with city-size. Relative advantages, in turn, would dictate the direction of migratory streams. The superimposition of Figures 5(b) and 8 is shown in Figure 9. Following Casetti (1980), the intersection between $\bar{u}[N]$ and $\bar{v}[1-N]$ marks an equilibrium partition of population between the urban and the alternative sector. It is obvious now that the structure of $\bar{u}[N]$ in this case cannot produce sudden urban growth. To see this observe that both curves in Figure 9 shift upwards with the development of technology. Then, if one asserts that the city is more responsive to such development than the alternative sector, one may obtain a growing city. Nevertheless, under smooth technological improvements, growth will be smooth, whereas the structure of sudden urban growth clearly suggests some discontinuity. Such discontinuity is obtained, for example, with a $\bar{u}[N]$ structure as in Figure 6(b). This is shown in Figure 10 which also follows Casetti (1980). Whereas both $\bar{u}$ and $\bar{v}$ shift upwards with the development of technology the city is more responsive to such development than the alternative sector. Figure 10(a) indicates shifts of $\bar{u}$ *relative* to $\bar{v}$. Until period three, smooth changes in technology cause smooth changes in the equilibrium population partition. Period three however marks a critical point beyond which a small relative upward shift in $\bar{u}$ creates a discontinuity in the equilibrium urban population path [Figure 10(b)]. This, together with a statement that the rate of migration is proportional to the difference between $\bar{u}$ and $\bar{v}$, is sufficient to account for sudden urban growth. One may imagine the non-discernible equilibrium path in Figure 10(b) as guiding the urban population trajectory. At t_4, $\bar{u}-\bar{v}$ is small and so is the rate of migration with time; however, the rate of migration increases with $\bar{u}-\bar{v}$ and then declines until the new equilibrium is reached [Figure 10(a)]. One therefore may expect that the structure of the urban population trajectory will be logistic under these circumstances.

It is now possible to answer within this framework questions akin to those posed at the introduction. Why do some cities suddenly "explode?" Because the combination of individual characteristics, production technology, and the such, associated with these cities are expressed in a combination of constants k_i and m_i that generate utility variation paths as, for example, in Figures 6(a) or 7(a). Notice also that Figures 5(a) and 6(b) suggest the possibility of cities suddenly created, not merely of existing cities suddenly changing size. Why do most cities fail to "explode?" Because the combination of circumstances

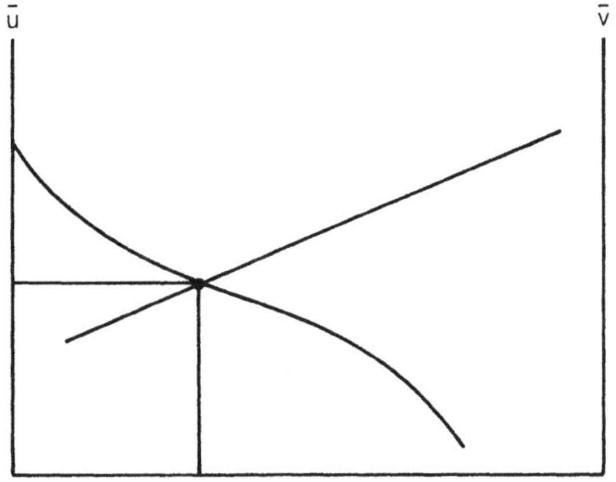

Fig. 9. Superimposition of Figures 5(b) and 8.

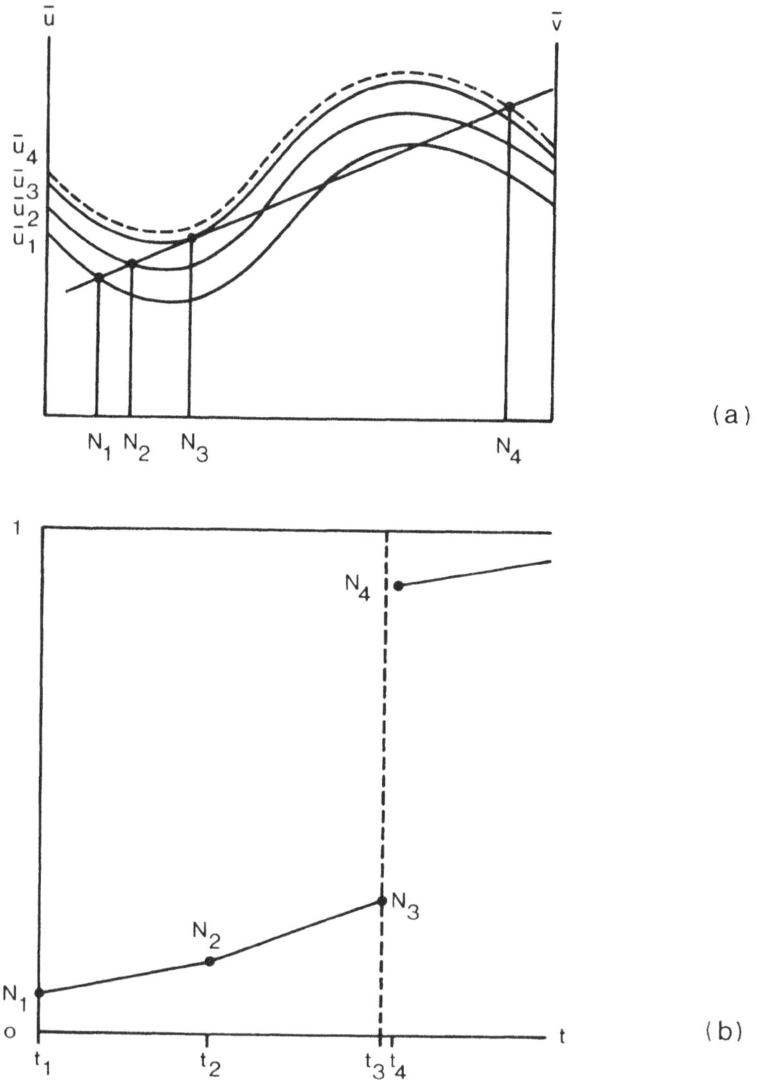

Fig. 10. Shifts in $\bar{u}$ and $\bar{v}$ with the Development of Technology

characterizing such cities generate utility variation paths as, for example, in Figure 7(b). The possibility for an unbounded maximum $\bar{u}[N]$, as in Figure 7(a), does not affect the path toward urbanization. The only consequence is a fleeting moment of infinite bliss that vanishes through migration as suddenly as it has appeared, because if the rate of migration is proportional to the difference between $\bar{u}$ and $\bar{v}$, then the migration-adjustment in the neighborhood of this unbounded maximum will be close to instantaneous. Is there a possibility for prediction? In principle yes, because the constants k_i and m_i are known functions of individual characteristics such as marginal utilities, and of observables such as the opportunity cost of land. Prediction however would require a theory concerning $\bar{v}[1-N]$ at the same level of detail as the one proposed for $\bar{u}[N]$, and a theory about how $\bar{v}$ shifts relative to $\bar{u}$ through time. Simply to assert that agricultural societies enjoy a lower standard of living than urbanized ones and that the city is more responsive to the development of technology than the alternative sector, although both reasonable and sufficient to account for sudden urban growth, is nevertheless too abstract to be used for drawing conclusions about particular cities. What are the associated difficulties? The problems of measuring some of the entities that compose the constants k_i and m_i are well-known. This however does not imply some inherent micro-theoretical inadequacy. It rather stems and reflects the enormous complexity of the systems at hand.

Recall that the arguments in this section pertain to constant returns to scale in land as a factor of production. When this assumption is relaxed, the right-hand side of (2.7) becomes a third-degree polynomial. This implies an additional change in the direction of $\bar{u}[N]$, and hence an additional equilibrium configuration.

GENERALIZATIONS

Consider Figure 11. It contains two stable (N_1,N_3) and one unstable (N_2) equilibrium population positions. N_1 corresponds to an underdeveloped, low-urbanized state while N_3 corresponds to a developed, high-urbanized state. Given population homogeneity, this tends to suggest that all is well with development because N_3 corresponds to higher welfare for everyone. If however the population is heterogeneous, Figure 11 should be interpreted as indicating average welfare so that the issue of welfare distribution becomes relevant. What can be said about this issue now? Hartwick, et al. (1976) may provide us with some insight into the complexities involved. In the case of population heterogeneity they have discovered that increasing the size of a class, with everything else being constant, decreases the utility of all classes, whenever increasing the income of a class, with everything else being held constant, increases the utilities of the poorer

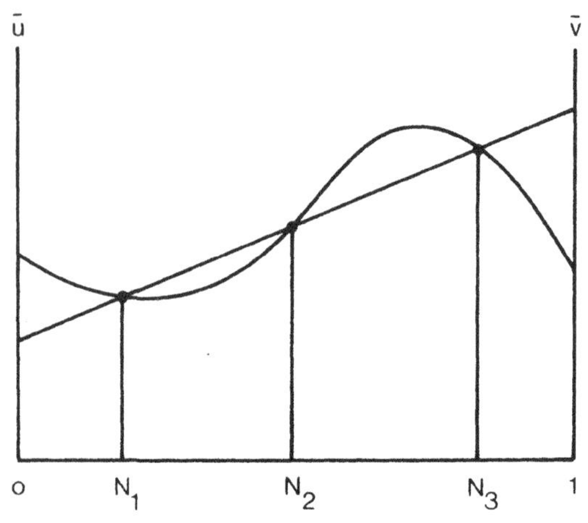

Fig. 11. Equilibrium Population Positions

classes and decreases the utilities of the richer classes. Here, that both urban population and utility have increased as the system moves from its underdeveloped to its developed state implies a positive overall income effect average on utility stronger than the corresponding negative population effect. But if the income of the poorer half has mainly increased then the utility of the richer half has decreased, which together with the urban population increase tends to suppress average utility as the system moves from N_1 to N_3. Intuitively this suggests that development may increase social inequility in laissez-faire situations.

What may happen in the case of a system of cities? Then $\bar{u}_i$ relates to a particular city i and $\bar{v}_i$ to the rest of the economy relative to that city. For smaller cities, where the economy is relatively larger, the slope of $\bar{v}_i$ must flatten. For very large central place systems, migration to and from city i cannot affect the general level of utility because it involves a relatively insigificant part of the total population: $\bar{v}_i$ becomes horizontal over the population range of city i as determined by $\bar{u}_i$.

Complexity multiplies if one allows for the interdependencies between cities. Hence the position of $\bar{u}_i$ relative to $\bar{v}_i$ remains unknown. For some cities, it may well be that discontinuity is imminent, and regional policies aimed at altering, even slightly, the relative configuration of advantages in the system may be prone to unexpectedly catastrophic population shifts unless they are supported by effective controls on migration. Furthermore, once sudden urban growth is well underway, it is very difficult to stop it by means other than direct controls on migration. This happens because, according to this theory, a fast rate of urban growth implies a large positive difference between $\bar{u}$ and $\bar{v}$. Thus, in order to stop sudden urban growth and at the same time avoid direct controls on migration, one has to eliminate this large difference either by substantially reducing urban opportunities, or by substantially increasing the opportunities in the rest of the economy, or both, within a prohibitively short period. The case of direct controls on migration is also difficult to support. This happens because to maintain a large difference between $\bar{u}$ and $\bar{v}$ is to create a (probably unmanageable) political problem. As it is also morally untenable to support such institutionalized inequality, the consequent political pressure will eventually neutralize any imposed direct controls. Hence, following Doxiadis (1967), I am led to believe that the challenge of the future, far from being to stop sudden urban growth, is rather to accept it and to plan for it.

REFERENCES

Casetti, E., 1980. Equilibrium Population Positions between Urban and Agricultural Occupations, _Geograhhical Analysis_, 12:47-54.

Doxiadis, C., 1967. Ecumonopolis: The Coming World-City, in: _Cities of Destiny_, ed. by A. Toynbee, London: Thomas and Hudson.

Hartwick, J., _et al._, 1976. Comparative Statics of a Residential Economy with Several Classes, in: _Mathematical Land Use Theory_, ed. by G. Papageorgiou, Lexington, Mass.: Lexington Books, 55-78.

Papageorgiou, G., 1980. On Sudden Urban Growth, _Environment and Planning A_, 12: 1035-1050.

STABILITY AND CHANGE IN TWO-DIMENSIONAL FLOWS

Tönu Puu

Department of Economics
Umeå University

This presentation concerns structural stability and structural change in a continuous space economy. The continuous flow model, in addition to its intuitive geometrical representation of space, has the particular advantage that we can apply the generic theory of differential equations and catastrophe theory to the study of stability and change. In order to make the flow, the properties of which we study, more substantial, I will start by presenting a very simple continuous model of the space economy. The particular model chosen is, however, not important, as the reasoning is applicable equally well to a wide variety of models.

Suppose first that there is available a production technology represented by a simple Cobb-Douglas function:

$$Q = K^{\alpha} L^{\beta} M^{\gamma} , \tag{1}$$

where K, L, and M represent capital, labor and land inputs, Q represents output, and the exponents sum to unity. Due to the linear homogeneity of the function we can divide through by M and obtain:

$$q = k^{\alpha} l^{\beta} , \tag{2}$$

where the lower case symbols $q = Q/M$, $k = K/M$ and $l = L/M$ now denote areal densities of output and inputs. If we assume these areal densities to be functions of the space coordinates (x,y), we have the picture of a spatially extended production economy.

Due to traditional economic theory, the optimum of the individual producers at every location would be determined by the conditions:

$$rk/\alpha = wl/\beta = g/\gamma = pq \quad , \tag{3}$$

where r denotes interest, w wages, g land rent, and p product price. The two first equations are optimum conditions, whereas the last one simply serves to determine land rent. Equations (3) together with (2), of course, determine inputs k and l as well as output q, as soon as the set of prices r, w and p are known.

As a condition of long-run equilibrium allocation of capital we assume interest r to be constant over space. Product price p and wages w will, as we shall discover, vary over space, depending on the patterns of labor and commodity flows between industrial and residential areas. Land rent, finally, is determined residually from the local profitability of production as determined by real wages, reflecting the spatial closeness to market places and labor supply.

We are now ready to introduce the flows. We define two vector fields:

$$\phi = [\phi_1(x,y), \phi_2(x,y)] \quad , \text{ and} \tag{4}$$

$$\psi = [\psi_1(x,y), \psi_2(x,y)] \quad , \tag{5}$$

representing the flows of goods and labor, respectively. These vector fields associate with every point in space (x,y) vectors, the norms of which represent the quantities transported and the directions of which represent the directions of flows.

Due to Gauss' integral theorem the divergence of a vector field is interpreted as the local change of flow due to sources and sinks. If we now take q', the local demand of goods, and l', the local supply of labor, as given data, mainly determined by the distribution of population and, if desired, dependent on the local prices, we can define the excess supplies and write

$$\text{div } \phi = (q - q') \quad , \text{ and} \tag{6}$$

$$\text{div } \psi = -(l - l') \ . \tag{7}$$

These are the continuity equations or the divergence laws.

The gradient laws can be formulated as

$$f\phi/|\phi| = \text{grad } p \quad , \text{ and} \tag{8}$$

$$f\psi/|\psi| = \text{grad } w \quad , \tag{9}$$

where the scalar field $f(x,y)$ expresses the local (for simplicity, isotropic) cost of hauling goods or labor across the location. The cost function can be made equal for both flows by a suitable choice of unit quanta. Relations (8) and (9) simply tell us that the flows are in the direction of steepest price or wage increase, and that in these directions price or wage changes with the rate of local transportation cost.

In principle, the equations written down so far allow us to solve the whole flow and spatial organization problem given certain boundary conditions. By taking squares of equations (8) - (9) we get partial differential equations for the product price and wage rate. Solving these we directly obtain the unit flow fields. Next, knowing the unit flow fields, equations (6) - (7) furnish partial differential equations in flow intensities or the norms of the flow fields. This is the logical way of proceeding to obtain an explicit solution to the problem. Rather than proceeding in this way, we will put additional structure into the model by introducing stability considerations. To make these particularly simple, however, we first introduce an assumption of polarity of the flows:

$$\phi/|\phi| + \psi/|\psi| = 0 \ . \tag{10}$$

This assumption simply means that the two flow fields have opposite direction, labor flowing from residences to industry, and finished goods in the opposite direction. This assumption allows us to deal with only one set of flow lines and one potential, denoted $\lambda(x,y)$. By this:

$$p = \bar{p} + \lambda(x,y) \quad , \text{ and} \tag{11}$$

$$w = \bar{w} - \lambda(x,y) \ . \tag{12}$$

All these simplifying conditions have been introduced in order to make the model as simple as possible. Most of them are not difficult to relax. Before treating the subject of structural stability I would like to refer to the "correspondence principle" of traditional economics. The philosophy is the following: As changes in an economic system always occur, an equilibrium would be without interest if it is unstable. On the other hand the general equilibrium theory of economics contains very little information of the comparative statics type. By making a simple dynamization of the model, and assuming that the dynamic system is asymptotically stable, it is possible to draw conclusions of a comparative static character that would be otherwise impossible, and this puts a lot more structure into the model.

The concept of structural stability is being used here in exactly the same manner, with the difference that we need not make any explicit dynamization that may seem arbitrary, and that we actually obtain even more information by assuming stability. The philosophy, however, is exactly the same. A flow pattern that would be thrown into something qualitatively completely different by the slightest disturbance would not persist in the real world, and therefore would be without interest to us.

Of course, the disturbances we now consider do not concern displacements of the initial state with a given set of differential equations. Rather, we deal with deformations of the differential equations themselves. The differential equations we are concerned with are those for the flow lines. As we see from equations (8) - (9) these flow lines are ultimately determined by the local transportation cost function $f(x,y)$. There is, however, no harm in making the mathematical trick of assuming the resulting potential function $\lambda(x,y)$ as known instead. As the flow lines are gradient to the potential, we can choose a suitable parameterization of the flow lines so that they are solutions to the differential equations:

$$dx/ds = \lambda_x(x,y) \text{ , and} \tag{13}$$

$$dy/ds = \lambda_y(x,y) \text{ .} \tag{14}$$

A deformation of these differential equations, or a perturbation, is defined as another set of differential equations:

$$dx/ds = \mu_x(x,y) \text{ , and} \tag{15}$$

$$dy/ds = \mu_y(x,y) \text{ ,} \tag{16}$$

such that in some sense it is close to the previous one.

Closeness is understood in the sense of a C^1 topology, so that $|\lambda_i - \mu_i| < \varepsilon$ and $|\lambda_{ij} - \mu_{ij}| < \varepsilon$, where i and j range over x and y. Observe that the condition of the derivatives of the differential equations, i.e., on the second derivatives of the potential, are necessary, as otherwise isolated singularities could always be converted into accumulations of singular points, and vice versa. By using the weaker C^0 topology, we therefore could not catch the intuitive meaning of structural stability.

Structural stability can now be defined. The solution curve portrait for the λ-system of differential equations is said to be structurally stable, provided that we can define a homeomorphism mapping trajectories into trajectories (of the same direction), and singularities onto singularities (of the same type) in the corresponding portrait for the λ-system whenever the latter is an ε-perturbation of the former. This mathematical definition makes precise and catches the intuitive meaning of the concept of structural stability. By this we may rely on well established results from the generic theory of differential equations, and can use Peixoto's very powerful theorem to obtain a surprisingly precise characterization of the structurally stable flow.

For a gradient flow like ours Peixoto's theorem states that:

(1) The flow consists almost everywhere of regular points, i.e., points through which only one trajectory passes.

(2) Except for these regular points, there is in a bounded area only a finite number of isolated singular points of a hyperbolic type. At a singular point we have $\lambda_x = \lambda_y = 0$. The singularities are hyperbolic when the Hessian of λ is non-zero, i.e., $\lambda_{xx}\lambda_{yy} \neq \lambda_{xy}^2$. Hence, at the singular point there is no unique direction of flow defined, so that several trajectories may be incident. When the Hessian is non-zero, the eigenvalues of the corresponding linear system are non-zero and real, and the singularities are either nodes (stable or unstable) or simple saddle points.

(3) In addition to the above local characteristics of a stable flow there is the global result that a trajectory joins saddle points. This last condition helps us to characterize the whole set of global patterns of a structurally stable flow--up to topological equivalence.

Before considering the global arrangements we shall stop to consider the spatial organization around a singular point of any one of the two admitted classes: nodes and saddles.

All the trajectories in a basin of attraction (or repulsion) are incident to the singular point itself. The simplest way of drawing the spatial organization around a node is by a von Thünen-like organization, with the trajectories being radial straight lines and the orthogonal isopotential curves being concentric circles. The organization depicted in Figure 1 is hence one of the concentric rings. As the isopotentials, according to (11) - (12), represent loci of constant price and wage rate, and hence, according to (3), lead to the choice of a certain production density and a certain choice among capital- and labor-intensive technologies, these curves obviously become the boundaries between subregions with certain characteristics. Obviously production is concentrated in industrial areas, even though it is distributed everywhere in space at various degrees of concentration, and is capital-intensive in industrial areas and labor-intensive in residential areas. A node in this model would represent a center of one of two kinds: industrial or residential. In the former case goods flow outwards and labor upwards; in the latter case the directions are reversed.

The organization around a saddle point is illustrated in Figure 2. To the saddle point only four trajectories, one pair ingoing and one pair outgoing, are actually incident. All the other trajectories have the look of hyperbolas and miss the singular point itself. The orthogonal isopotential curves, that are boundaries in the spatial organization pattern, in the present

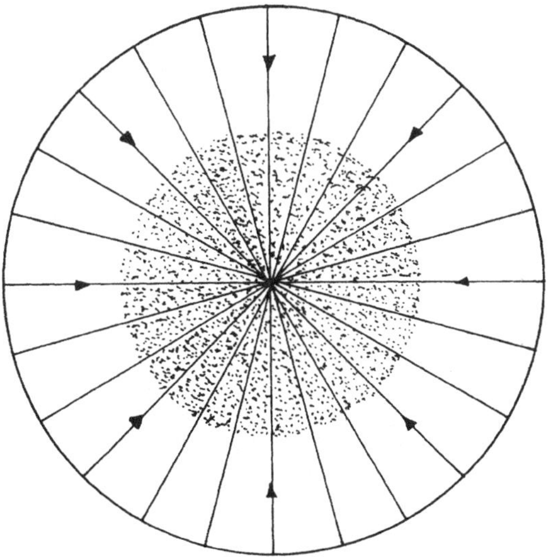

Fig. 1. Flow and spatial organization around a node singularity.

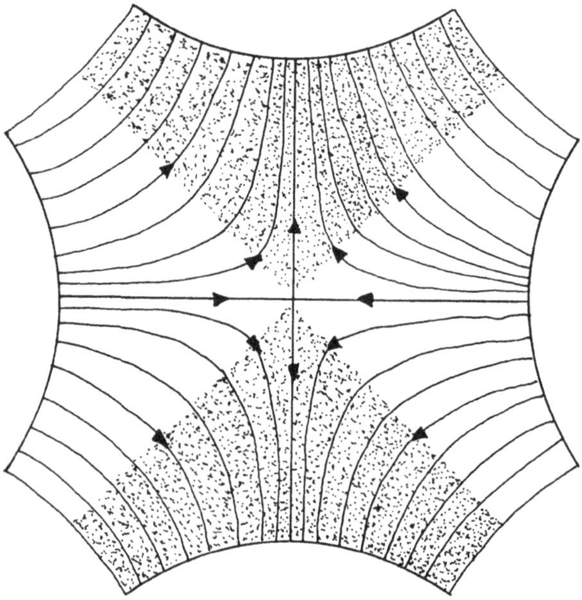

Fig. 2. Flow and spatial organization around a saddle singularity.

case look exactly like the trajectories themselves, only rotated by an angle of 45°. The spatial organization is now sectoral rather than ring shaped. In this particular model we have considered pairs of industrial and residential sectors. Otherwise the comments about the chracteristics of the various regional subareas and the flows between them apply. We will see that in the global arrangement the saddle singularities simply fit in between "empty space" between industrial and residential areas. We can imagine that, due to particularly good transportation facilities, the optimal routes tend to be attracted towards these saddle points.

The global pattern is most easy to arrange if for the moment we make the simplifying assumption that no pair of outgoing or ingoing trajectories from any saddle point is incident to the same node. As we know, there are no saddle connections in a structurally stable flow. So we can quite easily arrange the global patterns on a square grid made up by trajectories incident to saddle points alone. The singular points are the intersection points in this grid.

Starting out from any saddle point we can orient the whole graph. The initial saddle is surrounded by two stable and two unstable nodes. Continuing this orientation, we must conclude that in the diagonal directions from the initial saddle there are again saddle points, because there are both ingoing and outgoing trajectories to them. This result is shown in Figure 3. The spatial organization is a chessboard pattern of quadratic areas of industrial and residential character, as indicated in the diagram. What we have seen in Figures 1 and 2 is simply the local organization around a node and a saddle point, picked out from this global picture.

The corresponding potential surface or "price landscape" is shown in Figure 4. We can imagine labor as flowing downhill and goods as flowing uphill in gradient direction. Varying production intensities and choices of labor- or capital-intensive techniques depend on altitude, and belong to different "climatic zones." It should be emphasized that this characterization has been of a topological nature. The whole pattern should be imagined as drawn on a perfectly elastic rubber sheet, and it actually

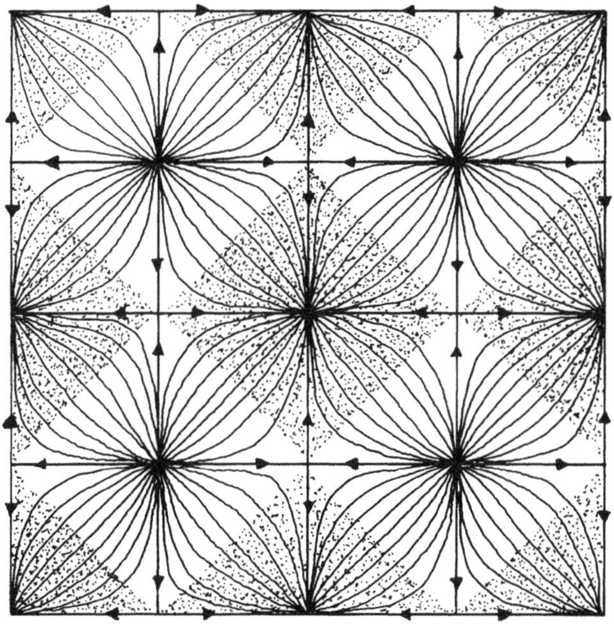

Fig. 3. Global picture of a structurally stable flow.

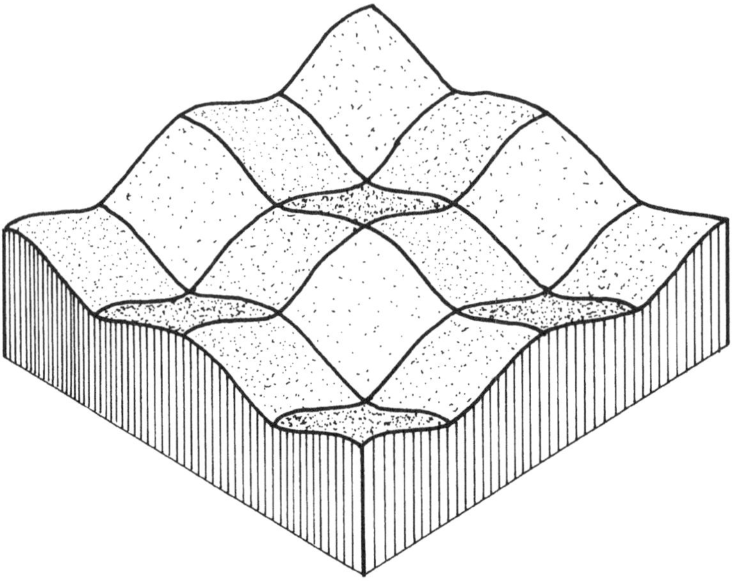

Fig. 4. The potential function or "price landscape."

represents all the patterns that can be obtained from the one depicted by any stretching without tearing.

One possibility was omitted, namely that a pair of ingoing or outgoing trajectories from a saddle point could end up at the same unstable or stable node. Both pairs could not be joined in this manner, as this could not be arranged without intersecting trajectories. What happens is that a circuit is set up delimiting an isolated trade area with a singularity of the node type inside. As shown in Figure 5, such circuits can be nested inside each in any number of levels. By these considerations we have, again up to topological equivalence, exhausted all possible features of a structurally stable flow. In my opinion this gives as astonishingly precise description of the flow pattern merely by assuming structural stability.

We have argued above that structurally unstable flows are in themselves not interesting because they cannot persist in the real world of constant perturbation. There is, however, an interesting feature in the transition from the state of flow just before, to the state just after the momentary passage of the unstable state. Both before and after the transition we deal with structurally stable flows. Compared, they will,

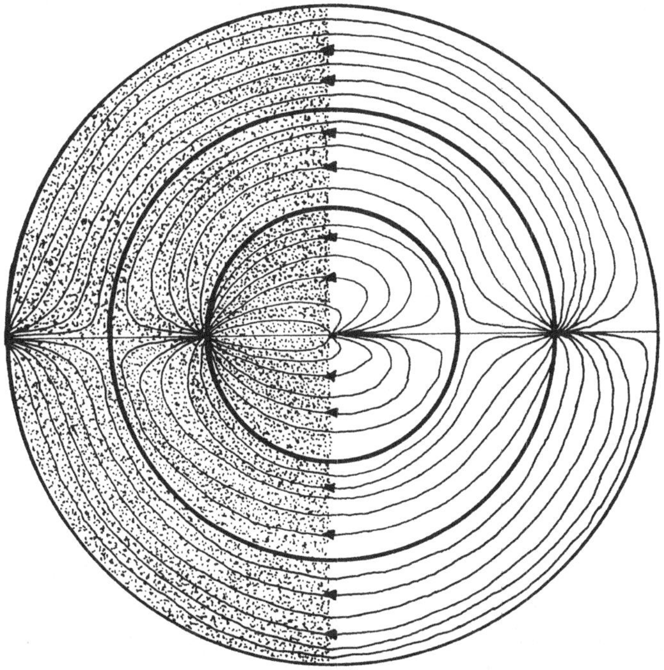

Fig. 5. Saddle trajectories joined to the same node and nested trade areas.

however, be qualitatively different, involving phenomena of sudden rotations of the whole trajectory portraits, emergence, disappearance and multiplication of singular points. So, it would certainly be interesting to study structural change. Like the generic theory of differential equations was a tool for the study of structural stability, catastrophe theory is a tool for the systematic study of structural change. For this we rely on Thom's classification theorem that deals with structural stability for families of functions depending on certain numbers of parameters. Introducing such exogeneous parameters makes it possible to obtain structurally unstable flows for certain parameter combinations, even though the families of functions always remain structurally stable.

The parameters we need to deal with in the present context are the factors influencing the spatial distribution of transportation costs. It therefore seems reasonable to single out three such factors:

(1) road constructions and repair;

(2) the load of traffic; and,

(3) fuel prices.

These seem to constitute three independent and important factors influencing f(x,y), and therefore the potential $\lambda(x,y)$.

Let us denote local road capacity, traffic and fuel price by the triple (u,v,w). This puts the potential in the form $\lambda(x,y,u,v,w)$, and the differential equations become:

$$dx/ds = \lambda_x(x,y,u,v,w) \text{ , and} \tag{17}$$

$$dy/ds = \lambda_y(x,y,u,v,w) \text{ .} \tag{18}$$

As we deal with a two-dimensional gradient field, dependent on three parameters, we know from Thom's classification theorem that we only need to consider the canonical forms of the elliptic and hyperbolic umblics, namely

$$\lambda = x^3 - 3xy^2 + w(x^2 + y^2) - ux - vy \text{ , and} \tag{19}$$

$$\lambda = x^3 + y^3 + wxy - ux - vy \text{ .} \tag{20}$$

By studying all the phenomena that can occur with the gradient fields to these λ-functions when we let the triple (u,v,w) take on all possible value combinations, we obtain, again up to topological equivalence, a complete description of the structural changes that can occur in a structurally stable family of functions.

In Figures 6 and 7 we illustrate the canonical forms of the elliptic and hyperbolic umblic catastrophes. On top of each figure is the bifurcation manifold in parameter space. Below are the flow fields connected with various portions of parameter space. As long as the parameter combination moves in the (u,v,w)-space without crossing the bifurcation manifold, we deal with smooth transformations of structurally stable flow fields of the types depicted in Figures 6 and 7. As soon as this manifold is crossed we deal with a sudden change of flow pattern. Then singular

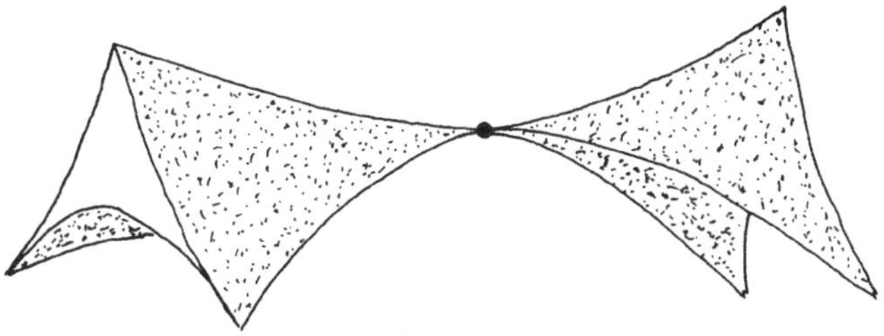

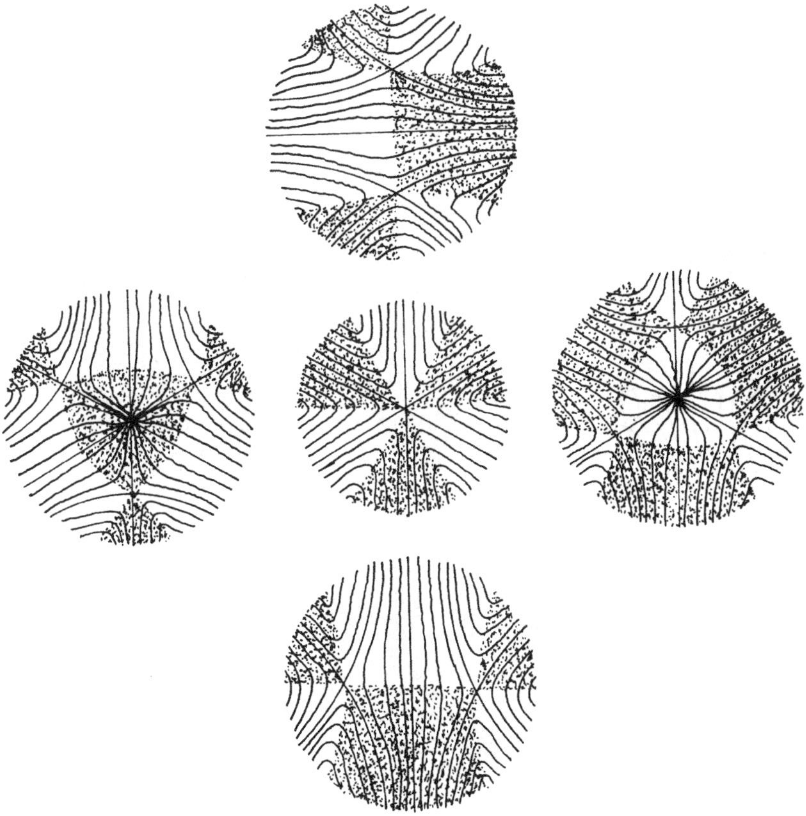

Fig. 6. Changes of pattern with the elliptic umblic catastrophe.

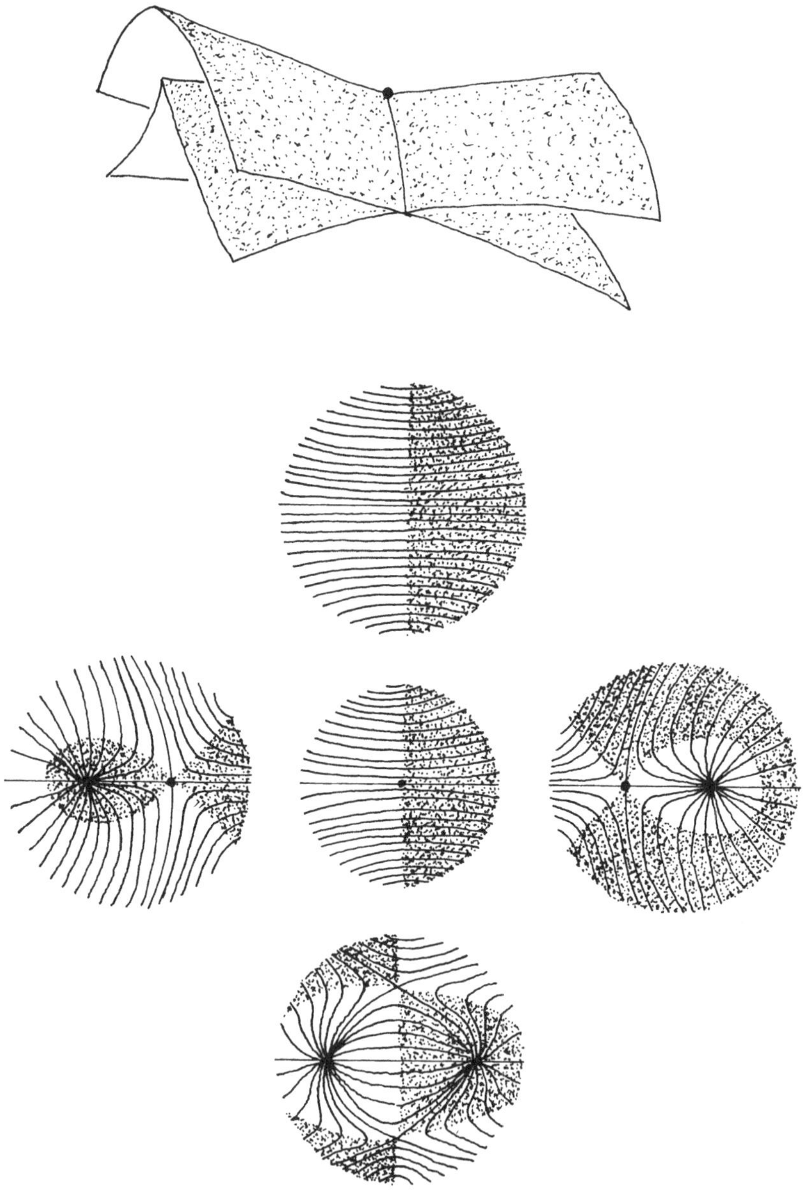

Fig. 7. Changes of pattern with the hyperbolic umblic catastrophe.

points collapse, multiply, emerge, or disappear, or there occur sudden rotations of the whole diagram. In the small disks are illustrated the most pathological cases, corresponding to the parameter combinations u = v = w = 0, and marked by a dot on the bifurcation manifolds. From that point all possible patterns depicted are obtainable. What is interesting is that, by these general considerations, we have obtained not only a complete description of the structurally stable flow, but also the transitions from one structurally stable flow to another. Observe, however, that the latter description is only local, not global.

The characterization has been only topological, in contrast to the Euclidean characterization in traditional central place and land use theory. Nevertheless, there are conclusions that are, at least superficially, in contradiction with this traditional theory. One issue is that a structurally stable flow never stagnates on the boundary of a market area. Another is that the spatial organization connected with structurally stable flows tends to be quadratic, rather than triangular or hexagonal.

REFERENCES

Angel, S. and Hyman, G.M., 1976. Urban Fields - A Geometry of Movement for Regional Science, Pion, London.

Beckmann, M.J., 1952. A Continuous Model of Transportation, Econometrica 20: 643.

Beckmann, M.J., 1953. The Partial Equilibrium of a Continuous Space Market, Welwirtschaftliches Archiv 71:73.

Beckmann, J.J. and Püu, T., 1980. Continuous Flow Modelling in Regional Science - State of the Arts and New Perspectives (forthcoming).

Hirsch, M.J. and Smale, S., 1974. Differential Equations, Dynamical Systems, and Linear Algebra, Academic, New York.

Marsden, J.E. and Tromba, A.J., 1976. Vector Calculus, Freeman, San Francisco.

Peixoto, M.M., 1977, Generic Properties of Ordinary Differential Equations, in: Studies in Ordinary Differential Equations, J. Hale (ed.), Studies in Mathematics, The Mathematical Association of America, 14: 52.

Poston, T. and Stewart, I., 1978. Catastrophe Theory and Applications, Pitman, London.

Puu, T., 1979. The Allocation of Road Capital in Two-Dimensional Space--A Continuous Approach, North-Holland, Amsterdam.

Puu, T., 1979. Regional Modelling and Structural Stability, Environment and Planning A, 11: 1431.

STATISTICAL ANALYSIS OF SPACE-TIME SERIES

A statistical theme is pursued in this section. The overriding question here is whether statistical analysis can extract useful, illuminating information from a space-time data series or manifestations of its underlying processes. Approaches to obtaining an answer for this question utilize the statistical method from multivariate analysis, quadrat analysis, spectral analysis and econometrics.

Griffith investigates the relationship between time series and spatial series, and finds that merely attaching subscripts to denote a point in time or a location in space is insufficient. Space and time are intertwined, and space-time data can not be reduced to two independent components corresponding to these two dimensions. A similar contention has been presented already by Beguin and Thisse. Consequently, this inseparability property implies that on the average neither serial correlation nor spatial autocorrelation indices reveal much about space-time processes. Because of this finding a space-time autocorrelation index is formulated, and formulae for its expected value and standard error are presented for the special case of normally distributed intra-areal unit variates. The interpretation of this new index furnishes a rich topic for subsequent investigations.

Haining is concerned about deriving process inferences from map patterns. The two features of maps he focuses on are distributional or frequency counts, and arrangement or the geographic pattern of frequency counts. One could assume that some map pattern is the manifestation of any of an infinite number of processes. Constraints upon the set of possibilities are supplied by characteristics of the corresponding observed frequency distribution and geographic arrangement. From this reduced set one wishes to select the most likely map pattern generator. Such a selection must refer to the underlying sampling distribution, which arises because of stochastic attributes combining with phenomena in space and time. An exciting area for further research concerns processes that generate clustered map patterns in some systematic manner.

Marchand analyzes a sample of 23 urban areas for Quebec by means of spectral and cross spectral methods. She applies the extremal approach of the maximum entropy method to the spectral analysis of employment time series for each of these urban places in an attempt to verify the theoretical assumptions of traditional models of change. The respective importance of two types of demand fluctuation is measured, namely (1) a Juglar cycle that is found to dominate the employment variations of all cities in the set, and

(2) a series of minor industrial cycles that is found to appear in various combinations in certain urban areas. Cross spectral results further confirm the complexity of demand variations in time and space for this sample. Application of the maximum entropy method of spectral analysis to space-time series constitutes a fruitful area for future research, especially since it can handle successfully relatively small time series and does not need a stationarity assumption.

The final paper in this section, by Paelinck and Ancot, reports on work that has been carried out at the Netherlands Economic Institute during the past decade. The primary goal of this work has been to achieve a better match between statistical data and spatial model construction. Although much of this paper is devoted to static, econometric analysis, the derivations are obtained in order to say something about the threshold theory perspective for modeling interregional growth of phenomena.

These four papers suggest that statistical techniques indeed can extract useful, enlightening information from space-time data series or manifestations of their underlying processes in terms of the trajectories, or transformation from one spatial distribution to another. Because of problems peculiar to these types of data, however, classical techniques often must undergo considerable modification, a theme to which much attention is devoted in these papers.

INTERDEPENDENCE IN SPACE AND TIME: NUMERICAL AND INTERPRETATIVE CONSIDERATIONS

Daniel A. Griffith

Department of Geography
State University of New York at Buffalo

Historically econometricians have paid attention to serial correlation whereas spatial analysts have been concerned with spatial autocorrelation. The purpose of this exposition, which spans these two problem areas, is to investigate dimensions of autocorrelation for space-time series. Much research has been done during the past decade on such topics as spatial time series (Bennett, 1979), the space-time autoregressive model (Haggett, Cliff and Frey, 1977), spatial comparisons of time series (Cliff, Ord, Bassett and Davies, 1975), and the space-time moving average model (Curry, 1971). This inquiry departs from these pursuits, first by outlining a space-time autocorrelation index, and second by relating to it the traditional indices of serial and spatial autocorrelation.

BACKGROUND

Bennett and Haining (1976) have provided an excellent survey of modelling in space and/or time. They state that for space-time studies a valuable preliminary step is to conduct a simple time series analysis of individual areal unit time series, thus treating these series initially as independent of one another. Next the nature and degree of spatial autocorrelation for these time series results can be studied for various spatial infrastructures. Then space-time correlograms are to be computed, where the cross-covariance or autocovariance at lag K between region i at time t and region j at time t-τ is estimated by

$$C_{ij}(K) = \frac{1}{T} \sum_{s=0}^{s=T-\tau} (y_{i,s} - \bar{y}_i)(y_{j,s+k} - \bar{y}_j) \quad , \qquad (1)$$

while the associated space-time autocorrelation is estimated by

$$R_{ij}(K) = C_{ij}(K)/C_{ij}(0) \quad . \tag{2}$$

This perspective raises several noteworthy questions. Can time-series results be properly interpreted when the underlying trend is a space-time one? Is spatial autocorrelation a manifestation of serial trends? Granger (1975) maintains that the principal answer to this latter question is "yes." Similarly, can space series results be properly interpreted when the underlying trend is a space-time one? Or, is serial correlation a manifestation of spatial trends?

Hepple (1978) expands upon this modelling theme by casting the solution of a space-time series problem as a "seemingly unrelated regression equations" problem. Each areal unit time series $\underline{y}_i$ is defined as a function of the time series for K variates, denoted $\underline{x}_i$. A normally distributed random error time series vector $\underline{u}_i$ is assumed, yielding K regression coefficients to be determined. The resulting model becomes

$$\begin{pmatrix} \underline{y}_1 \\ \underline{y}_2 \\ \cdot \\ \cdot \\ \cdot \\ \underline{y}_N \end{pmatrix} = \begin{pmatrix} \underline{x}_1 & & 0 & \underline{\beta}_1 \\ & \underline{x}_2 & & \underline{\beta}_2 \\ & & \cdot & \cdot \\ & & \cdot & \cdot \\ & & \cdot & \cdot \\ 0 & & \underline{x}_N & \underline{\beta}_N \end{pmatrix} + \begin{pmatrix} \underline{u}_1 \\ -2 \\ \cdot \\ \cdot \\ \cdot \\ \underline{u}_N \end{pmatrix} \quad . \tag{3}$$

One concern about equation (3) is the presence of autocorrelation in the disturbances. A general space-time autocorrelation parameter estimate is presented as

$$\hat{\rho} = \frac{\sum_{i=1}^{i=N} \sum_{t=2}^{t=T} \hat{u}_{i,t} \, \hat{u}_{j,t-1}}{\sum_{i=1}^{i=N} \sum_{t=2}^{t=T} \hat{u}^2_{i,t-1}} \quad . \tag{4}$$

At least two noteworthy questions are raised here. First, is $\hat{\rho}$ a maximum likelihood estimate since the Jacobian of the transformation has been ignored? Ord (1975) suggests that equation (4) would render results that are inconsistent, even if they prove to be sufficient and unbiased. Second, should the denominator of equation (4) ignore $u_{i,t}$ terms? The Durbin-Watson statistic includes them.

Hordijk and Nijkamp (1976) work with the Moran Coefficient to develop the following generalized form for space-time series:

$$m = \frac{\sum_{t=2}^{t=T} \sum_{i=1}^{i=n} \sum_{j=1}^{j=n} w_{ij} (y_{i,t} - \bar{y})(y_{j,t-1} - \bar{y})}{\sum_{t=2}^{t=T} \sum_{i=1}^{i=n} (y_{i,t} - \bar{y})^2} , \qquad (5)$$

where $\sum_{j=1}^{j=n} w_{ij} = 1$, and

$$\bar{y} = \sum_{t=2}^{t=T} \sum_{i=1}^{i=n} y_{it} .$$

This specification eventually led to the formulation of a modified space-time Moran Coefficient that is in keeping with the Durbin-Watson statistic.

Griffith (1978) has noted similarities between these two indices. Traditionally the Moran Coefficient is written as

$$MC = \frac{n}{\sum_{i=1}^{i=n} \sum_{j=1}^{j=n} c_{ij}} \quad \frac{\sum_{i=1}^{i=n} \sum_{j=1}^{j=n} c_{ij} (y_i - \bar{y})(y_j - \bar{y})}{\sum_{i=1}^{i=n} (y_i - \bar{y})^2} . \qquad (6)$$

where $c_{ij} = \begin{cases} 1 \text{ if areal units i and j are juxtaposed, and} \\ 0 \text{ otherwise; and} \end{cases}$

$$\bar{y} = \sum_{i=1}^{i=n} y_i/n .$$

Meanwhile the traditional Durbin-Watson statistic is written as

$$D\text{-}W = \frac{\sum_{t=2}^{t=T} (y_t - y_{t-1})^2}{\sum_{t=1}^{t=T} (y_t - \bar{y})^2} , \qquad (7)$$

where $\bar{y} = \sum_{t=2}^{t=T} y_t/T$.

Synthesizing equations (6) and (7) yields a space-time auto-correlation index that may be written as

$$MC_T = \frac{(T-1)n}{\sum_{t=2}^{t=T}\sum_{i=1}^{i=n}\sum_{j=1}^{j=n} c_{ijt-1}} \frac{\sum_{t=2}^{t=T}\sum_{i=1}^{i=n}\sum_{j=1}^{j=n} c_{ijt-1}(y_{it}-\bar{y})(y_{jt-1}-\bar{y})}{\sum_{t=1}^{t=T}\sum_{i=1}^{i=n} (y_{it}-\bar{y})^2} , \qquad (8)$$

where $c_{ijt} = \begin{cases} 1 \text{ if areal units i and j are juxtaposed during time t-1, and} \\ 0 \text{ otherwise; and} \end{cases}$

$$\bar{y} = \sum_{t=1}^{t=T}\sum_{i=1}^{i=n} y_{it}/nT .$$

Because of the relative ease with which equation (6) can be expressed in matrix notation, it together with equation (8) is used in subsequent analyses.

It is, however, desirable to spell out an explicit relationship between equations (6) and (7). First, an affine transformation links the Moran Coefficient with the Geary Ratio. Moreover,

$$GR = \frac{n-1}{2n} \frac{\sum_{i=1}^{i=n}\sum_{j=1}^{j=n} c_{ij}(y_i-\bar{y})^2}{\sum_{i=1}^{i=n}\sum_{j=1}^{j=n} c_{ij}(y_i-\bar{y})(y_j-\bar{y})} MC . \qquad (9)$$

Further, a spatially adjusted Durbin-Watson statistic may be rewritten in terms of a temporally modified Geary Ratio as

$$GR_T = \frac{(T-1)(n-1)}{2\sum_{t=2}^{t=T}\sum_{i=1}^{i=n}\sum_{j=1}^{j=n} c_{ijt-1}} D\text{-}W_S . \qquad (10)$$

Therefore, temporally adjusted Moran Coefficients and Geary Ratios produce results that are equivalent to those from a spatially adjusted Durbin-Watson statistic.

THE FUNDAMENTAL THEOREM OF SPACE-TIME INTERDEPENDENCE

Does the presence of serial correlation spuriously enhance the importance of spatial relatedness when temporal relatedness is overlooked? What about the converse? What is the impact of space-time covariation? These and other questions allude to a fundamental theorem of space-time interdependence, which may be stated as follows:

> THEOREM: Given a discrete space-time process characterized by the system of partial difference equations
>
> $$\underline{y}_{t+1} - \underline{A}\,\underline{y}_t = \underline{\phi}_t\ , \qquad (11)$$
>
> where matrix $\underline{A}$ depicts space-time covariation, and the level of spatial auto- or serial correlation contained in vector $\underline{y}_t$ or $\underline{y}_i$, respectively, is a function of properties of matrix $\underline{A}$.

An algebraic proof of the spatial autocorrelation portion of this theorem appears in Table I. As Table I indicates, the proof of this theorem involves eight cases. Those cases dealing with stationary processes will be discussed first. Case I depicts a situation where zero space-time covariation is present, the rate of change through time is constant and is the same for each areal unit, and the forcing function is constant. These three descriptions respectively correspond to $a_{ij} = 0$ $(i \neq j)$, $a_{ij} = a$ $(i = j)$, and $\phi_i = b$. Consequently the numerical value produced by equation (6) for this type of data is constant through time. In other words, the presence of serial correlation does not spuriously enhance the importance of spatial relatedness when temporal relatedness is overlooked.

Case II preserves the stationarity property while introducing non-zero space-time covariation. These two descriptions respectively correspond to $a_{ij} = a$ $(i = j)$ and $\phi_i = b$, and $a_{ij} \neq 0$ for some $i \neq j$. Therefore the numerical value yielded by equation (6) for this type of data will vary through time. The path and attributes of this time-series are functions of the connectivity matrix $\underline{C}$ and prevailing space-time covariations. Hence, the numerical value of a spatial autocorrelation index will be due in part to spatial relatedness and in part to space-time covariations.

TABLE I

PROOF OF THE FUNDAMENTAL THEOREM OF SPACE-TIME INTERDEPENDENCE

PROOF: The solution to the difference equation $\underline{y}_{t+1} - \underline{A}\,\underline{y}_t = \underline{\phi}_t$ is $\underline{y}_t = \underline{A}^t \underline{y}_0 + \sum_{k=0}^{k=t-1} \underline{A}^j \underline{\phi}_{t-k-1}$.

The Moran Coefficient at time t may be written, using matrix notation, as

$$MC_t = \frac{n\{(\underline{y}_t - \bar{\underline{y}}_t)^T \underline{C}(\underline{y}_t - \bar{\underline{y}}_t)\}}{\underline{1}^T_{n\times 1}\underline{C}\,\underline{1}_{n\times 1}(\underline{y}_t - \bar{\underline{y}}_t)^T(\underline{y}_t - \bar{\underline{y}}_t)}, \qquad \bar{\underline{y}}_t = \frac{1}{n}\underline{1}_{n\times n}\underline{y}_t$$

$$= \frac{n\{[\underline{A}^t \underline{y}_0 + \sum_{k=0}^{k=t-1} \underline{A}^k \underline{\phi}_{t-k-1} - \frac{1}{n}\underline{1}_{n\times n}(\underline{A}^t \underline{y}_0 + \sum_{k=0}^{k=t-1} \underline{A}^k \underline{\phi}_{t-k-1})]^T \underline{C}[\underline{A}^t \underline{y}_0 + \sum_{k=0}^{k=t-1} \underline{A}^k \underline{\phi}_{t-k-1} - \frac{1}{n}\underline{1}_{n\times n}(\underline{A}^t \underline{y}_0 + \sum_{k=0}^{k=t-1} \underline{A}^k \underline{\phi}_{t-k-1})]\}}{\underline{1}^T_{n\times 1}\underline{C}\,\underline{1}_{n\times 1}[\underline{A}^t \underline{y}_0 + \sum_{k=0}^{k=t-1} \underline{A}^k \underline{\phi}_{t-k-1} - \frac{1}{n}\underline{1}_{n\times n}(\underline{A}^t \underline{y}_0 + \sum_{k=0}^{k=t-1} \underline{A}^k \underline{\phi}_{j-k-1})]^T[\underline{A}^t \underline{y}_0 + \sum_{k=0}^{k=t-1} \underline{A}^k \underline{\phi}_{t-k-1} - \frac{1}{n}\underline{1}_{n\times n}(\underline{A}^t \underline{y}_0 + \sum_{k=0}^{k=t-1} \underline{A}^k \underline{\phi}_{t-k-1})]\}}$$

CASE I: $\forall_i\ \forall_j\ a_{ij} = 0,\ i \neq j\ ;\ \forall_i\ \forall_j\ a_{ij} = a,\ i = j\ ;\ \forall_i\quad \phi_i = b$

$$\underline{A}^t \underline{y}_0 = a^t \underline{I}\,\underline{y}_0 = a^t \underline{y}_0$$

$$\underline{\phi}_{t-k-1} = [b] = b\,\underline{1}_{n\times 1}$$

$$MC_t = \frac{n\{[a^t\underline{y}_0 + \underline{1}_{n\times 1}\, b \sum_{k=0}^{k=t-1} a^k - \frac{1}{n}\underline{1}_{n\times n}(a^t\underline{y}_0 + \underline{1}_{n\times 1}\, b \sum_{k=0}^{k=t-1} a^k)]^T \underline{C}[a^t\underline{y}_0 + \underline{1}_{n\times 1}\, b \sum_{k=0}^{k=t-1} a^k - \frac{1}{n}\underline{1}_{n\times n}(a^t\underline{y}_0 + \underline{1}_{n\times 1}\, b \sum_{k=0}^{k=t-1} a^k)]\}}{\underline{1}^T_{n\times 1}\underline{C}\,\underline{1}_{n\times 1}[a^t\underline{y}_0 + \underline{1}_{n\times 1}\, b \sum_{k=0}^{k=t-1} a^k - \frac{1}{n}\underline{1}_{n\times n}(a^t\underline{y}_0 + \underline{1}_{n\times 1}\, b \sum_{k=0}^{k=t-1} a^k)]^T[a^t\underline{y}_0 + \underline{1}_{n\times 1}\, b \sum_{k=0}^{k=t-1} a^k - \frac{1}{n}\underline{1}_{n\times n}(a^t\underline{y}_0 + \underline{1}_{n\times 1}\, b \sum_{k=0}^{k=t-1} a^k)]}$$

$$\frac{1}{n}\underline{1}_{n\times n}\underline{1}_{n\times 1} = \frac{1}{n}[\sum_{j=1}^{j=n} 1] = \frac{1}{n}[n] = \underline{1}_{n\times 1}$$

$$a^t \underline{y}_0 + \underline{1}_{n\times 1}\, b \sum_{k=0}^{k=t-1} a^k - \frac{1}{n}\underline{1}_{n\times n}\,(a^t \underline{y}_0 + \underline{1}_{n\times 1}\, b \sum_{k=0}^{k=t-1} a^k) = a^t(\underline{I} - \frac{1}{n}\underline{1}_{n\times 1})\,\underline{y}_0 + b \sum_{k=0}^{k=t-1} a^k(\underline{1}_{n\times 1} - \frac{1}{n}\underline{1}_{n\times n}\underline{1}_{n\times 1}) = a^t(\underline{I} - \frac{1}{n}\underline{1}_{n\times 1})\,\underline{y}_0$$

$$MC_t = \frac{n\{[a^t(\underline{I} - \frac{1}{n}\underline{1}_{n\times 1})\underline{y}_0]^T \underline{C}[a^t(\underline{I} - \frac{1}{n}\underline{1}_{n\times 1})\underline{y}_0]\}}{\underline{1}^T_{n\times 1}\underline{C}\,\underline{1}_{n\times 1}[a^t(\underline{I} - \frac{1}{n}\underline{1}_{n\times 1})\underline{y}_0]^T[a^t(\underline{I} - \frac{1}{n}\underline{1}_{n\times 1})\underline{y}_0]} = \frac{a^{2t}\, n\{[(\underline{I} - \frac{1}{n}\underline{1}_{n\times 1})\underline{y}_0]^T \underline{C}[(\underline{I} - \frac{1}{n}\underline{1}_{n\times 1})\underline{y}_0]\}}{a^{2t}\,\underline{1}^T_{n\times 1}\underline{C}\,\underline{1}_{n\times 1}[(\underline{I} - \frac{1}{n}\underline{1}_{n\times 1})\underline{y}_0]^T(\underline{I} - \frac{1}{n}\underline{1}_{n\times 1})\underline{y}_0}$$

$\therefore MC_t = MC_0$

CASE II: $\forall_i\ \phi_i = b\,;\ \forall_i\ \forall_j\ a_{ij} = a,\ i = j;$

$\exists_i\ \exists_j \ni\ a_{ij} \neq 0,\ i \neq j$

$\underline{A}^t = \underline{M}\underline{\Lambda}^t\underline{M}^{-1}$

$$MC_t = \frac{n\{[\underline{M}\underline{\Lambda}^t\underline{M}^{-1}\underline{y}_0 + \underline{1}_{n\times 1}\, b\sum_{k=0}^{k=t-1}\underline{M}\underline{\Lambda}^k\underline{M}^{-1} - \frac{1}{n}\underline{1}_{n\times n}(\underline{M}\underline{\Lambda}^t\underline{M}^{-1}\underline{y}_0 + \underline{1}_{n\times 1}\, b\sum_{k=0}^{k=t-1}\underline{M}\underline{\Lambda}^k\underline{M}^{-1})]^T\underline{C}[\underline{M}\underline{\Lambda}^t\underline{M}^{-1}\underline{y}_0 + \underline{1}_{n\times 1}\, b\sum_{k=0}^{k=t-1}\underline{M}\underline{\Lambda}^k\underline{M}^{-1} - \frac{1}{n}\underline{1}_{n\times n}(\underline{M}\underline{\Lambda}^t\underline{M}^{-1}\underline{y}_0 + \underline{1}_{n\times 1}\, b\sum_{k=0}^{k=t-1}\underline{M}\underline{\Lambda}^k\underline{M}^{-1})]\}}{\underline{1}^T_{n\times 1}\underline{C}\,\underline{1}_{n\times 1}[\underline{M}\underline{\Lambda}^t\underline{M}^{-1}\underline{y}_0 + \underline{1}_{n\times 1}\, b\sum_{k=0}^{k=t-1}\underline{M}\underline{\Lambda}^k\underline{M}^{-1} - \frac{1}{n}\underline{1}_{n\times n}(\underline{M}\underline{\Lambda}^t\underline{M}^{-1}\underline{y}_0 + \underline{1}_{n\times 1}\, b\sum_{k=0}^{k=t-1}\underline{M}\underline{\Lambda}^k\underline{M}^{-1})]^T[\underline{M}\underline{\Lambda}^t\underline{M}^{-1}\underline{y}_0 + \underline{1}_{n\times 1}\, b\sum_{k=0}^{k=t-1}\underline{M}\underline{\Lambda}^k\underline{M}^{-1} - \frac{1}{n}\underline{1}_{n\times n}(\underline{M}\underline{\Lambda}^t\underline{M}^{-1}\underline{y}_0 + \underline{1}_{n\times 1}\, b\sum_{k=0}^{k=t-1}\underline{M}\underline{\Lambda}^k\underline{M}^{-1})]}$$

$$\underline{M}\underline{\Lambda}^t\underline{M}^{-1}\underline{y}_0 + \underline{1}_{n\times 1}\, b\sum_{k=0}^{k=t-1}\underline{M}\underline{\Lambda}^k\underline{M}^{-1} - \frac{1}{n}\underline{1}_{n\times n}(\underline{M}\underline{\Lambda}^t\underline{M}^{-1}\underline{y}_0 + \underline{1}_{n\times 1}\, b\sum_{k=0}^{k=t-1}\underline{M}\underline{\Lambda}^k\underline{M}^{-1}) = (\underline{I} - \frac{1}{n}\underline{1}_{n\times n})\underline{M}\underline{\Lambda}^t\underline{M}^{-1}\underline{y}_0 + (\underline{1}_{n\times 1} - \frac{1}{n}\underline{1}_{n\times n}\underline{1}_{n\times 1})\, b\sum_{k=0}^{k=t-1}\underline{M}\underline{\Lambda}^k\underline{M}^{-1} =$$

$$(\underline{I} - \frac{1}{n}\underline{1}_{n\times n})\underline{M}\underline{\Lambda}^t\underline{M}^{-1}\underline{y}_0$$

$$MC_t = \frac{n\{[(\underline{I} - \frac{1}{n}\underline{1}_{n\times n})\,\underline{M}\,\underline{\Lambda}^t\,\underline{M}^{-1}\,\underline{y}_0]^T\underline{C}[(\underline{I} - \frac{1}{n}\underline{1}_{n\times n})\,\underline{M}\,\underline{\Lambda}^t\,\underline{M}^{-1}\,\underline{y}_0]\}}{\underline{1}_{n\times 1}^T\,\underline{C}\,\underline{1}_{n\times 1}[(\underline{I} - \frac{1}{n}\underline{1}_{n\times n})\,\underline{M}\,\underline{\Lambda}^t\,\underline{M}^{-1}\,\underline{y}_0]^T[(\underline{I} - \frac{1}{n}\underline{1}_{n\times n})\,\underline{M}\,\underline{\Lambda}^t\,\underline{M}^{-1}\,\underline{y}_0]}$$

$$= \frac{n\{\underline{y}_0^T\,\underline{M}^{-T}\,\Lambda^t\,\underline{M}^T(\underline{I} - \frac{1}{n}\underline{1}_{nxn})\underline{C}(\underline{I} - \frac{1}{n}\underline{1}_{nxn})\underline{M}\,\underline{\Lambda}^t\,\underline{M}^{-1}\,\underline{y}_0\}}{\underline{1}_{n\times 1}^T\,\underline{C}\,\underline{1}_{n\times 1}\,\underline{y}_0^T\,\underline{M}^{-T}\,\underline{\Lambda}^t\,\underline{M}^T(\underline{I} - \frac{1}{n}\underline{1}_{n\times n})^2\,\underline{M}\,\underline{\Lambda}^t\,\underline{M}^{-1}\,\underline{y}_0}$$

Because in general matrix multiplication is not commutative, MC_t does not necessarily reduce to MC_0 .

$\therefore\ MC_t \neq MC_0, t \neq 0$

CASE III: $\forall_i\ \forall_j\ a_{ij} = 0,\ i \neq j\ ;\ \exists_i\ \exists_k \ni a_{ii} \neq a_{kk},\ i \neq k\ ;\ \ \forall_i\ \phi_i = b$

$\underline{A}^t = \underline{M}\,\underline{\Lambda}^t\,\underline{M}^{-1}$ has $\underline{M} = \underline{I}$

$\underline{A}^t = \underline{\Lambda}^t$

From Case II MC_t reduces to

$$\frac{n\{\underline{y}_0^T\,\underline{\Lambda}^t(\underline{I} - \frac{1}{n}\underline{1}_{nxn})\underline{C}(\underline{I} - \frac{1}{n}\underline{1}_{nxn})\,\underline{\Lambda}^t\,\underline{y}_0\}}{\underline{1}_{n\times 1}^T\,\underline{C}\,\underline{1}_{n\times 1}\,\underline{y}_0^T\,\underline{\Lambda}^t(\underline{I} - \frac{1}{n}\underline{1}_{n\times n})^2\,\underline{\Lambda}^t\,\underline{y}_0}$$

Because matrix multiplication is not commutative when one of the two matrices is a diagonal matrix, MC_t does not necessarily reduce to MC_0.

$\therefore\ MC_t \neq MC_0, t \neq 0$

CASE IV: $\forall_i \forall_j\ a_{ij} = 0,\ i \neq j;\ \forall_i \forall_j\ a_{ij} = a,\ i = j;\ \exists_i \exists_j \ni \phi_i \neq \phi_j,\ i \neq j$

From Case I, $\underline{A}^t \underline{y}_0 = a^t \underline{y}_0$

$$a^t \underline{y}_0 + \sum_{k=0}^{k=t-1} a^k \underline{\phi}_{t-k-1} - \frac{1}{n} \underline{1}_{n\times n} (a^t \underline{y}_0 + \sum_{k=0}^{k=t-1} a^k \underline{\phi}_{t-k-1}) =$$

$$a^t (\underline{y}_0 - \bar{\underline{y}}_0) + \sum_{k=0}^{k=t-1} a^k (\underline{\phi}_{t-k-1} - \bar{\underline{\phi}}_{t-k-1})$$

$$MC_t = \frac{n\{[a^t (\underline{y}_0 - \bar{\underline{y}}_0) + \sum_{k=0}^{k=t-1} a^k (\underline{\phi}_{t-k-1} - \bar{\underline{\phi}}_{t-k-1})]^T \underline{C} [a^t (\underline{y}_0 - \bar{\underline{y}}_0) + \sum_{k=0}^{k=t-1} a^k (\underline{\phi}_{t-k-1} - \bar{\underline{\phi}}_{t-k-1})]\}}{\underline{1}_{n\times 1}^T \underline{C}\, \underline{1}_{n\times 1} [a^t (\underline{y}_0 - \bar{\underline{y}}_0) + \sum_{k=0}^{k=t-1} a^k (\underline{\phi}_{t-k-1} - \bar{\underline{\phi}}_{t-k-1})]^T [a^t (\underline{y}_0 - \bar{\underline{y}}_0) + \sum_{k=0}^{k=t-1} a^k (\underline{\phi}_{t-k-1} - \bar{\underline{\phi}}_{t-k-1})]}$$

Because a^t is no longer common to all terms, it does not vanish.

$\therefore MC_t \neq MC_0,\ t \neq 0$

CASE V: $\exists_i \exists_j \ni a_{ij} \neq 0,\ i \neq j;\ \exists_i \exists_j \ni \phi_i \neq \phi_j,\ i \neq j$

From Cases II and IV, $MC_t \neq MC_0, t \neq 0$

CASE VI: $\forall_i \forall_j\ a_{ij} = 0,\ i \neq j;\ \exists_i \exists_j \ni a_{ii} \neq a_{kk},\ i \neq k;\ \exists_i \exists_j \ni \phi_i \neq \phi_j$

From Cases III and IV, $MC_t \neq MC_0, t \neq 0$

CASE VII: $\exists_i \exists_j \ni a_{ij} \neq 0,\ i \neq j;\ \exists_i \exists_k \ni a_{ii} \neq a_{kk},\ i \neq k;\ \forall_i\ \phi_i = b$

From Cases II and III, $MC_t \neq MC_0,\ t \neq 0$

CASE VIII: $\exists_i \exists_j \ni a_{ij} \neq 0,\ i \neq j;\ \exists_i \exists_k \ni a_{ii} \neq a_{kk},\ i \neq k;\ \exists_i \exists_i \ni \phi_i \neq \phi_j,\ i \neq j$

From Cases II, III and IV, $MC_t \neq MC_0,\ t \neq 0$

Therefore, the level of spatial autocorrelation contained in $\underline{y}_t$ is a function of matrix $\underline{A}$, Q.E.D.

The six remaining cases pertain to non-stationary situations. In practice one could assume that random perturbations occur in which the expected value is a stationary situation. Nevertheless, Case III is a modified version of Case I in which the rate of change is constant through time, but is allowed to vary between regions. These descriptions respectively correspond to $a_{ij} = 0$ $(i \neq j)$ and $\phi_i = b$, and $a_{ii} \neq a_{jj}$ $(i \neq j)$. Once again the numerical values produced by equation (6) will vary through time. In this case, though, it is detecting the spatial autocorrelation present in temporal rates of change. Thus the path and attributes of this time-series are functions of the connectivity matrix $\underline{C}$ as well as the spatially differentiated rates of change through time. Moreover, it is not the presence of serial correlation that spuriously enhances the importance of spatial relatedness when temporal relatedness is overlooked, but rather the spatial interdependence of a geographically varying temporal process.

Case IV is another version of Case I, where now the forcing function becomes a variable. These descriptions respectively correspond to $a_{ij} = 0$ $(i \neq j)$ and $a_{ij} = a$ $(i = j)$, and $\phi_i \neq \phi_j$ $(i \neq j)$. Again the numerical value rendered by equation (6) will vary through time. Here, however, it is detecting the spatial autocorrelation present in this perturbation term. Consequently, the path and attributes of this time series are functions of both the connectivity matrix $\underline{C}$ and the spatial allocation of ϕ_i's to areal units. As such the time series of each ϕ_i term is not affecting the numerical value of MC, but rather the spatial autocorrelation latent in each of the T geographic distributions of $\underline{\phi}$ effect MC.

Case V is a variation of Cases II and IV, where non-zero space-time covariation coexists with a constant rate of change through time and a spatially variant forcing function. These descriptions respectively correspond to $a_{ij} \neq 0$ $(i \neq j)$, $a_{ij} = a$ $(i = j)$ and $\phi_i \neq \phi_j$ $(i \neq j)$. Clearly the numerical value yielded by equation (6) will fluctuate with the passing of time. Here the path and attributes of this time-series are functions of the connectivity matrix $\underline{C}$ as well as both the nature and degree of spatial autocorrelation found in the geographic distribution of the ϕ_i's and prevailing space-time covariations. Untangling effects attributable to these two components is essential to a proper interpretation of MC.

Case VI is a combination of Cases III and IV, where zero space-time covariation is present coupled with a geographically varying stationary temporal process and a spatially related perturbation term. These descriptions respectively correspond to $a_{ij} = 0$ $(i \neq j)$, $a_{ii} \neq a_{jj}$ $(i \neq j)$ and $\phi_i \neq \phi_i$. Once again a numerical value calculated with equation (6) will fail to be constant through time. Now the path and attributes of this time-series depend upon the connectivity matrix $\underline{C}$ in addition to that spatial autocorrelation embedded in the temporal rate of change. Separating effects due to these two components also is necessary to a proper interpretation of MC.

Case VII is a convolution of Cases II and III, where non-zero space-time covariation is concomitant with a geographically non-stationary temporal process. These descriptions respectively correspond to $a_{ij} \neq 0$ $(i \neq j)$, $a_{ii} \neq a_{jj}$ $(i \neq j)$ and $\phi_i = b$. Obviously the numerical value yielded by equation (6) suffers from a compounding of those inflictions associated with Cases V and VI.

Similarly, the final case, namely VIII, collects aspects from Cases II, III and IV. It is described by $a_{ij} \neq 0$ $(i \neq j)$, $a_{ii} \neq a_{jj}$ $(i \neq j)$ and $\phi_i \neq \phi_j$. And, equation (6) simultaneously measures all three effects for data affiliated with this situation.

These eight scenarios outline the possibilities for spatial autocorrelation. Equations (9) and (10) imply that a relaxation of the stationarity assumption for temporal processes results in equivalent conclusions.

To summarize, because most operational models assume stationarity, Cases I and II are of prime interest. The other six cases are useful only in uncovering implications when stationarity is not present. If random deviations from stationarity are supposed, then the expectations reduce to these same two cases. A compendium for this taxonomy appears in Table II.

THE SPACE-TIME AUTOCORRELATION INDEX MC_T

In this section Cases I and II are assumed to be representative of the process in question. In the investigation of the dimensions of autocorrelation for space-time series, two additional questions emerge at this point. First, is a new index necessary, or can the same information be acquired from the conventional indices? The fundamental theorem of space-time interdependence suggests that a new index is indeed necessary, and that the conventional autocorrelation indices are not suitable for examining space-time phenomena. Second, how powerful is the index MC_T?

In order to answer these two questions, as well as to see whether or not MC_T could be partitioned into a geographical and temporal component, numerous simulation experiments conforming to Cases I and II were conducted. (The computer program upon which these simulations are based appears in Appendix A.)

TABLE II

SOURCES OF SPATIAL AUTOCORRELATION DETECTED AT SOME POINT IN TIME

Case	Non-Stationary Term: Space-Time Covariation	Non-Stationary Term: Temporal Process	Non-Stationary Term: Perturbation	Interpretation Problem
I	No	No	No	No
II	Yes	No	No	No
III	No	Yes	No	Yes
IV	No	No	Yes	Yes
V	Yes	No	Yes	Yes
VI	No	Yes	Yes	Yes
VII	Yes	Yes	No	Yes
VIII	Yes	Yes	Yes	Yes

Eighteen experiments consisting of three ideal geographic configurations, twenty-five time periods, and roughly 100 repetitions were undertaken. The simulation equation used was

$$y_{it} = y_{it-1} + \rho \sum_{j=1}^{j=n} w_{ijt-1} y_{jt-1} \quad , \quad t \geq 2 \quad , \tag{12}$$

where y_{i0} was randomly selected, and $\sum_{j=1}^{j=n} w_{ijt-1} = 1.$

Accordingly $\forall_i \; a_{ii} = 1$, $\forall_i \phi_i = 0$ and $\exists a_{ij} \neq 0 \; (i \neq j)$. The autocorrelation parameter ρ took on values of $\pm.9$, $\pm.5$ and $\pm.1$. Furthermore, 3-by-3, 6-by-6 and 9-by-9 lattices were employed. The various combinations of these two sets yielded eighteen experiments. Meanwhile, sets of y_{*0} were drawn from a random normal distribution, and then a random allocation of these values was searched for $-(n-1)^{-1} - .025 \leq MC \leq -(n-1)^{-1} + .025$. The value of .025 was derived from a consideration of the standard errors of GR and MC. Consequently, the temporal process was stationary in both space and time, the forcing function was constant, and the initial vector of values was both normally distributed and spatially independent.

Two approaches were utilized to help determine whether or not a new index is needed, and to ascertain whether or not this index can be factored into a spatial component and a temporal component. First regression equations were constructed, in which the space-time index (i.e., STI) was cast as a function of the Durbin-Watson statistic calculated for the time series of Geary Ratios (i.e., DWGR) and of Moran Coefficients (i.e., DWMC), and the Moran Coefficient and Geary Ratio for the spatial distribution of Durbin-Watson statistics (i.e., MCDW and GRDW, respectively). Results of this regression analysis appear in Table III.

Generally speaking four basic inferences can be drawn from these regression results. Foremost is a rejection of the idea that traditional autocorrelation indices provide sufficient information about space-time processes. Although several multiple correlation coefficients are very close to unity (e.g., .99995 and .99872), these coefficients widely vary in value, in some instances reaching lows of .38692 and .32688. Most coefficients are moderate, falling in the .5-.6 category. Another finding is that as the number of areal units increases, the predictability of STI tends to decrease. While a similar pattern may be expected as the autocorrelation parameter ρ of equation (12) approaches zero, except for very small numbers of areal units this particular relationship seems to be somewhat constant. Meanwhile the only consistently significant term is the constant $\hat{\alpha}$. Strong evidence is provided here to reject the null hypothesis H_0: $\alpha = 0$. In addition, the sign patterns suggest the $|STI|$ tends to be overpredicted by these four independent variables. Finally, a tendency appears to be present for STI to be more closely affiliated with a time series of spatial autocorrelation index values. This implies that to some degree observed spatial autocorrelation values are more sensitive to underlying space-time covariation than are serial correlations.

The second approach involved an attempt to discriminate between the six different levels of space-time covariation on the basis of DWGR, DWMC, GRDW and MCDW. This task was tackled by subjecting the simulated data to a discriminant function analysis. Results of this endeavor appear in Tables IV and V.

Two basic discriminant functions were extracted for all three geographic configurations. One function represents spatial trends in serial correlation indices while the other represents temporal trends in spatial autocorrelation indices. In each case their combination accounted for at least 80% of the variance. Unfortunately these pairs of functions are relatively unsuccessful in allocating observations to their correct autocorrelation parameter classes. While success does tend to improve with an

TABLE III

STEP-WISE REGRESSION MODELS (FINAL STEP) BASED UPON SIMULATION RESULTS FOR THE SPACE-TIME AUTOCORRELATION INDEX

	9 Areal Units				36 Areal Units				81 Areal Units			
Autocorrelation Parameter	Variable	Regression Coefficient	Standard Error	Multiple Correlation	Variable	Regression Coefficient	Standard Error	Multiple Correlation	Variable	Regression Coefficient	Standard Error	Multiple Correlation
+.9	Constant	.5668***	$.508 \times 10^{-4}$	.99995***	Constant	.5193***	$.836 \times 10^{-2}$	.54363***	Constant	.5337***	$.880 \times 10^{-2}$	.32688***
	DWGR	-.1374***	$.143 \times 10^{-3}$		GRDW	-.0191*	$.106 \times 10^{-1}$		GRDW	-.0738	$.526 \times 10^{-1}$	
					DWGR	-.0656	$.853 \times 10^{-2}$		MCDW	.0359	$.481 \times 10^{-1}$	
					MCDW	-.0034	$.887 \times 10^{-2}$		DWMC	-.0136	$.942 \times 10^{-2}$	
					DWMC	-.0020	$.898 \times 10^{-2}$		DWGR	-.0139	$.982 \times 10^{-2}$	
+.5	Constant	.7843***	$.750 \times 10^{-2}$	.99874***	Constant	.6798***	$.122 \times 10^{-1}$	.68051***	Constant	.7039***	$.120 \times 10^{-1}$	.57381***
	DWGR	-.6590***	$.867 \times 10^{-2}$		DWGR	-.0406***	$.120 \times 10^{-1}$		GRDW	-.7302***	.256	
	DWMC	.1832***	$.928 \times 10^{-2}$		GRDW	-.2295***	$.630 \times 10^{-1}$		DWMC	-.0389***	$.121 \times 10^{-1}$	
	MCDW	-.0013**	$.582 \times 10^{-3}$		MCDW	.1227**	$.517 \times 10^{-1}$		DWGR	-.0314***	$.119 \times 10^{-1}$	
	GRDW	.0012*	$.643 \times 10^{-3}$		DWMC	-.0243*	$.134 \times 10^{-1}$		MCDW	.2822	.235	
+.1	Constant	.4810***	$.921 \times 10^{-1}$	.54246***	Constant	.4187***	$.731 \times 10^{-1}$	.38692***	Constant	.8079***	.103	.53086***
	MCDW	37.1015***	$.773 \times 10^{1}$		GRDW	10.9963	$.952 \times 10^{1}$		MCDW	-46.6620*	$.260 \times 10^{2}$	
	GRDW	-29.4124***	$.776 \times 10^{1}$		MCDW	-3.2281	$.860 \times 10^{1}$		DWGR	.0737*	$.433 \times 10^{-1}$	
	DWMC	-.6380***	.135		DWGR	-.2254	$.673 \times 10^{-1}$		GRDW	23.4177	$.254 \times 10^{2}$	
	DWGR	-.4067***	.107		DWMC	-.2806	$.945 \times 10^{-1}$		DWMC	.0129	$.622 \times 10^{-1}$	

TABLE III
(continued)

Autocorrelation Parameter	9 Areal Units				36 Areal Units				81 Areal Units			
	Variable	Regression Coefficient	Standard Error	Multiple Correlation	Variable	Regression Coefficient	Standard Error	Multiple Correlation	Variable	Regression Coefficient	Standard Error	Multiple Correlation
-.1	Constant	-.5752***	.125	.60152***	Constant	-.4532***	.115	.54082***	Constant	-1.2465	.212	.61799***
	DWMC	.4261***	.124		MCDW	-31.7865***	$.954 \times 10^{1}$		GRDW	27.6178	$.219 \times 10^{2}$	
	DWGR	.3224***	.116		DWGR	.2183**	$.976 \times 10^{-1}$		DWGR	.0738	$.805 \times 10^{-1}$	
	MCDW	-53.4723***	$.112 \times 10^{2}$		GRDW	17.1614**	$.838 \times 10^{1}$		MCDW	16.3273	$.306 \times 10^{2}$	
	GRDW	40.3001***	$.892 \times 10^{1}$		DWMC	.1515	.131					
-.5	Constant	-.8438***	$.313 \times 10^{-1}$	.95418***	Constant	-.7167***	$.336 \times 10^{-1}$	.54208***	Constant	-.7152***	$.332 \times 10^{-1}$	.44680***
	DWGR	.3492***	$.333 \times 10^{-1}$		DWGR	.0989***	$.321 \times 10^{-1}$		DWGR	.1149***	$.323 \times 10^{-1}$	
	DWMC	.1253***	$.385 \times 10^{-1}$		GRDW	.0655	.161		DWMC	.1075***	$.360 \times 10^{-1}$	
	GRDW	.1388	.108		DWMC	.0633	$.388 \times 10^{-1}$		MCDW	-1.1776**	.563	
	MCDW	-.1464	.117		MCDW	.0425	.211		GRDW	.8789*	.470	
-.9	Constant	-.2955***	$.105 \times 10^{-1}$	.99872***	Constant	-.5522***	$.134 \times 10^{-1}$	.80778***	Constant	-.5520***	$.196 \times 10^{-1}$	.50349***
	DWMC	-.3808***	$.140 \times 10^{-1}$		DWGR	.0592***	$.131 \times 10^{-1}$		DWGR	.0592***	$.187 \times 10^{-1}$	
	DWGR	-.0949***	$.965 \times 10^{-2}$		GRDW	.0054	$.314 \times 10^{-1}$		DWMC	.0402*	$.226 \times 10^{-1}$	
	GRDW	-.0018	$.385 \times 10^{-2}$		DWMC	.0271*	$.149 \times 10^{-1}$		MCDW	.0574	$.449 \times 10^{-1}$	
	MCDW	.0013	$.421 \times 10^{-2}$		MCDW	.0263	$.388 \times 10^{-1}$					

*Denotes a significant value at the 10% level of significance.

**Denotes a significant value at the 5% level of significance.

***Denotes a significant value at the 1% level of significance.

TABLE IV

SIGNIFICANT DISCRIMINANT FUNCTIONS BASED UPON SIMULATION RESULTS FOR THE AUTOCORRELATION PARAMETER GROUPS

Number of Areal Units	Variable	Varimax Rotated Discriminant Function Coefficient			
		DF 1	DF 2	DF 3	DF 4
9	χ^2 after function removed	207.48***	3.65		
	DWGR	.5089*	-.0156		
	DWMC	-.5101*	.0106		
	GRDW	-.0159	.5038*		
	MCDW	-.0107	.5046*		
Per Cent of Variance Accounted For		67.16	15.37		
36	χ^2 after function removed	166.01***	38.27***	10.59	0.00
	DWGR	-.0450	.5227*	3.0339	-.1164
	DWMC	.0641	-.5084*	3.0570	-.0416
	GRDW	.5116*	-.0621	-.0449	2.3051
	MCDW	.5307*	-.0459	.1614	-2.2879
Per Cent of Variance Accounted For		48.86	42.90	5.13	3.09
81	χ^2 after function removed	404.35***	60.31***	12.72***	0.00
	DWGR	.5128*	-.0295	2.5560	-.0216
	DWMC	-.5123*	.0305	2.5567	-.0320
	GRDW	-.0296	.5081*	-.0171	3.4895
	MCDW	-.0310	.5076*	.0221	3.4909
Per Cent of Variance Accounted For		64.05	29.39	3.57	2.99

*Denotes a significant correlation between the variable and the discriminant function.

***Denotes a significant difference from zero at the .01 level of significance.

TABLE V

CLASSIFICATION OF SIMULATION RESULTS
BY THE EXTRACTED DISCRIMINANT FUNCTIONS

Number of Areal Units	Actual Autocorrelation Parameter	Predicted Autocorrelation Parameter					
		-.9	-.5	-.1	.1	.5	.9
9	-.9	12	40	1	0	33	13
	-.5	9	74	2	2	9	0
	-.1	0	10	39	48	0	0
	.1	0	4	42	53	0	0
	.5	11	44	0	0	17	25
	.9	17	10	0	0	14	57
Percent of cases correctly classified = 43.0 (n = 586)							
36	-.9	31	23	2	0	27	17
	-.5	7	48	4	14	25	2
	-.1	0	0	64	36	0	0
	.1	0	2	42	56	0	0
	.5	5	13	0	0	76	6
	.9	5	0	0	0	19	76
Percent of cases correctly classified = 58.50 (n = 600)							
81	-.9	68	13	1	0	10	8
	-.5	6	49	13	16	12	4
	-.1	0	0	92	8	0	0
	.1	0	2	23	75	0	0
	.5	1	4	1	3	89	2
	.9	2	0	0	0	20	78
Percent of cases correctly classified = 75.17 (n = 600)							

increase in the number of areal units, approximately 24% of the cases are classified as belonging to adjacent classes, while an additional 17% are mis-allocated at random. No apparent trend can be observed as $\rho \to 0$, $\rho \to -1$ or $\rho \to +1$.

Consequently unambiguous spatial and temporal autocorrelation components of STI could not be uncovered in these experiments. The regression results are counter to Haining and Bennett's speculation that time series should be investigated before space-time modelling takes place. Furthermore results presented in Table III, IV and V demonstrate a clear need for space-time autocorrelation indices.

Next the power of MC_T will be explored. This exploration has been achieved through analysis of variance. Results of this technique appear in Table VI. For all three geographic landscapes the calculated F-ratios were thousands of times greater than a 1% level of significance critical value. These values must be interpreted with caution, however. Robustness tests for these analysis of variance results are summarized in Table VII. Generally speaking, STI appears to come from a normal distribution. Hence the Bartlett/Box homogeneity of group variances F-ratios are interpretable, and clearly imply variance inhomogeneity. The Scheffe test consistently indicates that all group means are significantly different. Therefore, the F-ratio values presented in Table VI are due in part to variance inhomogeneity. More than likely their enormous magnitudes reveal true group differences. This inference indicates that STI is a good discriminator and would permit cases to be properly

TABLE VI

ANALYSIS OF VARIANCE FOR SIMULATION RESULTS OF THE SPACE-TIME AUTOCORRELATION INDEX

Number of Areal Units	Source	Sum of Squares	Degrees of Freedom	Mean Squares	F-Ratio
9	Between Autocorrelation Parameters	183.797	5	36.795	5845.597*
	Within Autocorrelation Parameters	3.647	580	.006	
	Total	187.444	585		
36	Between Autocorrelation Parameters	193.448	5	38.690	13032.834*
	Within Autocorrelation Parameters	1.763	594	.003	
	Total	195.211	599		
81	Between Autocorrelation Parameters	162.575	5	32.515	13186.710*
	Within Autocorrelation Parameters	1.465	594	.003	
	Total	164.040	599		

*Denotes a significant difference from unity at the .01 level of significance, $F_{.01,df} = 3.05$.

TABLE VII

ROBUSTNESS MEASURES FOR THE ANALYSIS OF VARIANCE RESULTS

Number of Areal Units	Autocorrelation Parameter	Kolmogorov-Smirnov Statistic	Bartlett/Box F-Ratio	Scheffe Test Groups (α = .05)
9	+.9	.449***	244.605***	all
	+.5	.375***		all
	+.1	.036		all
	-.1	.056		all
	-.5	.273***		all
	-.9	.490***		all
36	+.9	.050	180.728***	all
	+.5	.031		all
	+.1	.029		all
	-.1	.031		all
	-.5	.040		all
	-.9	.030		all
81	+.9	.050	192.162***	all
	+.5	.050		all
	+.1	.041		all
	-.1	.023		all
	-.5	.019		all
	-.9	.050		all

*** Denotes a significant difference at the .01 level of significance.

allocated to their corresponding classes.

MOMENTS OF MC_T

One key theorem permits the moments of MC_T to be evaluated under the assumption of normality. It is due to Pitman (1937) and Koopmans (1942), and states if x_j $(i = 1,2,\ldots,n)$ are independent identically distributed normal variates, then any scale-free function $g(x_{ij};\ i = 1,2,\ldots,n)$ is distributed independently of $\sum_{i=1}^{i=n} x_i^2$. Therefore from equation (8)

$$E(STI) = E\left[\frac{(T-1)n}{\sum_{t=2}^{t=T}\sum_{i=1}^{i=n}\sum_{j=1}^{j=n} c_{ijt-1}} \; \frac{\sum_{t=2}^{t=T}\sum_{i=1}^{i=n}\sum_{j=1}^{j=n} c_{ijt-1}(y_{it}-\bar{y})(y_{jt-1}-\bar{y})}{\sum_{t=1}^{t=T}\sum_{i=1}^{i=n}(y_{it}-\bar{y})^2}\right]$$

$$= \frac{(T-1)n}{\sum_{t=2}^{t=T}\sum_{i=1}^{i=n}\sum_{j=1}^{j=n} c_{ijt-1}} \; \frac{-\frac{\sigma^2}{nT}\sum_{t=2}^{t=T}\sum_{i=1}^{i=n}\sum_{j=1}^{j=n} c_{ijt-1}}{nT(1-\frac{1}{nT})\sigma^2} \tag{13}$$

$$= -\frac{(T-1)}{T(nT-1)} \; .$$

Hence the mean of the sampling distribution of STI is $-(T-1)/T(nT-1)$.

Following Cliff and Ord (1973), the second moment derived from equation (8) is

$$E(STI^2) = E\left[\frac{(T-1)^2 n^2}{(\sum_{t=2}^{t=T}\sum_{i=1}^{i=n}\sum_{j=1}^{j=n} c_{ijt-1})^2} \; \frac{\{\sum_{t=2}^{t=T}\sum_{i=1}^{i=n}\sum_{j=1}^{j=n} c_{ijt-1}(y_{it}-\bar{y})(y_{jt-1}-\bar{y})\}^2}{\{\sum_{t=2}^{t=T}\sum_{i=1}^{i=n}(y_{it}-\bar{y})\}^2}\right]$$

$$= \frac{(T-1)^2 n^2}{(\sum_{t=2}^{t=T}\sum_{i=1}^{i=n}\sum_{j=1}^{j=n} c_{ijt-1})^2} \; \frac{\frac{\sigma^4}{n^2T^2}[2n^2T^2\sum_{t=2}^{t=T}\sum_{i=1}^{i=n}\sum_{j=1}^{j=n} c_{ijt-1}^{\;2} - 4nT\sum_{i=1}^{i=n}(\sum_{t=2}^{t=T}\sum_{j=1}^{j=n} c_{ijt-1})^2 + 3(\sum_{t=2}^{t=T}\sum_{i=1}^{i=n}\sum_{j=1}^{j=n} c_{ijt-1})^2]}{(nT-1)(nT+1)\sigma^4} \tag{14}$$

$$= \frac{(T-1)^2[2n^2T^2\sum_{t=2}^{t=T}\sum_{i=1}^{i=n}\sum_{j=1}^{j=n} c_{ijt-1}^{\;2} - 4nT\sum_{i=1}^{i=n}(\sum_{t=2}^{t=T}\sum_{j=1}^{j=n} c_{ijt-1})^2 + 3(\sum_{t=2}^{t=T}\sum_{i=1}^{i=n}\sum_{j=1}^{j=n} c_{ijt-1})^2]}{T^2(nT-1)(nT+1)(\sum_{t=2}^{t=T}\sum_{i=1}^{i=n}\sum_{j=1}^{j=n} c_{ijt-1})^2}$$

Therefore the standard error of the sampling distribution of STI may be determined by combining equations (13) and (14) such that

$$\sigma_{STI} = \{E(STI^2) - [E(STI)]^2\}^{\frac{1}{2}} \; .$$

Neither equation (13) nor equation (14) require any parameters to be estimated in order to calculate σ_{STI}. Further, since nT will need to be reasonably large, where $T \geq 20$ and $n \geq 4$, then the significance test will be based upon the standard normal deviate z. Thus the test statistic becomes

$$z = \frac{MC_T + (T - 1)/T(nT - 1)}{\sigma_{STI}}$$

CONCLUSIONS AND IMPLICATIONS

Three sequential goals were set at critical points in this exposition. The first objective was to investigate dimensions of autocorrelation for space-time series. The second objective was to outline a space-time autocorrelation index, and explore its relations with the traditional spatial auto- and serial correlation indices. The third objective was to provide a fundamental theorem of space-time interdependence. As each of these goals was set, ancillary questions were raised and then answers to these questions were sought. The more important questions together with their answers will be recapitulated at this time, collated with a drawing of some prominent inferences.

One question connected with the first objective asks whether or not the first step of a space-time study should be an investigation of individual areal unit time-series. According to simulation results a somewhat strong tendency is prevalent for STI to be linked only to a time series of spatial autocorrelation index values. Presumably, then, measures of spatial autocorrelation are more sensitive to space-time covariation than are ones of serial correlation. This finding contradicts Bennett and Haining's contentions, and in a modified way corroborates Granger's postulate. An important inference that can be drawn from it suggests that a time-series of spatial autocorrelation index values will illuminate the path of space-time modelling ventures more brightly than a cursory inspection of areal unit time series. In addition interpretations attached to conventional time series analyses may be misleading. Space-time covariation apparently cannot be reduced to a spatial component and a temporal component. Hence the degrees of freedom problem in numerical work may not be as severe as Hepple suggests.

Another question affiliated with this first objective asks whether or not spatial autocorrelation is a manifestation of temporal trends. Case I of the proof of the fundamental theorem is a definitive demonstration that the answer to this question is "no." In contrast, for space-time covariations the answer is "yes." Once again interpretative and numerical considerations focus on space-time rather than temporal interdependencies.

Three closely related questions were raised with respect to the second objective. Is a new index strictly dealing with space-time phenomena necessary? What are the properties of MC_T? How powerful is MC_T? In response to the first of these questions, the fundamental theorem as well as the simulation experiments lend credibility to a need for space-time indices. While the geographic distribution of Durbin-Watson statistics offers very little information about space-time series, the time series of Geary Ratios fails to correlate highly with STI in a consistent manner. Likewise, the conspicuous patterns of covariation displayed by spatial series of serial correlation and time series of spatial autocorrelation are not very successful predictors of the latent space-time covariation. In response to the second question, simulation experiments indicate that STI is an extremely good predictor, however. Consequently numerical results derived from traditional autocorrelation statistics cannot be properly interpreted. But equations (13) and (15) specify sufficient properties of MC_T for rendering meaningful numerical results, answering the third question.

The third objective alludes to questions concerning material and interpretative considerations for stationary versus non-stationary processes. A fundamental theorem of space-time interdependence was spelled out and proven for spatial situations, mathematical transformations from spatial to temporal autocorrelation indices were sketched, the remaining part of the proof being self-evident. Outcomes for stationary situations are that in the absence of space-time covariation autocorrelation indices are insensitive to trends in their spatial or temporal counterparts. The opposite is true when non-zero space-time covariation exists. The same conclusions hold when quasi-stationary situations are observed for which random deviations from a stationary state occur. Here the expectation is one of stationarity. Non-stationary situations introduce complications when the forcing function is autocorrelated and/or the areal unit temporal processes are autocorrelated.

In conclusion dimensions of space-time autocorrelation are not straight-forward extensions of concomitant spatial and temporal autocorrelation measures. Embedded complexities mean numerical considerations must include a distinct space-time autocorrelation index, whereas interpretation considerations include ascribing some portion of a spatial autocorrelation index value to space-time covariation, or attaching some meaning to serial correlation values that distinguishes between temporal processes and space-time processes.

APPENDIX A: COMPUTER PROGRAM FOR SIMULATING SPACE-TIME DATA SETS

The following computer program may be used to generate space-time data sets where each $x_{it=0}$ is drawn from a normal distribution, and the spatial distribution of $x_{it=0}$ $(i=1,2,\ldots,n)$ displays no noticeable geographical relatedness. This program includes seven subroutines. INIT selects the initial spatial distribution. It utilizes ITIMER to select random numbers. This second subroutine retrieves single-digit milli-seconds from the computer clock. Ten of these numbers are summed, and the total is inputed to IMSL subroutine MDNRIS which returns a normal deviate. Because uniform random variates are being added, the seed inputed to MDNRIS tends to be normally distributed, helping to insure that the frequency distribution of the initial selection of $x_{it=0}$ $(i=1,2,\ldots,n)$ will conform to a normal distribution.

Next subroutine SPAUTO is called to determine the nature and degree of chance spatial autocorrelation displayed by the geographic distribution of $x_{it=0}$ $(i=1,2.,,,.n)$. Then if the latent level exceeds some predescribed amount, subroutine SWITCH is employed to randomly permute $x_{it=0}$ $(i=1,2,\ldots,n)$ values in search of some distribution containing a negligible amount of spatial autocorrelation. This algorithm was modelled after Goodchild (1979). In the simulation of data sets, this step tends to be extremely time consuming. For relatively small sets of areal units a desirable geographical distribution is difficult to find, resulting in the inspection of a large number of permutations. As n increases, the probability becomes somewhat small that a random allocation of $x_{it=0}$ $(i=1,2,\ldots,n)$ is achieved in such a fashion that sizeable spatial autocorrelation can be detected. However, interchanging pairs of values tends to produce small gains. In the simulations carried out in this investigation, searches for n=9 and n=81 were more time consuming than those searches for n=36.

GEN takes the initial spatial distribution transferred from SWITCH and generates a space-time series. TMAUTO calculates the Durbin-Watson statistic for individual time series. And, SPATH calculates the space-time index STI for a space-time series.

```
      PROGRAM SPTMSM(IN,IND,INPUT,OUTPUT,OUT,FILE,DATA,TAPE3=IN,TAPE4=
     CIND,TAPE5=INPUT,TAPE6=OUTPUT,TAPE7=OUT,TAPE8=FILE,TAPE9=DATA)
C
C     PROGRAM TO GENERATE SPACE-TIME SERIES
C
C     INITIAL SPATIAL DISTRIBUTION IS RANDOM, AND THE VALUES CAN BE
C          DRAWN FROM A NORMAL DISTRIBUTION
C
C     M = NO. OF TIME PERIODS (MAX = 25)
C     N = NO. OF AREAL UNITS (MAX = 81)
C     NS = NO. OF SIMULATIONS TO BE RUN
C     ISP = NO. OF AUTOCORRELATION PARAMETERS TO BE EVALUATED
C     ISTART = NO. OF RANDOM PERMUTATIONS ALLOWED IF A RESTART IS
C              NECESSARY WHEN ESTABLISHING THE INITIAL SPATIAL
C              DISTRIBUTION
C     IN = NUMBER OF RANDOM NUMBERS IN A PERMUTATION ROW OF INDX(IN,ISTART)
C     SDM = THE PART OF A STANDARD DEVIATION SPATIAL DISTRIBUTIONS WILL
C           BE PERMITTED TO VARY FROM THE EXPECTED VALUES OF 1 AND -1/(N-1)
C     IFLAG = 0:  INITIAL VALUES GENERATED
C           = OTHERWISE:  INITIAL VALUES SUPPLIED EXTERNALLY
C     FMT:  FORMAT OF (NF1.0) FOR CONNECTIVITY MATRIX
C     SPAT:  VECTOR OF AUTOCORRELATION PARAMETERS
C
C     FILE 3 HAS THE FOLLOWING STRUCTURE:
C          CARD 1 - M,N,NS,ISP,ISTART,IN,SDM,IFLAG   (6I3,F3.0,I3)
C          CARD 2 - FMT   (8A10)
C          CARD 3 - FMT2   (8A10)
C          CARD 4 - SPAT   (16F5.0)
C          CARD 5+ - DATA TO BE READ IN, ONE VALUE PER CARD, IF IFLAG IS
C                    NOT EQUAL TO 0   (F5.0)
C
C     FILE 4 HAS THE FOLLOWING  STRUCTURE:
C          TEN ROW MATRIX OF RANDOM NUMBERS BETWEEN 1 AND N   (FMT2) -
C                                                       INTEGER FORMAT
C
C     FILE 5 HAS THE FOLLOWING STRUCTURE:
C          CONNECTIVITY MATRIX   (FMT) - FLOATING POINT FORMAT
C
C     OUTPUT:  FILE 7 = OUT - PERMUTATIONS FOR INITIAL VALUES
C                           - SIMULATED SPACE-TIME SERIES
C              FILE 8 = FILE - SPACE-TIME STATISTICS IN THE COLUMN ORDER:
C                              (1) AUTOCORRELATION PARAMETER
C                              (2) SIMULATION NO.
C                              (3) SPACE-TIME MORAN COEFFICIENT
C                              (4) DURBIN-WATSON STAT. FOR GEARY RATIO TIME
C                                  SERIES
C                              (5) DURBIN-WATSON STAT. FOR MORAN COEFFICIENT
C                                  TIME SERIES
C                              (6) GEARY RATIO FOR DURBIN-WATSON SPATIAL
C                                  DISTRIBUTION
C                              (7) MORAN COEFFICIENT FOR DURBIN-WATSON SPATIAL
C                                  DISTRIBUTION
C              FILE 9 = DATA - SPACE-TIME SERIES
C
C     WRITTEN BY - DANIEL A. GRIFFITH
C                  DEPARTMENT OF GEOGRAPHY
C                  STATE UNIVERSITY OF NEW YORK AT BUFFALO
C
C     IMPLEMENTED - JUNE, 1980
C
C     NOTES:  COLUMN 1 OF FORMAT STATEMENTS THAT ARE OPTIONAL IS MARKED WITH
C                  A "C"
C             CDC CYBER 173 SYSTEMS CONTROL CARDS:
C                  OLD,NATO
```

```
C                   ATTACH,OPLC/UN=LIBRARY
C                   FTN,I=NATO,L=0
C                   REWIND,LGO
C                   GET,PARM,RAND,CONMTX
C                   UCSLIB,IMSL7.
C                   ENTER./LDSET,LIB=IMSL7./LGO,PARM,RAND,CONMTX.
C
      REAL GR1,MC1,XINT(81),SPAT(16),MC(25),GR(25),DW(81),MCT,MCDW
      INTEGER INDX(10,81)
      COMMON XDATA(81,25),SPADT(81,81)
      READ(3,100) M,N,NS,ISP,ISTART,IN,SDM,IFLAG
      READ(3,104) (SPAT(I),I=1,ISP)
      READ(3,102) FMT
      READ(3,102)  FMT2
      READ(4,FMT2) ((INDX(I,J),J=1,IN),I=1,ISTART)
      READ(5,FMT) ((SPADT(I,J),J=1,N),I=1,N)
      DO 6 IA=1,ISP
      SP = SPAT(IA)
      WRITE(8,106) SP
      DO 5 ICOUNT=1,NS
      WRITE(7,103) ICOUNT
      CALL INIT(IN,SDM,ISTART,N,IFLAG,XINT,GR1,MC1,INDX)
      GR(1) = GR1
      MC(1) = MC1
      CALL GEN(XINT,M,N,SP)
      WRITE(9,1000) SP,ICOUNT
      WRITE(7,1002) SP
      DO 30 I=1,N
   30 WRITE(9,1001) (XDATA(I,J),J=1,M)
      CALL TMAUTO(M,N,DW)
      DO 4 J=2,M
      DO 3 K=1,N
    3 XINT(K) = XDATA(K,J)
      CALL SPAUTO(XINT,N,GR1,MC1,VARX)
      GR(J) = GR1
    4 MC(J) = MC1
      CALL SPATM(M,N,MCT)
      CALL SPAUTO(DW,N,GR1,MC1,VARX)
      DWGR = GR1
      DWMC = MC1
      N2 = 2
      DO 20 K=1,M
      XDATA(1,K) = GR(K)
   20 XDATA(2,K) = MC(K)
      CALL TMAUTO(M,N2,DW)
      GRDW = DW(1)
      MCDW = DW(2)
      WRITE(8,105) ICOUNT,MCT,DWGR,DWMC,GRDW,MCDW
    5 CONTINUE
    6 CONTINUE
      STOP
  100 FORMAT(6I3,F3.0,I3)
  102 FORMAT(8A10)
  103 FORMAT(1X,17HSIMULATION NUMBER,I5,///)
  104 FORMAT(16F5.0)
  105 FORMAT(1X,I3,5X,5(F12.7,5X))
  106 FORMAT(1X,42HTHE SPACE-TIME AUTOCORRELATION PARAMETER =,F5.3,//)
 1000 FORMAT(1X,27HAUTOCORRELATION PARAMETER =,F5.3,5X,
     C17HSIMULATION NUMBER,I5,///)
 1001 FORMAT(1X,4(11X,6F15.5,/),11X,F15.5)
 1002 FORMAT(/,1X,27HAUTOCORRELATION PARAMETER =,F5.3,//)
      END
C
      SUBROUTINE INIT(IN,SDM,ISTART,N,IFLAG,XINT,GR1,MC1,INDX)
      REAL GR1,MC1,XINT(N),RSUM(81)
      INTEGER IR(81),RAN,IXT(10),INDX(10,81)
```

```
      COMMON XDATA(81,25),SPADT(81,81)
      IF(IFLAG.NE.0) GO TO 5
      DO 4 I=1,N
      IX = 0
      DO 55 II=1,10
      CALL ITIMER(RAN)
      IXT(II) = RAN/10
      IXT(II) = IXT(II)*10
      IXT(II) = RAN - IXT(II)
   55 IX = IX + IXT(II)
      IR(I) = IX
      P  = FLOAT(IX)/100
      CALL MDNRIS(P,Y,IER)
      XINT(I) = -Y
    4 CONTINUE
      WRITE(7,102) (IR(L),L=1,N)
      GO TO 50
    5 DO 49 IK=1,N
   49 READ(3,104) XINT(IK)
   50 CALL SPAUTO(XINT,N,GR1,MC1,VARX)
      WRITE(7,100) (XINT(L),L=1,N)
      WRITE(7,2000) GR1,MC1
      SUMC = 0.0
      DO 10 I=1,N
      RSUM(I) = 0.0
      DO 10 J=1,N
      RSUM(I) = RSUM(I) + SPADT(I,J)
   10 SUMC = SUMC + SPADT(I,J)
      SUMCL1 = 0.0
      DO 15 I=1,N
   15 SUMCL1 = SUMCL1 + RSUM(I)*(RSUM(I) - 1.0)
      STEGR = ((SUMC + SUMCL1/2)*(N-1)-(SUMC**2)/2)/((N+1)*(SUMC**2/4))
      STEMC = (2*(N**2)*SUMC - 4*N*(SUMC+SUMCL1)+3*(SUMC**2))/((SUMC**2)
     C*(N**2 - 1)) - 1.0/((N - 1)**2)
      CVRMC = SDM*SQRT(STEMC)
      CVRGR = SDM*SQRT(STEGR)
      WRITE(6,105) CVRGR,CVRMC
      CALL SWITCH(IN,ISTART,XINT,N,VARX,CVRGR,CVRMC,GR1,MC1,INDX)
      WRITE(7,103) (XINT(L),L=1,N)
      WRITE(7,2001) GR1,MC1
  100 FORMAT(5X,15HINITIAL VECTOR:,10(5X,13F7.4/)/)
 2000 FORMAT(//,25X,12HINITIAL GR =,F5.3,5X,12HINITIAL MC =,F5.3,//)
  101 FORMAT(1X,F10.5,I5)
  102 FORMAT(1X,51HTHE SET OF RANDOM NUMBERS INPUTED TO MDNRIS(IMSL7):
     C,//,4(5X,31I3,/))
  103 FORMAT(///,5X,13HFINAL VECTOR:,10(5X,13F7.4/)/)
 2001 FORMAT(25X,10HFINAL GR =,F5.3,5X,10HFINAL MC =,F5.3,//)
  104 FORMAT(F5.0)
  105 FORMAT(1X,19HGR CRITICAL VALUE =,F7.5,5X,19HMC CRITICAL VALUE =
     C,F7.5)
      RETURN
      END
C
      SUBROUTINE SPAUTO(XINT,N,GR1,MC1,VARX)
      REAL XINT(N),GR1,MC1
      COMMON XDATA(81,25),SPADT(81,81)
      SUMX = 0.0
      SUMX2 = 0.0
      DO 10 I=1,N
      SUMX = SUMX + XINT(I)
   10 SUMX2 = SUMX2 + XINT(I)**2
      XBAR = SUMX/N
      VARX = SUMX2 - (SUMX**2)/N
      SUMC = 0.0
      SUMMC = 0.0
      SUMGR = 0.0
```

```
      DO 20 I=1,N
      DO 19 J=1,N
      IF(SPADT(I,J).EQ.0.0) GO TO 19
      SUMC = SUMC + 1.0
      SUMMC = SUMMC + (XINT(I) - XBAR)*(XINT(J) - XBAR)
      SUMGR = SUMGR + (XINT(I) - XINT(J))**2
   19 CONTINUE
   20 CONTINUE
      MC1 = (N*SUMMC)/(SUMC*VARX)
      GR1 = ((N-1)*SUMGR)/(2*SUMC*VARX)
      RETURN
      END
C
          IDENT  ITIMER      RETURN MILLI-SECONDS SINCE DEADSTART.
          ENTRY  ITIMER

* GET REAL TIME USING *RTIME* MACRO FROM *COMCSYS* CMMON DECK.
*
* WRITTEN BY - ANDREW RAKOWSKI
*              DEPT OF GEOGRAPHY
*              STATE UNIVERSITY OF NEW YORK AT BUFFALO
*
*IMPLEMENTED - 23-MAY-1980
*
*EQUIPMENT   - CONTROL DATA CYBER 174
*
*
* NOTES :
*
* THIS IS A FORTRAN 4 CALLABLE ROUTINE. MAY BE CALLED EITHER AS
*         A FUNCTION OR A SUBROUTINE. IF USED AS A FUNCTION, IT
*         MUST HAVE A DUMMY PARAMETER.
*
* THE LIBRARY 'OPLC' MUST BE ATTACHED BEFORE COMPILING
*         THIS ROUTINE. FOR EXAMPLE :
*
* ATTACH,OPLC/UN=LIBRARY. GET OPLC LIBRARY
*
* SHOULD APPEAR IN YOUR JCL PRIOR TO COMPILATION.
*

ITIMER    BSS    1           ENTRY/EXIT
          SA7    X1          GET ADDRESS OF PARAMETER.
          RTIME  A7          GET RTIME AND PLACE IT INTO PARAMETER.
          SA2    A7          PLACE RTIME FIELDS INTO X2.

* DISCARD THE SECONDS FIELD, LEAVING JUST MILLI-SECONDS SINCE
*         DEADSTART IN X2 AFTER LOGICAL 'AND' WITH MASK.

          MX3    36          36 BIT MASK FIELD IN X3.
          LX3    36          LEFT SHIFT MASK INTO LOWER 36 BITS.
          BX2    X2*X3       LOGICAL PRODUCT TO EXTRACT MILLI-SECONDS.

* RETURN VALUES FOR FORTRAN 4 ROUTINES CALLING ITIMER.
*
          BX7    X2          PLACE VALUE INTO X7 FOR TRANSFER.
          BX6    X7          PUT VALUE INTO X6 FOR FUNCTION RETURN.
          SA7    A7          PUT TIME INTO PARAMETER.
          EQ     ITIMER      AND RETURN TO CALLER.

OPLC      XTEXT  COMCSYS
          END
C
      SUBROUTINE SWITCH(IN,ISTART,XINT,N,VARX,CVRGR,CVRMC,GR1,MC1,INDX)
      INTEGER INDX(10,81)
      REAL XIT(81),MC1,XINT(N)
```

```
      KL = 0
      CALL SPAUTO(XINT,N,GR1,MC1,VARX)
      DIFF1 = ABS(GR1 - 1.0)
      IF(DIFF1.GT.CVRGR) GO TO 80
      EX = 1.0/(N - 1)
      DIFF2 = ABS(MC1 + EX)
      IF(DIFF2.LT.CVRMC) GO TO 12
   80 TGR = GR1
      TMC = MC1
      DO 11 K=1,ISTART
      IK = K - 1
      IF(K.GT.1) GO TO 90
      WRITE(7,103) K,(INDX(K,LL),LL=1,N)
   90 DO 10 I=1,IN
      DO 9 J=1,IN
      IF(XINT(INDX(K,I)).EQ.XINT(INDX(K,J))) GO TO 9
      Y = XINT(INDX(K,I))
      XINT(INDX(K,I)) = XINT(INDX(K,J))
      XINT(INDX(K,J)) = Y
      CALL SPAUTO(XINT,N,GR1,MC1,VARX)
      KL = KL + 1
      EX = 1.0/(N - 1)
      DIFF1 = ABS(GR1 - 1.0)
      IF(DIFF1.GT.CVRGR) GO TO 19
      DIFF2 = ABS(MC1 + EX)
      IF(DIFF2.GT.CVRMC) GO TO 19
      GO TO 12
   19 IF(ABS(TGR-1.0).LT.ABS(GR1-1.0)) GO TO 20
      IF(ABS(TMC-EX).LT.ABS(MC1-EX)) GO TO 20
      GO TO 8
   20 Y = XINT(INDX(K,I))
      XINT(INDX(K,I)) = XINT(INDX(K,J))
      XINT(INDX(K,J)) = Y
    8 TGR = GR1
      TMC = MC1
C     WRITE(7,100) KL,(XINT(L),L=1,N)
C     WRITE(7,3000) GR1,MC1
    9 CONTINUE
   10 CONTINUE
      L = K + 1
      WRITE(7,101) L,(INDX(L,LL),LL=1,N)
   11 CONTINUE
  100 FORMAT(5X,6HPERM =,I5,13(5X,10F7.4/))
 3000 FORMAT(25X,4HGR =,F5.3,5X,4HMC =,F5.3)
  101 FORMAT(//,5X,46HANALYSIS RESTARTED WITH RANDOM PERMUTATION NO.,
     C5X,I5,/,5X,11HPERMUTATION,4(5X,31I3,/),/)
  103 FORMAT(//,5X,42HANALYSIS BEGAN WITH RANDOM PERMUTATION NO.,
     C9X,I5,/,5X,11HPERMUTATION,4(5X,31I3,/),/)
  102 FORMAT(//5X,51HTOTAL NO. OF PERMUTATIONS THROUGH FINAL ITERATION =
     C,I5,5X,23HTOTAL NO. OF RESTARTS =,I5/)
   12 WRITE(7,102) KL,IK
      RETURN
      END
C
      SUBROUTINE GEN(XINT,M,N,SP)
      DIMENSION XINT(N),RSUM(81)
      COMMON XDATA(81,25),SPADT(81,81)
      DO 10 I=1,N
      RSUM(I) = 0.0
      DO 9 J=1,N
      IF(SPADT(I,J).EQ.0.0) GO TO 9
      RSUM(I) = RSUM(I) + 1.0
    9 CONTINUE
   10 XDATA(I,1) = XINT(I)
      DO 21 K=2,M
      IK = K - 1
```

```
      DO 20 I=1,N
      XDATA(I,K) = 0.0
      DO 19 J=1,N
      IF(SPADT(I,J).EQ.0.0) GO TO 19
      XDATA(I,K) = XDATA(I,K) + SP*XDATA(J,IK)/RSUM(I)
   19 CONTINUE
      XDATA(I,K) = XDATA(I,K) + XDATA(I,IK)
   20 CONTINUE
   21 CONTINUE
      RETURN
      END
C
      SUBROUTINE TMAUTO(M,N,DW)
      DIMENSION DW(N)
      COMMON XDATA(81,25),SPADT(81,81)
      DO 30 K=1,N
      SUMY = 0.0
      SUMY2 = 0.0
      SUMDW = 0.0
      DO 10 I=1,M
      SUMY = SUMY + XDATA(K,I)
   10 SUMY2 = SUMY2 + XDATA(K,I)**2
      YBAR = SUMY/M
      VARY = SUMY2 - (SUMY**2)/M
      DO 20 I=2,M
      J = I - 1
   20 SUMDW = SUMDW + (XDATA(K,I) - XDATA(K,J))**2
      DW(K) = SUMDW/VARY
   30 CONTINUE
      RETURN
      END
C
      SUBROUTINE SPATM(M,N,MCT)
      COMMON XDATA(81,25),SPADT(81,81)
      REAL MCT
      SUMX = 0.0
      SUMX2 = 0.0
      SUMCT = 0.0
      SUMMCT = 0.0
      DO 10 J=1,M
      DO 10 I=1,N
      SUMX = SUMX + XDATA (I,J)
   10 SUMX2 = SUMX2 + XDATA(I,J)**2
      XBAR = SUMX/(N*M)
      VARX = SUMX2 - (SUMX**2)/(N*M)
      DO 20 K=2,M
      L = K - 1
      DO 18 I=1,N
      DO 17 J=1,N
      IF(SPADT(I,J).EQ.0.0) GO TO 17
      SUMCT = SUMCT + 1.0
      SUMMCT = SUMMCT + (XDATA(I,K) - XBAR)*(XDATA(J,L) - XBAR)
   17 CONTINUE
   18 CONTINUE
   20 CONTINUE
      MCT = ((M - 1)*N*SUMMCT)/(SUMCT*VARX)
      RETURN
      END
```

REFERENCES

Bennett, R., 1979. Spatial Time Series, Pion, London.
Bennett, R. and Haining, R., 1976. Space-Time Models: An Introduction to Concepts, Occasional Papers No. 28, Department of Geography, University College, London.
Cliff, A. and Ord, J., 1973. Spatial Autocorrelation, Pion, London.
Cliff, A., Haggett, P., Ord, J., Bassett, K. and Davies, R., 1975. Elements of Spatial Structure, Cambridge University Press, New York.
Curry, L, 1971. Applicability of Space-Time Moving-Average Forecasting, in: Regional Forecasting, M. Chisholm, A. Frey and P. Haggett (eds.), Butterworth, London, 11-24.
Goodchild, M., 1979. Simulation of Autocorrelation for Aggregate Data, Environment and Planning 12A: 1073-1081. University of Western Ontario.
Granger, C., 1975. Aspects of the Analysis and Interpretation of Temporal and Spatial Data, The Statistician, 24: 197-210.
Griffith, D., 1978. The Impact of Configuration and Spatial Autocorrelation on the Specification and Interpretation of Geographical Models, unpublished doctoral dissertation, Department of Geography, University of Toronto.
Haggett, P., Cliff, A. and Frey, A., 1977. Locational Analysis in Human Geography, Edward Arnold, London, Vol. 2.
Hepple, L., 1978. The Econometric Specification and Estimation of Spatio-Temporal Models, in: Time and Regional Dynamics, T. Carlstein, D. Parkes and N. Thrift (eds.), Wiley, New York, 66-80.
Hordijk, L. and Nijkamp, P., 1976. Dynamic Models of Spatial Autocorrelation, Research Memorandum, No. 57, Department of Economics, Free University, Amsterdam.
Koopmans, T., 1942. Serial Correlation and Quadratic Forms in Normal Variables, Annals of Mathematical Statistics, 13: 14-33.
Ord, K., 1975. Estimation Methods for Models of Spatial Interaction, Journal of the American Statistical Association, 67: 120-126.
Pitman, E., 1937. The "Closest Estimates" of Statistical Parameters, Proceedings, Cambridge Philosophical Society, 33: 212-222.

AN APPROACH TO THE STATISTICAL ANALYSIS OF CLUSTERED MAP PATTERN

Robert Haining

Department of Geography
University of Sheffield

INTRODUCTION

This paper is concerned with the analysis of spatial pattern where the map data are in the form of quadrat counts. Properties of such maps may be divided into two categories: distributional properties, which relate to such things as frequency counts, and arrangement properties which relate to the pattern of frequency count values on the map itself. Whereas the former is essentially non-spatial, the latter recognizes the order or organizational attributes of maps. It is obvious, for example, that n distinct quadrat counts (computed for a region partitioned into n cells) can be permutted to create n! different arrangements of those counts. Although the frequency distribution naturally must remain the same over the n! cases, the arrangement characteristics of the map change.

Excepting the use of non-parametric forms of modelling such as spectral analysis, geographers have traditionally treated distributional and arrangement properties of maps separately. Yet not only do many geographical problems suggest the importance of both properties, they also suggest the need for simultaneous analysis of these properties. As a result of developments in statistical theory over the last ten years or so methods now exist for a program of simultaneous analysis via parametric modelling.

The particular objective of this paper is to help provide a focus for studying these properties, and it is not the intention here to describe general classes of models. We shall be analyzing rural settlement patterns as examples of spatial location processes for which distributional and arrangement properties are both important. In particular, this emphasis allows us to concentrate on clustered or contagious map patterns.

Philosophically we approach map analysis recognizing three levels underlying any study. At the deepest level (level I) is some (spatial-temporal) process, by which is meant a set of rules that define how the map at one point in time becomes the map at a later point in time. The process is assumed to contain chance or stochastic attributes so that the process induces (level II) a probability distribution both for the spatial-temporal development of the process and for spatial cross sections. With these convenient assumptions, probability theory and in particular multivariate distributions provide models for map patterns which may include both distributional and arrangement properties.

To give an example, consider the multivariate normal density function $MVN(\underline{\mu},\underline{\Sigma})$ as a model of a map pattern (the subscript on variates $X_1,X_2,X_3,\ldots,X_n$ is used to identify each area or site on the map). A study focusing only on distributional properties would assume $\underline{\Sigma}$ to be diagonal--perhaps a scalar, σ^2, times the identity matrix. Thus, quadrat measures would be assumed to be n independent samples from a normal distribution with mean μ and variance σ^2. The n! permutations, within this modelling strategy, would be indistinguishable. However, a study focusing on distributional and arrangement properties might model the arrangement properties via the covariance matrix $\underline{\Sigma}$, permitting off diagonal entries to be non-zero. Although constraints on the behavior of $\underline{\Sigma}$ would be necessary to ensure that the model could be specified, the general point is that the covariance matrix provides a way of modelling the arrangement properties of maps [see for example Streitberg (1978)]. A usual constraint is to assume homogeneity in the map pattern with estimable inter-cell associations dependeng only on distance or lag separation.

An individual realization of the process (or single drawing from the multivariate probability distribution) generates the observed map pattern (level III). This is the only level to which we have direct access.

Map analysis may focus on any one of the above levels: description of map pattern, inferences concerning the nature of the appropriate spatial probability model or even process specification in terms of real physical events. Map pattern analysis may sometimes be used to infer model or process, but is better used to test hypotheses from an a priori specification of the process.

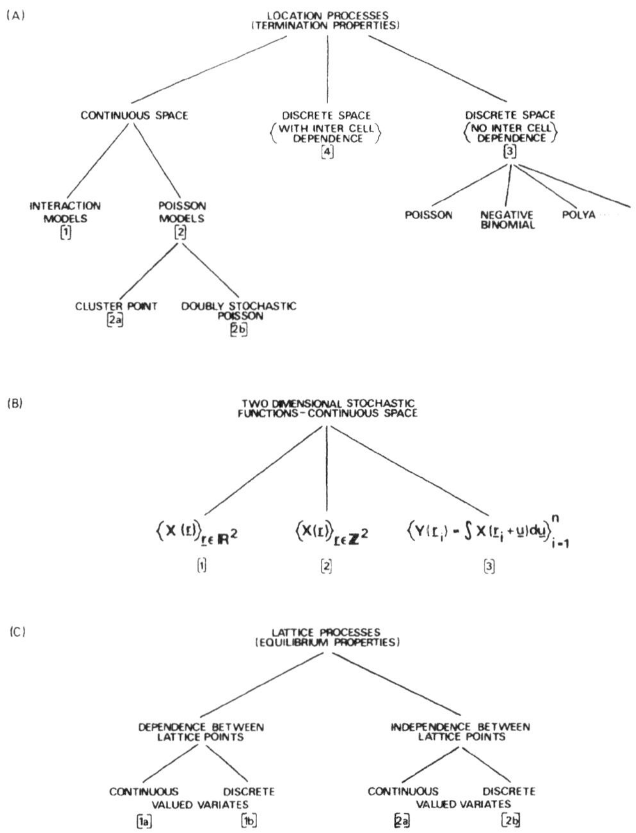

Fig. 1. A Taxonomy of Spatial Processes

A CLASSIFICATION OF SPATIAL PROCESSES

The spatial modelling in this paper can be put in context using the taxonomy presented in Figure 1 which is based on that in Bartlett (1974). He distinguishes three types of processes:

A. Location (Point) Processes:

All these processes involve the location of points in a region under different sets of assumptions. Interest focuses on properties of the induced probability distribution when the process has terminated.

(1) Interaction Process. Locate n objects, where location of the kth object $(1 < k \leq n)$ is influenced by the location of the objects $1,2,\ldots,k-1$. The kth object is either attracted to the areas of earlier points (contagion models) or repelled (inhibitory models).

(2) Poisson Process. Locate n objects according to a two-dimensional Poisson process with density parameter λ. The cluster-point or center-satellite variant of the Poisson process now assigns to each point the realization of a random variable which denotes the number of offspring or satellites associated with these n objects. The offspring are then distributed about the cluster centers according to a two-dimensional random dispersal function. The doubly stochastic variant of the Poisson is identical to the ordinary Poisson except that the density parameter λ is itself a random variable and thus varies over the region.

These processes are defined on continuous space (though for empirical analysis we may wish to know their properties when observed through an imposed partition of the region). However, these location processes should be distinguished from another group of processes (3) that are defined in terms of a specific partition. Each cell is like an independent replication of the process so that cell count values are assumed to be independent. Unlike the above processes, empirical analyses of these involve only distributional, not arrangement map properties. A good early review of this literature is provided by Skellam (1952) and the recent book by Getis and Boots (1978, Ch. 2 and 3) also provides a review with an emphasis on geographical applications. We are not directly interested in these models here because they explicitly rule out the need to specify arrangement properties--though they must satisfy tests for independence in order to be valid. It may be possible to extend these models to include inter-cell interaction (4).

B. Continuous Processes:

These are stochastic functions defined at every point of the two-dimensional surface (R^2). Three cases are distinguished--the first case where the entire surface configuration is recorded and analyzed, the second case where it is sampled (say at the integer points Z^2 of R^2) and the third case where aggregated values are recorded for n rectangular regions centered at $\{\underline{r}_i\}$. The auto-correlation function plays a particularly important role in the analysis of these types of surfaces.

C. Lattice Processes:

Consider the case where the variate is binary--site values are either zero or one. We start out with an initial distribution of occupied (one) and unoccupied (zero) sites. We may observe the changing occupance pattern as sites are occupied or vacated according to some stationary spatial-temporal process. The process may be described as a spatial "birth/death" process applicable, for example, to the study of the spread of infection (with recovery and possible reinfection) over a lattice of plants.

In these models the purely spatial dimension of the problem concerns the identification of the equilibrium properties of the lattice distribution. The process never terminates, but we are interested in the steady state pattern of zero's and ones. The classification in Figure 1 distinguishes between the "non-spatial problem" where variate values at a site are independent of neighboring site values from the "spatial problem" where there is dependence. Clearly this general class of models could include other discrete valued variables and continuous valued variables.

Given all these processes (with the exception noted) the analysis of map pattern realizations (inasmuch that quadrat methods rather than distance methods of analysis are used) must take into account not just frequency count properties--what are termed the aspatial properties of these maps--but also the arrangement properties, in particular those properties relating to the correlations between sites or areal values separated by fixed lags or distances.

CLUSTERING: SUBSTANTIVE AND STATISTICAL CONSIDERATIONS

Clustered patterns may result from at least two types of events, and should be distinguished at least for three different scales. Clustering may arise as a result of interactions between objects, which results in objects locating close to pre-existing objects. Objects are thus drawn to areas of existing concentrations. (The term interaction may also be extended to include diffusive processes, where new objects are "distance constrained" to locate close to parent locations). On the other hand clustering may not reflect any interactive process between objects, but be simply the result of objects locating, say, in a differentiated environment where the "carrying capacity" of subregions of the area differs. Thus clustering owes nothing to interaction between objects, simply to third party influences. It is the first type of clustering that is usually of most interest.

Clustering, as it is used in statistical contexts is a scale dependent term. We recognize three here:

(1) Intra-cell. Polya's distribution [Feller (1968, pp. 120-1)] provides a model of a clustered process at the intra-cell level (more usually called a contagious process). The probability of the kth object locating in the jth cell depends upon the location of objects 1,2,...,k-1. In particular the more objects (prior to the kth) that have located in the jth cell, the greater the probability of the kth object locating there too. For a fuller discussion, with an urn model of the process, the reader is referred to Feller, and to Dacey (1969) who considers the model in terms of a location process. Each cell is governed by the same probability rule and all cell counts are assumed mutually independent (for later reference, the negative binomial distribution is one limiting form of the Polya).

(2) Inter-cell. Clustering may also be an inter-cell characteristic, with cells containing a large number of objects being surrounded by cells that, for the most part, also contain a large number of objects. The process is thus spatially reinforcing, not just site reinforcing. Clearly the distinction between (1) and (2) may be only a feature of the scale of observation rather than a fundamental difference in terms of process.

The following processes may give rise to inter-cell clustering:

(a) Locate n objects one at a time. The probability of the kth object $(1 < k \leq n)$ locating in the jth subregion is a function of the number of objects already located in the jth subregion and in its neighbors. After proper specification we may be able to identify the probability distribution when the process has terminated.

(b) Given an initial distribution of objects over a region divided into cells, objects are then allowed to be born, or existing ones die according to a stationary process where transition probabilities may depend on the numbers of objects located both at each site plus neighboring sites. We can describe the distribution at equilibrium.

(c) Locate n "parents" at random over a group of cells. Generate a random number of offspring to each of the parents and distribute them around the parent cells, including neighboring cells. We can describe the distribution after the process has terminated.

(3) Inhomogeneity. In geographical situations we should recognize that clustering may reflect "third party" effects. Clustered patterns arise not only at inter- or intra-cell levels but as a result, say, of environmental variation. The doubly stochastic process could generate a clustered distribution with suitable variation in the density parameter λ. Alternatively an intrinsically non-stationary process could be used to create highly clustered map patterns. The point here is that clustering may be a large scale phenomenon, producing a differentiation between those areas of the map with objects and those without.

THE ANALYSIS OF RURAL SETTLEMENT MAPS

The theories of Bylund (1960) and Hudson (1969) provide the strongest impetus to the study of clustering in rural settlement maps. Map pattern may be present at three different scales. At the macro-scale, variations in environmental (biotope) characteristics lead to an irregular distribution of settlements over a region since this causes intra-regional variation in optimal farm size. This may give the appearance of macro-scale clustering. At a smaller scale, pattern is also induced by the morphology of offspring diffusion. The offspring of first generation colonizers set up their farms near the parent settlement and this process may extend to later generations. This also suggests clustering. However at a micro-scale there may be inhibitory patterns due to farmsteads being located at the center of each farm in order to minimize internal movement costs [these processes have analogues with a classification provided by Kershaw (1963) for clustering in plant species].

In addition to these considerations, Hudson's theory suggests the need for location processes of type A (Figure 1) to describe the early stages of colonization. These initial stages involve the spatial distribution of farmsteads whereas we are interested in the terminal properties of this allocation. However in the later stages inefficient producers are driven out (abandonment of farmsteads) while successful farmers expand (construction of new farmsteads near the centers of newly amalgamated farms). The final stage may, therefore, be more properly represented as the equilibrium state of a spatial birth/death process (type C of Figure 1). We shall consider this distinction again.

The empirical analysis in the next section treats pattern in terms of quadrat methods. There is no intention to discuss the relative merits of quadrat as opposed to distance methods; good discussions are available in, for example Dacey (1969), Ripley (1977, 1978). What has been an evident weakness in quadrat methods though has been an inability to model simultaneously the aspatial frequency count aspects of the data and the arrangement properties of these counts. Dacey's (1969) study of Puerto Rico settlement data is an example of this type of research and offers an appropriate point of departure for this discussion. In that paper he noted that while the negative binomial frequency function, for the most part, adequately described the frequency counts of settlements, there was strong evidence that neighboring quadrat counts were not independent. As a result he rejected the model as a description of the settlement pattern of Puerto Rico and in a concluding section discussed briefly models that might accommodate arrangement attributes of the maps.

The next section is an analysis of one of his maps. In a more complete analysis we might look at the data at a variety of different scales hoping to interpret scale shifts in parameter estimates and autocorrelation estimates (even frequency function fits) in terms of the earlier discussion of process. For this we would need to be able to vary the quadrat size.

MODELS OF DEPENDENCE FOR PUERTO RICO SETTLEMENT DATA

One of the maps used by Dacey (1969) was made available for study (Map 255). Table 1 provides a recapitulation of results given in that paper on the fitting of the negative binomial distribution to the count table. Table 1a gives results for the 1152 quadrat case, Table 1b gives results for the 288 quadrat case where cells are amalgamated by fours. As already indicated the negative binomial frequency function provides a good description of the aspatial frequency count but fails to account for the non-random arrangement of count values over the map.

0	1	2	1	0	0	0	0	0	0	0	0	0	0	0	0	0	0	0	0	3	0	0	0	0	0	0	0	0	0	0	0	0	0	0	0	0	0	0	2	0	0	0	0	0	0	0	0
0	0	0	0	0	0	1	3	0	0	1	0	0	0	0	0	0	0	0	0	0	0	0	0	0	0	0	0	0	0	0	0	0	0	0	0	0	0	0	6	0	0	0	0	0	0	0	0
0	0	1	0	1	2	1	1	1	0	0	0	0	0	0	0	0	0	0	0	0	0	0	0	0	0	0	0	0	0	0	0	0	0	0	0	0	0	0	7	0	0	0	0	3	0	0	0
0	0	0	0	0	0	0	0	0	0	0	0	0	0	0	0	0	0	0	0	0	0	0	0	0	0	0	0	0	0	0	0	0	0	0	0	0	0	0	0	0	0	0	0	1	0	0	0
0	0	0	0	0	0	0	1	0	0	0	0	0	0	0	0	0	0	0	0	0	0	1	0	0	0	0	0	0	0	1	2	0	0	0	0	0	0	0	0	0	0	0	0	0	0	0	0
0	0	0	0	0	0	0	0	0	0	0	0	0	0	1	0	0	0	0	0	0	0	0	0	0	0	0	0	0	0	1	2	0	0	0	0	0	0	0	0	0	0	0	0	0	0	0	0
0	0	0	0	0	0	0	1	1	0	0	0	0	0	0	0	0	0	0	0	0	0	0	0	0	0	0	0	0	0	0	0	0	0	0	0	0	0	0	0	0	0	0	0	0	0	0	0
0	0	0	0	0	1	1	1	0	0	0	1	1	0	1	0	0	0	0	0	0	0	0	0	0	0	0	0	0	0	0	0	0	0	0	0	0	0	0	0	0	0	0	0	0	0	0	0
0	2	5	0	0	0	0	2	1	1	0	0	1	2	0	0	0	0	0	0	0	0	0	0	0	0	0	0	0	0	0	0	0	0	0	0	0	0	0	0	0	0	0	0	0	0	0	0
0	0	0	1	0	5	2	1	1	3	0	0	1	0	0	0	0	0	0	0	0	0	0	0	0	0	0	0	0	0	0	0	0	0	0	0	0	0	0	0	0	0	0	0	0	0	0	0
0	2	1	0	0	1	1	4	1	0	0	2	0	1	0	0	0	0	0	0	0	0	0	0	0	0	1	0	0	1	0	0	0	0	0	0	0	1	0	0	0	0	0	0	0	0	0	0
2	0	1	3	1	1	4	0	0	0	1	1	1	2	0	0	0	0	0	0	0	0	0	0	0	0	0	0	0	0	0	0	1	0	0	0	2	4	0	4	3	4	1	0	0	0	0	0
3	0	1	3	4	3	3	0	0	0	1	0	3	1	1	0	0	0	0	0	0	0	0	0	0	0	0	0	0	0	0	0	0	0	0	2	2	1	3	4	2	0	1	1	1	0	0	0
3	1	0	1	3	3	2	0	1	0	1	0	6	2	0	0	0	0	0	0	0	0	0	0	0	0	0	0	0	0	0	0	0	0	0	0	1	0	0	0	0	0	0	1	1	0	0	0
1	2	0	1	1	0	0	2	3	0	1	1	2	1	0	0	0	0	0	0	0	0	0	0	0	0	0	0	0	0	0	0	0	1	2	0	0	0	0	0	0	1	0	0	1	0	0	0
1	1	0	1	1	1	0	1	0	1	2	3	4	2	0	0	0	0	0	0	0	0	0	0	0	0	0	0	0	0	0	0	0	5	0	0	0	1	1	0	1	2	0	0	0	1	0	0
0	0	4	0	1	1	0	0	0	0	0	0	0	0	0	0	0	0	0	0	0	0	0	0	0	0	0	0	0	0	0	0	0	2	0	0	0	0	0	0	0	0	0	0	0	0	0	1
0	0	1	2	2	0	0	0	0	0	0	1	0	0	0	0	0	0	0	0	0	0	0	0	0	0	0	0	0	0	0	0	0	0	0	0	0	0	0	0	0	0	1	0	0	0	0	0
0	0	0	0	2	1	0	0	0	0	0	1	0	0	0	0	0	0	0	0	0	0	0	0	0	0	0	0	0	0	0	0	0	0	0	0	0	0	0	0	0	0	0	0	0	0	0	0
0	0	0	0	2	2	0	0	0	0	0	1	0	0	1	0	0	0	0	0	0	0	0	0	0	0	0	0	0	0	0	0	0	0	0	0	0	0	0	0	0	0	0	0	0	0	0	0
0	0	0	0	0	0	0	0	0	1	0	1	0	1	5	7	0	0	0	0	0	0	0	0	0	0	0	0	0	0	1	6	0	0	0	0	0	0	0	0	0	0	0	0	0	0	0	0
0	0	0	0	0	0	0	3	3	1	2	2	1	0	1	2	0	0	0	0	0	0	0	0	0	0	0	0	0	0	0	0	1	0	0	0	0	0	0	0	0	0	0	0	0	0	0	0
0	0	0	0	0	0	0	0	0	2	2	2	5	0	0	0	0	0	0	0	0	0	0	0	0	0	0	0	0	0	0	0	0	0	0	0	0	0	0	0	0	0	0	0	0	0	0	0
0	0	0	0	0	0	0	0	0	2	1	0	3	0	0	0	0	0	0	0	0	2	1	0	0	0	0	0	0	0	0	0	0	0	0	0	0	0	0	0	0	0	0	0	0	0	0	0

Fig. 2. Quadrat Count Data: Map 255
(The author wishes to thank Professor Dacey for allowing him to use the Puerto Rico data.)

TABLE 1

(a) Frequency Count Data for Map 255 (1152 Cells)

Number of houses	Number of cells observed	Number of cells expected under negative binomial
0	972	971.35
1	101	105.55
2	40	38.73
3	20	17.55
4	9	8.70
5	5	4.54
6	3	2.45
7	2	1.35
≥ 8	0	1.78

Source: Dacey (1969)

(b) Frequency Count Data for Map 255 (288 cells)

Number of houses	Number of cells observed	Number of cells expected under negative binomial
0	197	195.07
1	29	34.11
2	12	17.47
3	9	10.91
4	12	7.42
5	9	5.30
6	4	3.91
7	9	2.94
≥ 8	7	10.85

Source: Dacey (1969)

Phase 1. Reduction to binary form.

The first analysis involved reducing the 1152 quadrat counts to binary form. Let Q_i denote the observed quadrat count in cell i; then

$$QB_i = \begin{cases} 1 & \text{if } Q_i > 0 \\ 0 & \text{if } Q_i = 0 \end{cases}$$

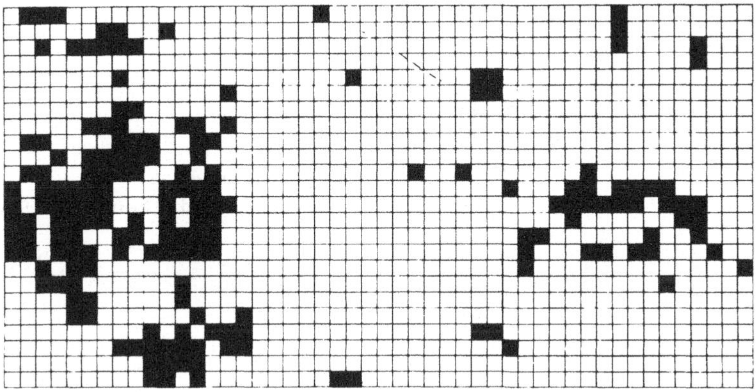

Fig. 3. Quadrat Count Data: Binary Form

The transformed map is presented here in Figure 3. A "rooks-move" join count test [Moran (1948) and Krishna Iyer (1949)] was then applied to the binary data. Results summarized in Dacey (1968, Table 1) were used for the test under the assumption of free sampling. Table 2 summarizes the results. We strongly reject the hypothesis of randomness and from the behavior of the three join count statistics may infer the presence of clustering as the cause of the rejection of the randomness hypothesis. [This is because the (1,1) and (0,0) joins are larger than expected and the (0,1) joins less than expected.]

We now turn to the question of modelling this pattern. Autocorrelation estimates were obtained using the $\{QB_i\}$ data and the results, summarized in Table 3, suggest that an isotropic model is probably acceptable (perhaps this fact should be partially discounted by the large number of zero counts on the map). Higher order auto-correlations were not computed. This is because although in many fields of spatial and temporal analysis correlation properties have an important part to play in model specification, in this particular field results (by which to judge empirical estimates) seem to be rather sparse. For a fuller discussion see Bartlett (1975, Section 2.2.1).

The next problem then, is the type of model to be fitted. Besag (1974) has identified consistent conditional probability models for describing lattice distributions with interaction. The model fitted here is the simplest one available for binary data--the

TABLE 2

Join Count Test Results for Map 255

Total number of houses located: 334

Total black (1) cells: 180 $p = 0.156$
Total white (0) cells: 972 $q = 0.844$

Join type	Observed	Expected (mean)	Expected (Standard Deviation)	Z-Score
BB	180	54.31	9.72	12.93
WW	1708	1589.93	40.90	2.88
BW	344	587.74	35.23	-6.91

TABLE 3

(a) Autocorrelations for Binary Form of Map 255

$$p_{1,1} = 0.321$$

$$p_{0,0} = 1.0 \qquad p_{1,0} = 0.422$$

$$p_{0,-1} = 0.421 \qquad p_{1,-1} = 0.326$$

(b) Autocorrelations for Original Count Data of Map 255

$$p_{1,1} = 0.211$$

$$p_{0,0} = 1.0 \qquad p_{1,0} = 0.358$$

$$p_{0,-1} = 0.349 \qquad p_{1,-1} = 0.218$$

first order isotropic autologistic. We now define that model.

Let x_{ij} denote the observed value in the (i,j)th cell. (Data are recorded on a rectangular grid.) This value is either zero or one. Then using the notation $P\{x_{ij}|\cdot\}$ to denote the conditional probability that $X_{ij} = x_{ij}$ where x_{ij} denotes the random variable associated with the (i,j)th cell, the first order isotropic autologistic defines:

$$P\{x_{ij} \mid \text{all other site values}\} = \frac{\exp\{(\alpha + \beta y_{ij})x_{ij}\}}{1 + \exp(\alpha + \beta y_{ij})}, \qquad (1)$$

where $y_{ij} = x_{i-1,j} + x_{i+1,j} + x_{i,j-1} + x_{i,j+1}$ and α and β are parameters. The β term is the interaction parameter and may assume positive values so that the model may describe clustered map patterns.

A fitting procedure described by Besag (1974) was followed. The lattice was first coded so that for each x_{ij} treated as a random variable, its four neighbors, as defined by y_{ij}, were treated as fixed variates. (A second coding can then be obtained by reversing roles and this allows a second analysis.) The model was then fitted by the simple expedient of maximizing the product of (1) over the coded sites with respect to α and β. A more sophisticated search algorithm employing first derivatives for example was not thought necessary at this stage. Results of fitting the model, that is parameter estimates and X^2 tables, are given in Table 4 for the two codings.

It is worth noting that $\hat{\beta}$, the interaction parameter estimate, is greater than zero; this is consistent with the join count test results. Chi-square goodness of fit values are 12.14 and 7.74 respectively after amalgamating the columns $y = 3$ and $y = 4$ because of small expected values. The fit is not very satisfactory. On the whole one is inclined to argue that the model does not provide a very good description of the count data in binary form. The map pattern does display clustering but this arrangement property is not well captured by the first order isotropic autologistic.

However the results should not be undersold. The description is far better than random. In Table 5, X^2 is used once again to compare observed counts with those that could be expected under the assumption of randomness. These two sets of counts are markedly different and hence the X^2 test implies rejecting the hypothesis of randomness. This X^2 test, suggested by Bartlett, has much to commend it as a simple test of randomness over the more powerful "custom made" join count tests. The latter are

TABLE 4

Fitting the Autologistic Model to Map 255

Coding 1 (506 observations)

$\hat{\alpha} = -3.37$ $\hat{\beta} = 1.52$

		y_{ij} 0	1	2	3	4	Total
x_{ij}	0	310 (304.5)	77 (82.0)	27 (31.9)	10 (6.9)	2 (0.7)	426
	1	5 (10.5)	18 (13.0)	28 (23.1)	20 (23.1)	9 (10.3)	80
Total		315	95	55	30	11	506

Chi-square = 12.14 Degrees of freedom = 3 Per cent level = 0.05

Coding 2 (506 observations)

$\hat{\alpha} = -3.08$ $\hat{\beta} = 1.43$

		y_{ij} 0	1	2	3	4	Total
x_{ij}	0	316 (310.8)	73 (78.8)	20 (23.2)	12 (8.7)	1 (0.5)	422
	1	9 (14.2)	21 (15.2)	22 (18.8)	26 (29.3)	6 (6.5)	84
Total		325	94	42	38	7	506

Chi-square = 7.73 Degrees of freedom = 3 Per cent level = 0.05

Figures in brackets are expected values under the hypothesis of an autologistic model.

TABLE 5

Chi-Square Randomness Test for Map 255

Coding 1 (506 observations)

		y_{ij} 0	1	2	3	4	Total
x_{ij}	0	310 (265.2)	77 (80.0)	27 (46.3)	10 (25.3)	2 (9.3)	426
	1	5 (49.8)	18 (15.0)	28 (8.7)	20 (4.7)	9 (1.7)	80
Total		315	95	55	30	11	506

Chi-square (combining y = 3 and y = 4) = 193.89   Degrees of freedom = 2

Coding 2 (506 observations)

		y_{ij} 0	1	2	3	4	Total
x_{ij}	0	316 (271.0)	73 (78.4)	20 (35.0)	12 (31.7)	1 (5.8)	422
	1	9 (54.0)	21 (15.6)	22 (7.0)	26 (6.3)	6 (1.2)	84
Total		325	94	42	38	7	506

Chi-square (combining y = 3 and y = 4) = 181.79   Degrees of freedom = 2

somewhat time consuming since they are not a direct part of the modelling process. The X^2 test on the other hand fits into the process of modelling as we have seen. There are drawbacks however. First the test is upset by small expected values, and thus can only be used on large data sets. Second the test assumes independence between columns in the table, but since coded sites share neighbors there is likely to be some non-independence.

In view of the obvious improvement over randomness analysis provided by the autologistic, it was extended to include a partial second-order scheme. Letting

$$\zeta_{ij} = x_{i-1,j-1} + x_{i-1,j+1} + x_{i+1,j-1} + x_{i+1,j+1}$$

the second order scheme fitted is given by

$$P\{x_{ij} \mid \text{all other site values}\} = \frac{\exp\{(a + \beta y_{ij} + \gamma\zeta_{ij})x_{ij}\}}{1 + \exp\{\alpha + \beta y_{ij} + \gamma\zeta_{ij}\}} . \quad (2)$$

There are now 4 codings, hence 4 separate analyses [Besag (1974)]. Parameter estimation is facilitated as before and in order to test for improvement in the goodness of fit (with respect to the first-order model) the first-order model is refitted using the 4 second-order codings. Results are summarized in Table 6.

Table 6 includes a likelihood ratio (L.R.) test (ψ^2). Let $f_p(x)$ denote the likelihood function after fitting $p = 2$ parameters. The $f_{p+q}(x)$ denotes the likelihood function after fitting $p+q$ $(q = 1)$ parameters [these functions are simply the maximized values of (1) and (2) respectively under the 4 separate codings]. Then

$$\psi^2 = (N - p - q) \frac{f_{p(x)}}{f_{p+q}(x)} ,$$

where N is the number of coded sites, and $N = 253$ for all four codings. ψ^2 is X^2 distributed with 1 degree of freedom and it is clear that the second-order model is not required (A X^2 test of the earlier type is inappropriate here because of the large number of entries in the table with small expected values).

Further attempts to improve the description of the arrangement properties of quadrat counts might proceed by disaggregating the columns of the X^2 table in terms of the particular arrangement of zero's and ones. Thus the $y = 3$ column could be disaggregated into the cases:

TABLE 6

Fitting a Second Order Autologistic to Map 255

N = Number of sites included in each of the 4 codings = 253

Coding	1st Order Parameter Estimates		2nd Order Parameter Estimates			Function Values f(x) 1st Order	Function Values f(x) 2nd Order	ψ^2
	$\hat{\alpha}$	$\hat{\beta}$	$\hat{\alpha}$	$\hat{\beta}$	$\hat{\gamma}$	$\hat{\alpha}$	$\hat{\beta}$	$\hat{\gamma}$
1	-3.40	1.54	-3.49	1.33	0.31	70.46	69.76	2.48 N.S.
2	-3.09	1.43	-3.20	1.22	0.30	74.68	73.93	2.52 N.S.
3	-3.07	1.44	-3.17	1.24	0.29	75.93	75.37	1.88 N.S.
4	-3.43	1.51	-3.39	1.38	0.18	69.23	68.97	0.92 N.S.

N.S. denotes not significant

1	1	0	1
1 x 1	0 x 1	1 x 1	1 x 0
0	1	1	1

Each should presumably appear with equal frequency, but it is already evident that the extra work is hardly justified in view of the earlier results.

One reason for the somewhat unsatisfactory fit of the autologistic may be that although the map pattern displays a certain amount of inter-cell clustering, there may also be inhomogeneity across the lattice, with this secondary component undermining model fits. It may be helpful therefore to partition the map and analyze each part separately. It could then be argued that if inhomogeneity is present this will be reflected in substantial differences in the interaction parameter β over the various partitions (and shifts from the β estimate previously obtained by analyzing the entire lattice). This inference is still very speculative however, since a homogeneous clustering process could still give the visual impression of inhomogeneity. We should not automatically conclude the presence of a second scale of pattern is consistent with it. There are no general statistical rules to follow here, but probably several different types of partitions should be analyzed.

Map 255 is made up of 24 rows by 48 columns of values. It is divided into three: columns 1 to 16 [map 255(1)], columns 17 to 32 [map 255(2)], columns 33 to 48 [map 255(3)]. Using join count tests all three display clear evidence of significant clustering. Table 7 gives first order autologistic parameter estimates for codings 1 and 2 on each of the three maps (N = 154 in all 3 maps). A X^2 test is given for 255(1) but not for the others because of the large number of entries in their X^2 tables with small expected values.

The results suggest possible heterogeneity over the map. The estimate of β on map 255 is inflated by the value of β on 255(2) and 255(3), sectors where there are fewer houses. It may be therefore that the β estimate is being forced to pick up two forms of clustering--inter-cell and inhomogeneity. Probably the estimate of β for 255(1) comes closest to a pure inter-cell measure of clustering.

Although this result is speculative, it perhaps does reflect a two fold spatial structuring of houses on the map. The first is a local intra-community distribution of houses, and the second is a regional distribution of the communities themselves.

TABLE 7

(a) Join Count Tests

255 (1) Total number of houses located: 232

Total black (1) cells = 129 $\hat{p} = 0.336$
Total white (0) cells = 255 $\hat{q} = 0.664$

Join type	Observed	Expected (mean)	Expected (Standard Deviation)	Z-Score
BB	143	82.18	13.30	4.57
WW	369	320.97	24.21	1.98
BW	216	324.84	16.70	-6.51

255 (2) Total number of houses located: 22

Total black (1) cells = 12 $\hat{p} = 0.031$
Total white (0) cells = 372 $\hat{q} = 0.969$

Join type	Observed	Expected (mean)	Expected (Standard Deviation)	Z-Score
BB	6	0.69	0.90	5.86
WW	692	683.56	12.58	0.67
BW	30	43.73	12.26	-1.11

255 (3) Total number of houses located: 80

Total black (1) cells = 39 $\hat{p} = 0.102$
Total white (0) cells = 345 $\hat{q} = 0.898$

Join type	Observed	Expected (mean)	Expected (Standard Deviation)	Z-Score
BB	31	7.57	3.38	6.93
WW	607	587.06	20.47	0.97
BW	90	133.36	18.68	-2.32

TABLE 7 (continued)

(b) Autologistic Model Fits

Map	Coding	Parameter $\hat{\alpha}$	Estimates $\hat{\beta}$
255 (1)	1	-2.19	0.97
255 (1)	2	-1.93	0.93
255 (2)	1	-4.38	2.85
255 (2)	2	-5.55	2.84
255 (3)	1	-4.52	2.46
255 (3)	2	-3.36	1.66

TABLE 7 (continued)

(c) Chi-square Tables for Autologistic Fit to Map 255 (1)

Coding 1

		y_{ij}					
		0	1	2	3	4	Total
x_{ij}	0	39 (36.9)	29 (31.6)	18 (18.5)	10 (9.4)	2 (1.5)	98
	1	2 (4.1)	12 (9.4)	15 (4.5)	19 (19.6)	8 (8.5)	56
Total		41	41	33	29	10	154

Chi-square Value = 2.407 Degrees of Freedom = 3

Coding 2

		y_{ij}					
		0	1	2	3	4	Total
x_{ij}	0	39 (38.4)	30 (30.0)	15 (16.5)	10 (8.9)	1 (1.0)	95
	1	5 (5.6)	11 (11.0)	17 (15.5)	20 (21.1)	6 (6.0)	59
Total		44	41	32	30	7	154

Chi-square Value = 0.603 Degrees of Freedom = 3

Phase 2. Modelling the original counts.

Conclusions reached in the previous analysis may to some extent be biased by the reduction of the data to binary form and for this reason it may be instructive to try and model the original count data. Besag (1974) has derived a number of "auto" models for discrete valued random variables including the autobinomial and autopoisson. Again, isotropic versions of these models would seem appropriate since autocorrelation values computed using the $\{Q_i\}$ data did not indicate the presence of directionality (Table 3).

In view of Dacey's results, a natural candidate for Map 255 would be the "auto" version of the negative binomial. Although this model is not explicitly discussed by Besag it may be derived using procedures given in the paper. Unfortunately, like the autopoisson, this "auto" model, to be referred to as autopascal, possesses an interaction parameter that is constrained to be less than zero. Thus it may only describe spatially inhibitory patterns and cannot describe spatial clustering. It is unfortunate that a model that is a natural candidate for describing intra-cell clustering is incompatible with a process that also possesses inter-cell clustering. It may be appropriate, therefore for describing the distribution of villages, where each is separated by a minimum distance, but not here where there is statistical evidence of clustering.

It was decided to fit the first order autobinomial. The reason for fitting this model is simply that it is one of the few discrete valued "auto" models that has an interaction parameter β, which may assume positive values and thus describe clustered distributions. There is no other justification, and indeed the binomial model itself provides a poor description of the aspatial frequency count. The expected number of cells with 0, 1 and 2 houses under a binomial hypothesis is 861.9, 250.1 and 36.1 respectively, which disagree considerably with the observed counts given in Table 1a and reflect a fit which is less satisfactory than the negative binomial. With these points in mind, define x_{ij} and y_{ij} as in the previous analysis, removing the restriction that the variate values be binary. Then

$$P\{x_{ij} \mid \text{all other site values}\} = {}_mC_{Nj} \left(\frac{\exp(\alpha+\beta y_{ij})}{1+\exp(\alpha+\beta y_{ij})}\right)^{x_{ij}} \times \left(\frac{1}{1+\exp(\alpha+\beta y_{ij})}\right)^{m-x_{ij}}, \qquad (3)$$

where m is the total number of objects located and β is the interaction parameter. Analysis involved coding the quadrats and then estimating α and β by maximizing (3) over each of the coded subsets. Table 8 summarizes the results and Table 9 gives the X^2 table for coding 1; because of the large number of small expected values no attempt is made to undertake a significance test. Generally, though, the fit is not impressive. The major source of discrepancy is the large number of $(x = 0, y = 0)$ counts and the small number of $(x = 0, y = 1)$ counts. This, again, is perhaps a reflection of the inhomogeneity of the lattice and the large numbers of zero cells. In view of earlier comments it may also be a reflection of the inadequacy of the binomial distribution in describing the aspatial frequency count. Indeed summing down the columns of the X^2 table in Table 9 as far as $y_{ij} = 7$, it is readily seen that observed and expected counts differ substantially.

To complete the analysis, Map 255 (1) was analyzed using the autobinomial. Results are summarized in Table 10 and further indicate the effects of pooling heterogeneity. The β estimate for 255(1) is less than the estimate for 255 and may be a better estimate of the strength of inter-cell interaction in the context of the autobinomial.

Analysis could be extended to consider a coarser scale of pattern, where we could look at the pattern amalgamated into blocks of four. The coarser the scale, the more we would expect interaction parameters to emphasize inter-community scale patterns. As already argued, interaction parameters might therefore start decreasing to zero, even becoming negative. Such an analysis is not undertaken here. The models discussed do not perform well on small data sets (particularly in view of the need to reduce the data further by adopting a coding procedure). Even with over 1000 quadrats it may be difficult to justify the use of these models to estimate interaction effects, with under 300 the results would be far too unreliable.

COMMENTS ON PROCESS MODELLING

The discussion in the previous section was only concerned with description. But it was also an aspect of this research field to identify the processes responsible for map patterns. Map patterns are the realizations of spatial-temporal processes, and the spatial models we fit may be deduced from spatial-temporal models. So a natural consistency check on the spatial models we have fitted that would enable us to assess their value is understanding the processes at work would be this: are the spatial-temporal generators for the spatial models consistent with what we know (or think we know) about the physical location process itself?

TABLE 8

Fitting the Autobinomial Count Data for Map 255

Coding	Parameter Estimates $\hat{\alpha}$	$\hat{\beta}$	Number of houses located (m)
1	-7.11	0.34	157
2	-7.20	0.36	148

TABLE 9

Chi-Square Table for Fitting Autobinomial to Coding 1

y_{ij} \ x_{ij}	0	1	2	3	4	5	6	Total
0	310 (277.2)	5 (35.3)	0 (2.2)	0 (0.1)	0 (0.0)	0 (0.0)	0 (0.0)	315
1	58 (57.6)	8 (10.4)	1 (0.9)	1 (0.0)	1 (0.0)	0 (0.0)	0 (0.0)	69
2	23 (27.9)	9 (7.1)	1 (0.9)	2 (0.0)	0 (0.0)	1 (0.0)	0 (0.0)	36
3	10 (14.6)	3 (5.3)	4 (0.9)	2 (0.1)	0 (0.0)	2 (0.0)	0 (0.0)	21
4	11 (12.6)	8 (6.4)	0 (1.6)	0 (0.3)	2 (0.0)	0 (0.0)	0 (0.0)	21
5	5 (5.8)	3 (4.2)	2 (1.5)	2 (0.4)	0 (0.0)	0 (0.0)	0 (0.0)	12
6	1 (1.8)	2 (1.8)	0 (0.9)	0 (0.3)	0 (0.0)	1 (0.0)	0 (0.0)	5
7	6 (2.8)	2 (4.1)	2 (3.0)	1 (1.4)	1 (0.5)	0 (0.1)	0 (0.0)	7
8	2	1	2	1	1	0	0	12
9	0	2	0	1	0	1	1	5
10	0	0	0	0	1	0	0	1
11	0	1	0	0	0	0	0	1
12	0	0	1	0	0	0	0	1
Total	426	43	15	11	5	5	1	506

TABLE 10

Fitting the Autobinomial to 255 (1)

Coding	Parameter Estimates $\hat{\alpha}$	$\hat{\beta}$	Number of houses located (m)
1	-5.84	0.22	105
2	-5.95	0.25	97

Fitting the Autobinomial to 255 (3)

Coding	Parameter Estimates $\hat{\alpha}$	$\hat{\beta}$	Number of houses located (m)
1	-6.07	0.47	43
2	-6.10	0.38	32

TABLE 10 (continued)

Chi-Square Table for Fitting Autobinomial to 255 (1) Coding 1

		x_{ij}						
		0	1	2	3	4	5	Total
	0	39 (30.2)	2 (9.2)	0 (1.4)	0 (0.1)	0 (0.0)	0 (0.0)	41
	1	23 (20.4)	6 (7.8)	0 (1.5)	0 (0.2)	1 (0.0)	0 (0.0)	30
	2	13 (13.6)	6 (6.5)	0 (1.6)	2 (0.2)	0 (0.0)	1 (0.0)	22
	3	7 (8.2)	2 (4.9)	3 (1.5)	2 (0.3)	0 (0.0)	1 (0.0)	15
	4	9 (8.0)	8 (6.1)	0 (2.3)	0 (0.5)	0 (0.0)	0 (0.0)	17
	5	5 (4.1)	3 (4.1)	2 (2.0)	1 (0.6)	0 (0.0)	0 (0.0)	11
y_{ij}	6	0	0	1	0	0	1	2
	7	0	1	1	1	0	0	3
	8	2	1	1	2	0	0	6
	9	0	2	0	1	0	1	4
	10	0	0	0	0	1	0	1
	11	0	1	0	0	0	0	1
	12	0	0	1	0	0	0	1
Total		98	32	9	9	2	4	154

We are, of course at a twofold advantage here. First, all our analyses are based on map properties deriving from an arbitrarily imposed grid. It is difficult to perform a full analysis of a location process defined in continuous space, but observed in this filtered form. Second process explanation requires detailed area analysis. In section 3 settlement clustering was recognized as arising possibly from two quite separate types of events--biotope variations and the location of offspring with respect to parent centers. It is unlikely that statistical arguments alone would ever fully distinguish between these two causes.

Notwithstanding these difficulties it is still useful to identify the types of spatial-temporal models that generate specific marginal spatial models. As we shall see even at this level we may be able to account for the relatively poor fits that were obtained in the last section. We now summarize results on the spatial-temporal processes that generate the autologistic and autobinomial models. These results are taken from Bartlett (1975, Ch. 2).

Consider the <u>linear</u> nearest neighbor spatial-temporal model defined by

$$\frac{d}{dt} X_{i,j;t} = -\lambda G_0(\Delta_i, \Delta_j)\, X_{i,j;t} + \frac{d}{dt} e_{i,j;t} \quad , \tag{4}$$

where the i,j subscripts reference locations on a rectangular spatial grid and the subscript t denotes time. Further $d\, e_{i,j;t}$ is uncorrelated with $X_{i,j;T}$ $(T \leq t)$, and $G_o\ (\Delta_i, \Delta_j)$ is a linear nearest neighbor spatial operator acting on $X_{i,j;t}$. Besag (1972) has shown that if X is binary there is <u>no</u> spatial nearest neighbor binary model which emerges as the spatial equilibrium form of the spatial-temporal model (4). In particular, therefore, the autologistic model cannot be justified from a process such as (4).

The autologistic and autobinomial models can, however, be derived from a <u>non-linear</u> nearest neighbor spatial-temporal model--in particular from non-linear Markov nearest neighbor birth-death processes in which probabilities of births or deaths as sites on the lattice are exponential functions of the values at nearest neighbor sites [Bartlett (1975, pp. 42-44)]. A slightly expanded derivation to that provided by Bartlett is given in Appendix I. It appears that both these auto models represent the marginal spatial distribution of a non-linear nearest neighbor spatial-temporal birth-death process at equilibrium.

Leaving aside the problem that these models are lattice models, the foregoing discussion suggests that these two auto models are only suitable for certain kinds of settlement maps. As is clear, these models describe equilibrium states of birth-death processes. We must assume therefore an initial distribution of houses and the process then evolves by additions and deletions to the housing stock.

Where do these observations leave us concerning the suitability of these auto models for describing settlement patterns? We refer back to Hudson's (1969) theory. After the colonization and spread phases are over, the theory suggests that some farmsteads will die while new ones will be built as a result of improving efficiency and amalgamation. Hence these models may be best suited to describing settlement patterns evolving through Hudson's third and final phase--a mature phase where settlement is of long standing and responding to a competitive economic environment [the β parameter ought then to be related to that birth/death process, though it remains to be shown in the birth-death process in the statistical model (Appendix I) can be related to any real settlement process].

For newly settled regions, or regions where the final competitive stage motivated by economic pressures has not taken place, these models would seem less suitable. In these regions models which express the terminal distribution of n houses distributed according to some location process would seem more appropriate; equilibrium distributions would only assume prominence in the last phase of the settlement history. Models for handling such terminal distributions, which also possess clustering properties, will be discussed in a later paper.

APPENDIX I

Spatial-Temporal Generators for the Auto Models in This Study

The following is simply an expansion of the argument in Bartlett (1975, p. 43) which may be helpful. Following Bartlett's notation:

> $\underline{x}$ is a vector of values associated with the n sites, and $\underline{y}_i$ is similarly a vector of values and $\underline{y}_i = \underline{x}$ except at site i where the value is y not x.

Now

$$P\{x_{i,t+dt} = y \mid \underline{x}_t\} = \lambda G_i(y,\underline{x}_t)dt \qquad (y \neq x_{i,t}) \quad .$$

So $G_i(y,\underline{x}_t)$ is an arbitrary transition function. Bartlett chooses the local solution of the equation that the limiting equilibrium distribution must satisfy, namely that

$$p(\underline{y}_i)G(x,\underline{y}_i) = p(\underline{x})G_i(y,\underline{x}) \quad .$$

Let

$$G_i(y,\underline{x}) = \begin{cases} (n-x_i)\gamma(\underline{x}) & y = x_i+1 \\ x_i\delta(\underline{x}) & y = x_i-1 \end{cases} ,$$

where

$$\gamma(\underline{x}) = \exp[\alpha + \sum_j \beta_j x_j] \quad ,$$

$$\delta(\underline{x}) = \exp[\alpha^1 + \sum_j \beta_j^1 x_j] \quad , \text{ and}$$

where summation is on the neighbors of site i.

Now

$$\frac{p(\underline{x})}{p(\underline{y}_i)} = \frac{p(x_i \mid \text{all other site values})}{p(y_i \mid \text{all other site values})} \equiv \frac{p(x_i \mid \cdot)}{p(y_i \mid \cdot)} \quad .$$

So

$$\frac{p(x_i \mid \cdot)}{p(\underline{y}_i \mid \cdot)} = \frac{G_i(x,\underline{y}_i)}{G_i(y,\underline{x}_i)}$$

$$= \frac{[n - (x_i - 1)]\gamma(\underline{y}_i)}{x_i \delta(\underline{x})}, \quad \text{for } y = x_i - 1.$$

But

$$\frac{p(x_i|\cdot)}{p(0|\cdot)} = \frac{p(x_i|\cdot)}{p(x_i-1|\cdot)} \cdot \frac{p(x_i-1|\cdot)}{p(x_i-2|\cdot)} \cdot \cdot \cdot \cdot \cdot \frac{p(1|\cdot)}{p(0|\cdot)}$$

$$= \frac{[n-(x_i-1)]}{x_i} \frac{\gamma(\underline{y}_i)}{\delta(\underline{x})} \quad \frac{[n-(x_i-2)]}{(x_i-1)} \frac{\gamma(\underline{y}_i)}{\delta(\underline{x}_i)} \cdots n \frac{\gamma(\underline{y}_i)}{\delta(\underline{x})}$$

$$= \frac{n!}{x_1!(n-x_i)!} \left(\frac{\gamma(\underline{y}_i)}{\delta(\underline{x})} \right)^{x_i}$$

$$= {}_nC_{x_i} [\beta(\underline{x})]^{x_i} ,$$

where $\beta(\underline{x}) = \exp [(\alpha-\alpha^1) + \sum_j (\beta_j-\beta_j^1)x_j]$ and $(\beta_j-\beta_j^1)$ has lateral symmetry in relation to x_i and x_j. Now this satisfies the autobinomial with parameters n and θ where $\theta = \{\beta(\underline{x})/[1+\beta(\underline{x})]\}$, since if X_i is autobinomial

$$p(x_i|\cdot) = {}_nC_{x_i} \theta^{x_i} (1 - \theta)^{n-x_i} ,$$

$$p(0|\cdot) = (1 - \theta)^n ,$$

and

$$\frac{p(x_i|\cdot)}{p(0|\cdot)} = {}_nC_{x_i} \left(\frac{\theta}{1 - \theta}\right)^{x_i} .$$

The case for $y = x_i + 1$ follows identical lines of reasoning. When $n = 1$ this model reduces to the autologistic.

REFERENCES

Bartlett, M., 1974. The Statistical Analysis of Spatial Pattern. Advances in Applied Probability, 6: 336-358.

Bartlett, M., 1975. The Statistical Analysis of Spatial Pattern. London: Methuen.

Besag, J., 1974. Spatial Interaction and the Statistical Analysis of Lattice Systems, Journal, Royal Statistical Society, B, 36: 192-236.

Bylund, E., 1960. Theoretical Considerations Regarding the Distribution of Settlement in Inner North Sweden, Geografiska Annaler, 42: 225-31.

Dacey, M., 1968. A Review of Measures of Contiguity for Two and k Color Maps, in Spatial Analysis: A Reader in Statistical Geography, ed. B.J.L. Berry and D.F. Marble, Englewood Cliffs: Prentice Hall, 479-495.

Dacey, M., 1969. An Empirical Study of the Areal Distribution of Houses in Puerto Rico, Transactions, Institute of British Geographers, 45: 51-69.

Feller, W., 1968. An Introduction to Probability Theory and Its Applications, New York: Wiley, Vol. 1.

Getis, A. and Boots, B., 1978. Models of Spatial Processes, Cambridge: University Press.

Hudson, J., 1969. A Location Theory for Rural Settlement, Annals, Association American Geographers, 59: 365-381.

Kershaw, K., 1962. Pattern in Vegetation and its Causality. Ecology, 44: 377-388.

Krishna Iyer, P., 1949. The First and Second Moments of Some Probability Distributions Arising from Points on a Lattice and their Application. Biometrika, 36: 135-141.

Moran, P., 1948. The Interpretation of Statistical Maps. Journal, Royal Statistical Society, B, 10: 243-251.

Ripley, B., 1977. Modelling Spatial Patterns, Journal, Royal Statistical Society, B, 39: 172-212.

Ripley, B., 1978. The Analysis of Geographical Maps, in Exploratory and Explanatory Statistical Analysis of Spatial Data, ed. C. Bartels and R. Ketellapper, Boston: Martinus Nijhoff, 53-72.

Skellam, J., 1952. Studies in Statistical Ecology I. Spatial Pattern. Biometrika, 39: 346-362.

Streitberg, B., 1978. Multivariate Models of Dependent Data: Applications in Spatial Analysis, in Exploratory & Explanatory Statistical Analysis of Spatial Data, eds. C. Bartels and R. Kettellapper, Boston: Martinus Nijhoff, 139-177.

MAXIMUM ENTROPY SPECTRAL AND CROSS SPECTRAL ANALYSIS OF ECONOMIC TIME SERIES IN A MULTIREGIONAL CONTEXT

Claude Marchand

Départment de Géographie
Université de Montréal

INTRODUCTION

There are fundamental problems in applying economic theory to study the short-term behavior of urban systems through time. Traditional formulations provide only partial approaches, focusing upon the special case of change in the metropolis [Alonso (1971, 1977); Thompson (1965, 1975)] or upon change in the resource supplying areas [North (1955, 1961); Watkins (1963)]. Relationships between urban areas are thus never fully defined and the urban system is reduced to one of its typical components. In addition, such formulations fail to specify the time scale of change, so that one author's definition of the short term often corresponds to another author's definition of the long term. As a result, many of the contradictions among theories of regional change--contradictions concerning the identification of generating processes (export services versus export industries)--pertain to the time scale of analysis.

In contrast to the single region approach adopted for theoretical purposes, modelling efforts have been successful in recognizing the behavior of urban or regional systems in their entirety. The differing economic performances of metropolis and periphery alike have been traced through a simultaneous regression analysis of their employment or unemployment levels. Specific economic models have been developed to evaluate the contribution of aggregate demand fluctuations to disparities in short term performances [Harvey (1956); Marchand (1981); Thirlwall (1966, 1975); Van Duijn (1975)]. More complete theoretical attempts incorporate the effects of both longer and shorter term variables such as supply, local demand and climate

[Brechling (1976); Jeffrey (1974); Jeffrey & Webb (1972); King, et al. (1974, 1978)]. All models however have perpetuated a basic ambiguity about the time scale of analysis as they can only specify the spatial dimension, national or local, or the explanatory variables considered. Indeed the method cannot analyze directly the time variations present in the employment series of each region, but estimates them indirectly through independent variables.

The usefulness of the models developed therefore relies upon a priori knowledge of both the causal mechanisms and the actual composition of the time series used as best estimator of regional employment variations. While differing theories of regional change provide significant guidance in the specification of empirical models, in particular with respect to the number and choice of causal variables, they cannot determine the exact nature of long swings and short term fluctuations. The latter knowledge, derived from empirical observation, has not been incorporated in theoretical formulations as the usual method of observation, namely visual inspection, is not based upon rigorous foundations, nor does it produce consistent results. The main thrust of such work has been carried by economists seeking to establish the turning points of short term fluctuations affecting various sectors of the economy. The nature of such fluctuations, let alone the timing of their successive phases, has however remained a matter of contention. Following Schumpeter (1935), many researchers admit to the occurrence of both a major and a minor business cycle where the major or Juglar cycle is of longer duration (8-12 years) and greater amplitude than the minor cycle (approximately 3 years). In contrast, Burns and Mitchell (1947) recognize the existence of a minor cycle only.

Because of the inconsistencies present in both theoretical and statistical work on the time scale of change and because the regression method itself can only be used effectively if the nature of short term change is known, and its regularity sufficient to be embodied in a general explanatory variable, regional econometric models are silent on the time dimension of the phenomenon examined. The time scale of aggregate demand fluctuations and the importance of the phenomenon as a source of variation remains to be determined, the more so as the variations have been discussed within the context of analyses of time series of various length without recognizing that this cycle has a definite time scale that may or may not be encompassed by the length of the time series. Moreover for analyses based on long records, the independent variable chosen provides an efficient estimator of aggregate demand fluctuations in the dependent series only if other short term variations are originally absent or have been properly filtered out.

Finally, in those models which assume the existence of demand fluctuations operating at two levels, the national and the regional, the nature of the later cycle has also remained ambiguous. This ambiguity is best expressed by the inability to specify the independent variable which could best estimate the local cycle. The effects of variations in the export base are thus assumed to generate residual fluctuations, whereas the interpretation of such residuals never makes it clear whether local cycles operate on a different time scale of shorter or longer duration than the national impulse. In addition, the possibility that more than one local cycle may influence the behavior of a region is not discussed, nor is the possibility that local cycles may operate on a time scale which varies from area to area.

These modelling efforts, although partially successful in their specification of the spatial dimension of change, are not based upon direct observations of time variations which could lead to a greater consistency in the formulation of theory. In contrast, the specific problem of determining the temporal and spatial scales of short term change in a multi-regional context is addressed here. Time series data are analyzed using spectral and cross-spectral methods. The spectral method is emphasized as it best resolves the problem of evaluating the nature, number and amplitude of recurring employment variations affecting urban economies. The properties of the method are presented and the estimation problems related to the traditional computation procedure are discussed. Application of this first estimation procedure to the employment series illustrates such problems as the level of resolution is unsatisfactory for the analysis of fluctuations whose period exceeds one year.

A second estimation technique which involves calculating the spectrum which is consistent with the maximum extropy in the data series is then introduced and its objectives and features outlined. This procedure is designed to provide better resolution to low frequencies in short records of data, and its application to the employment series validates the assumptions of the econometric models incorporating the effects of demand conditions at two spatial levels. The respective importance of two types of short term fluctuations is measured: a Juglar cycle is found to dominate the employment variations of all cities in the system, whereas a series of minor industrial cycles appears in various combinations in certain urban areas. These results suggest that the level of employment in an urban area is directly and uniquely tied to the level of its exports on a time scale which does not correspond to that of changes affecting the entire system.

These results are confirmed by cross-spectral analysis which allows us to trace similarities in response to change in aggregate demand conditions as measured by the Juglar cycle. Those economies most clearly integrated with the Montreal industrial complex do not cohere with the metropolis at the Juglar cycle. The similarities in response to the national cycle do not reflect the presence of a causal process whereby the metropolis diffuses fluctuations through export-import ties. Rather, similarities in behavior are determined by similarities in employment structure. The diversification of the local ecnomic base promotes the occurrence of a Juglar cycle whose regularity approaches that of Montreal.

TRADITIONAL AND MEM SPECTRAL ANALYSES

Since the factor base of urban economies is assumed to be expandable (or retractable) only over long intervals of time, short term changes in employment are attributed to variations in demand and in climate. The demand variations for the products and services of an urban area are assumed to be affected by the level of economic activity in the overall system as well as by local conditions. Annual change on the other hand is linked to climatic variables whose effects are separate from those attributable to supply and demand.

These different components of change may be summarized in an additive (or multiplicative) model:

$$X_j(t) = S_j(t) + A_j(t) + Q_j(t) + R_j(t) \quad , \qquad (1)$$

where X_j denotes the economic variable for the region, $S_j(t)$ its long term component, and $A_j(t)$, $Q_j(t)$ and $R_j(t)$ its short term components. When a regression method is employed, aggregate demand effects $A_j(t)$ are specified by a common variable, the national series, whose variations are assumed to be the best estimator of the national cycle for all regions (j). This is obviously only true when the regional performance corresponds closely to the national performance. Seasonal change, $Q_j(t)$, on the other hand, is evaluated by a best fit one year sinusoid specific to each region, whereas local demand effects are left unspecified.

Econometric models thus provide efficient estimations of seasonal variations (and trends) only. Indeed the regression method is not suited to the analysis of fluctuations which are not perfectly periodic. Only in the case of seasonal variations is change recurring in a sufficiently regular and uniform fashion that least square fitting of a best sinusoid remains a manageable and efficient estimation procedure. The major business cycle, however, has a period that may vary from 8 to 12 years. Similarly, minor cycles may extend over a period of 2 to 4 years. If visual inspection could provide the correct information, proper specification of these components for each region would destroy the very generality which regression analysis attempts to achieve.

Moreover, not only are short term components other than annual change only quasi-periodic in nature, but there is a sufficient complexity of them that the summary provided by equation (1) is adequate only for significant short term change. Under such conditions regression analysis proves to be both a cumbersome and an unsatisfactory method of decomposition, since instead of assessing the contribution of each component to total variations in the series, it averages over all oscillations.

The severe estimation problems encountered when regression analysis is employed to analyze cyclical variables leads us to emphasize the spectral and cross-spectral methods. The regional system analyzed here consists of the 23 largest urban areas in Quebec. For each of these regions, Statistics Canada published a monthly employment index. The time period covered by the data extends from January 1957 to December 1975, totalling 228 consecutive observations. The 19 year time span includes the economic recession of 1956-60, the 1961-65 expansion, the subsequent 1966-70 recession and the later recovery.

Spectral analysis is designed to provide an understanding of such recurring change. Its efficiency as an estimation procedure reflects the fact that it adapts itself to the characteristics of cyclical data, ascertaining the existence of various periodicities and assessing their importance as a source of variation. Moreover, it directly evaluates the structure of time series in a manner that presupposes minimal a priori knowledge of the phenomenon under investigation. It thus obviates one of the main difficulties linked to the use of the regression method, whose more ambitious objective explanation of the phenomenon subjects the analysis to a greater number of theoretical and statistical constraints.

The assumption of stationarity constitutes the major statistical constraint required for its application since the spectral method is suited to the analysis of quasi-periodic variables only. In effect, time series are decomposed into a sum of oscillations of frequency f whose respective amplitudes are independent:

$$X(t) = \sum_{i=1}^{T/2} (a_i \cos f_i t + b_i \sin f_i t) , \qquad (2)$$

where $a_i = A \cos \emptyset$ and

$b_i = A \sin \emptyset$

are spectral amplitudes, whereas f_i, $\emptyset_i$ and t are the angular frequency, the phase, and the argument of the oscillation, respectively. Also

$$a \cos ft + b \sin ft = A \cos (ft + \emptyset) , \quad (3)$$

indicating that any oscillation whose phase is arbitrary at $t = 0$ cannot be expressed by one periodic function only, but rather requires a mixture of sine and cosine components. The parameters of the Fourier series which is to represent X(t), namely a_i and b_i $(i = 1,\ldots,T/2)$, are normally distributed with mean μ and common variance σ^2; moreover a_i and b_i are orthogonal so that the variables are independently distributed.

A time series x_j can only be represented by a Fourier series, as in equation (3), if the process that generated it, X(t), is stationary with respect to the first and second moments. Similarly, because the method analyzes these two moments exclusively, its use is most appropriate with Gaussian processes, since the information about such processes is entirely contained in these moments. Because economic processes are not normal, and the construction of tests and confidence bands is based on this assumption, the method has not been employed for inferential purposes. The descriptive use of the method, however, does not require the assumption of normality and the spectra have been tested for stability. The behavior of a stochastic normal process x_j, sampled at discrete time intervals Δ, is thus entirely determined by its mean μ and its autocovariance function

$$A(\tau) = E[X(t + \tau)X(t)] - \mu^2 \quad , \quad \tau = s - t . \quad (4)$$

Since the autocovariance function is a weighted sum of the same components, neighboring values are highly correlated and the function A(k) conveys a view of the property of the process which is as distorted as that of the original sample data series itself. In case (a) of Figure 1, the correlation function is difficult to interpret beyond the obvious information already discernible in the original data series (i.e., the presence of a seasonal components). Case (b) provides an even more striking example, since two periodicities consisting of one year and six month cycles are present that can be observed in the data series x_j and the autocovariance function A(k). Only the spectrum, however, reveals the exact contribution of each component--a contribution equal in magnitude at the two periods even though the autocovariance function displays a more important peak at a period of 12 months.

The representation of the autocovariance function in the frequency domain is the power specturm. This representation partially overcomes the problem of correlated values since, following the properties of the Fourier series, the estimates of the variance are almost orthogonal between neighboring frequency bands. This representation for a discrete non-deterministic process is given in the following equation for the spectrum, employing Euler's formula:

$$S(f) = \Delta \sum_{k=-\infty}^{\infty} A(k) \exp(-i2\pi kf\Delta) \ , \ \frac{-1}{2\Delta} \leq f \leq \frac{1}{2} \Delta \ . \tag{5}$$

In the traditional procedure used to calculate the spectrum, the autocovariance function A(k) is first estimated from the sample data series x_j , with the results subsequently being Fourier transformed according to (5). Both examples in Figure 1 display these fundamental steps. The power spectrum illustrates the contribution of distinct components to total change in the data series, in successive frequency bands, from lowest frequency ($f = 0$) to the Nyquist frequency ($f = 1/2\Delta$). If a contribution is large then a sharp peak will be associated with the corresponding frequency band. Thus the sample spectrum for Shawinigan shows important sources of variation corresponding to cycles of 96, 12 and 6 months duration, respectively.

Given the length of the data record, it should also be possible to determine the existence and importance of fluctuations in the neighborhood of a 2-3 year period. However, although the traditional spectral method resolves the Juglar cycle for one third of the sample when 73 lags are employed (as in the case of 48 lags for Shawinigan), it fails to resolve the adjacent spectral lines corresponding to a series of minor cycles.

Indeed, a major estimation problem arises from the fact that the spectrum S(f) must be computed from a data series x_j of finite length n. The fact that a phenomenon X(t) is observed over a finite interval (i.e., through a rectangular data window) generates an important type of bias in the estimation of the spectrum. The operation of sampling a phenomenon during a finite interval is equivalent to a unit weighting of the computed autocovariance such that the estimated autocovariance function is actually A'(k), where

$$A'(k) = A(k) \cdot w(k) \quad ,$$

and where

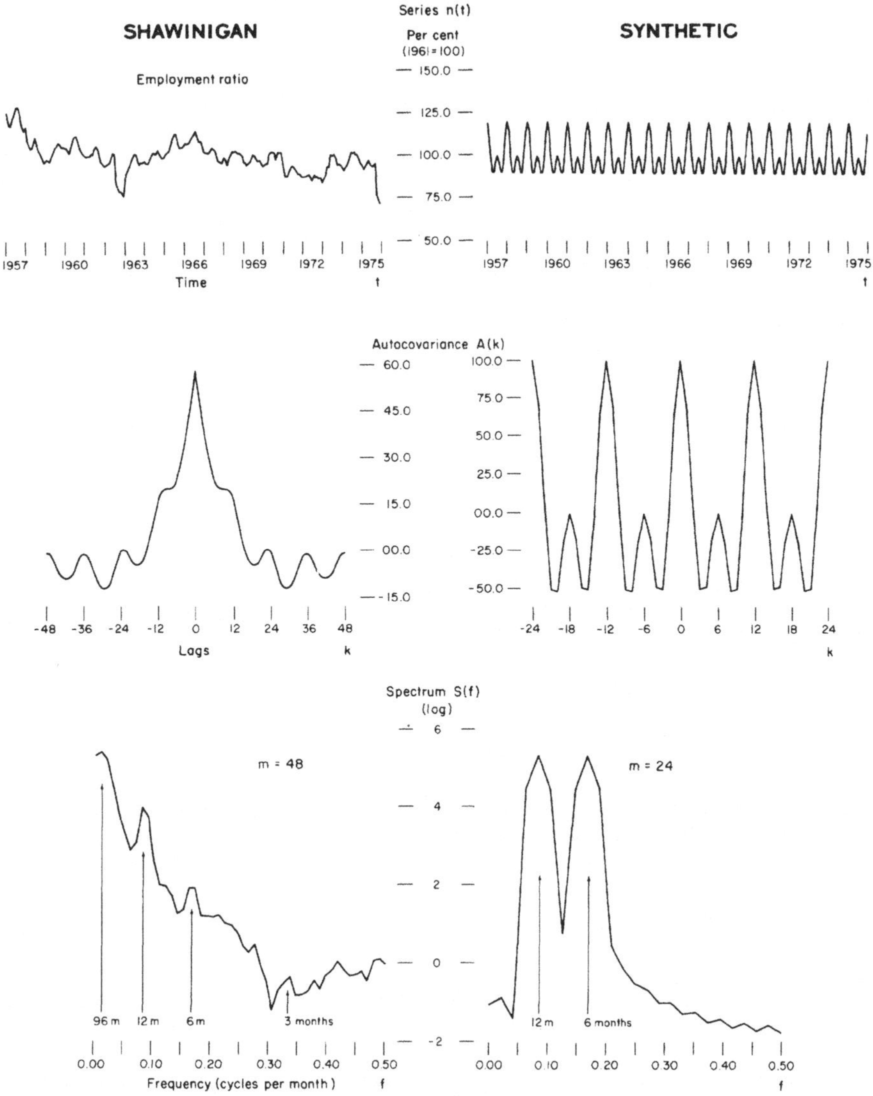

Fig. 1. Sample spectra estimated from the autocovariance function
(a) Per capita employment of Shawinigan
(b) Sum of two sinusoids of periodicities 12 and 6 months respectively

$$W(k) = \begin{cases} w(-k), & \\ 1, & k = 0,\pm1,\pm2,\ldots,\pm(n-1) \\ 0, & \text{otherwise,} \end{cases}$$

with transform $W(f) = \Delta \sum_{t=(n-1)}^{n-1} w(k) \exp(-i2\pi fk\Delta), \; -\frac{1}{2\Delta} \leq f \leq \frac{1}{2\Delta}$. (6)

The finite interval transform S'(f) of A'(k) becomes

$$S'(f) = \Delta \sum_{k=-(n-1)}^{n-1} w(k)A(k) \exp(-i2\pi fk\Delta), \; -\frac{1}{2\Delta} \leq f \leq \frac{1}{2\Delta} \;. \tag{7}$$

Substituting for A(k) in terms of its transform,

$$S'(f) = \Delta \sum_{k=-(n-1)}^{n-1} \left| w(k) \exp(-i2\pi fk\Delta) \int_{-\frac{1}{2}\Delta}^{\frac{1}{2}\Delta} S(g) \exp(i2\pi fk\Delta) \right| dg, \tag{8}$$

and rearranging yields

$$= \int_{-\frac{1}{2}\Delta}^{\frac{1}{2}\Delta} S(g) \left| \sum_{k=-(n-1)}^{n-1} w(k) \exp[i(2\pi f-g)k\Delta] \right| dg \tag{9}$$

$$= \int_{-\frac{1}{2}\Delta}^{\frac{1}{2}\Delta} S(g)\, W(f-g)\, dg \;, \tag{10}$$

that is, the estimated spectrum S'(f) results from the Fourier transformation of the autocovariance A(k) convolved with the Fourier transform of the data window w(k). The latter produces a transform or spectral window W(f), where

$$W(f) = \frac{\sin\pi fn}{\pi f} \;,$$

that is concentrated about f = 0 and dampens in numerous side lobes away from f = 0.

For data series of short duration n, the estimated spectrum S'(f) gives a very poor approximation of S(f), since the window S(f - g) is wide and values of S(g) far removed from g = f are leaked to other parts of the spectrum, resulting in a distortion of the estimates. As n increases, this distortion is somewhat reduced and the smoothed spectrum S'(f) converges to the true spectrum S(g). This convergence is slow, however, and convolution of the spectrum with this particular window does not guarantee good resolution.

Moreover, the estimation problem remains in its entirety whenever the amount of data available is limited. Traditionally, the problem is resolved by first controlling for window shape and subsequently experimenting with various truncation points m from which to estimate the autocovariance function.

Both the subjects of window shape and window closure are summarized in Jenkins and Watts (1968) and Fishman (1968), and Marchand (1981) provides a discussion of comparative performances as well as examples. In all cases window functions are devised independently of the data series x_j or of the properties of the process X(t) that is being analyzed. As a result, such windowing reduces the problem of bias at the expense of frequency resolution. In the present case, the spectrum was smoothed through application of a Hamming window. Although different truncation points corresponding to 24, 48 and 72 lags, respectively, were used, the latter procedure provides better resolution at high frequency only.

The need for long records and extremely regular fluctuations thus puts major limitations on the use of spectral analysis. The problem and its recognition, however, have by no means been confined to the fields of economics and geography. In effect, impetus for research into an alternative approach came within the field of geophysics with the first attempts at nuclear arms control. High resolution spectra of seismic signals became imperative if nuclear explosions were to be distinguished from natural earthquakes, and by the end of the 1960's data adaptive spectral methods had been developed. The use and reliability of two such methods, the maximum entropy method devised by Burg (1967) and the maximum likelihood estimator have been reviewed by Lacoss (1971) and Kanasewich (1973).

The maximum entropy method (MEM) constitutes a successful attempt at computing a high resolution spectrum through minimal modification of the data series x_j, and thus greater use of its information content. The assumption of a Gaussian, zero mean, process is the only constraint retained. Stationarity is not enforced on the data and the spectrum is estimated through a window which is unique to the data and as such does not modify it. The mean is the only information which is eliminated from the data, and the absence of smoothing guarantees that no extraneous information is added.

The objective of finding the spectrum which maximizes the information content of the data series x_j, as expressed by an entropy function H(x), is achieved with the design of a length n + 1 prediction error filter f_j+1. This filter, when convolved with the input series x_j, produces a white noise series ε_j. The optimum filter is the one which generates the greatest destruction of entropy in ε_j since it is the one best suited to the characteristics of the input series x_j. Filters of different lengths are calculated and, at each recursion step, the change in the entropy of the output series ε_j is computed. The minimal value in ΔH is obtained when the filter is of optimum length. The function ΔH is displayed in Figure 2 for the employment ratio series of Drummonville and Saint-Jean. The change in entropy (or its slowly varying envelope) reaches a minimum at the 50th step; accordingly, a 50-point prediction error operator is used to calculate the spectra.

The spectral computation consists of correcting the power density of the white noise spectra M_{n+1} (transform of ε_j) for the response of the filter. It can be shown that the output power spectrum M_{n+1} results from the Fourier transformation of the original series x_j multiplied by the power response of the filter F_{j+1}. The steps following the search for an optimum filter [i.e., the computation of the input power spectrum M(f) associated with it] thus involve a reverse operation, the division of the output M_{n+1} by the transform of the filter:

$$M(f) = \frac{M_{n+1}/f_n}{\left|1 + \sum_{j=1}^{n} F_{j+1} \exp(-i2\pi fn\Delta)\right|^2} \qquad (11)$$

where f_n is the Nyquist frequency.

Figure 2 displays the maximum entropy spectra M(f) for the employment ratio of two sample cities. This method resolves equally well the ten-year and one-year periodicities as well s minor cycles. The Drummondville employment series contains a three-year cycle and the Saint-Jean a weaker two-year fluctuation. Further examples are provided in Figure 3, where the performance of the Hamming estimator and of the MEM are displayed for two values of maximum lag and filter length F. The highest resolution (m = 49) afforded by the Hamming window gives approximately the same information as that provided by the application of MEM with a suboptimal prediction error filter. By far the most important contributions of this method of estimation, however, are the detection of cycles at low frequencies, and the complete resolution of adjacent spectra lines such as those corresponding to the minor cycles.

The application of the maximum entropy method to the employment series demonstrates that most urban areas in the sample region experience a fluctuation whose period is approximately 10 years. Shorter cycles of various periodicities also influence certain cities to a lesser degree. These results resolve the debate on the presence of cyclical fluctuations whose existence

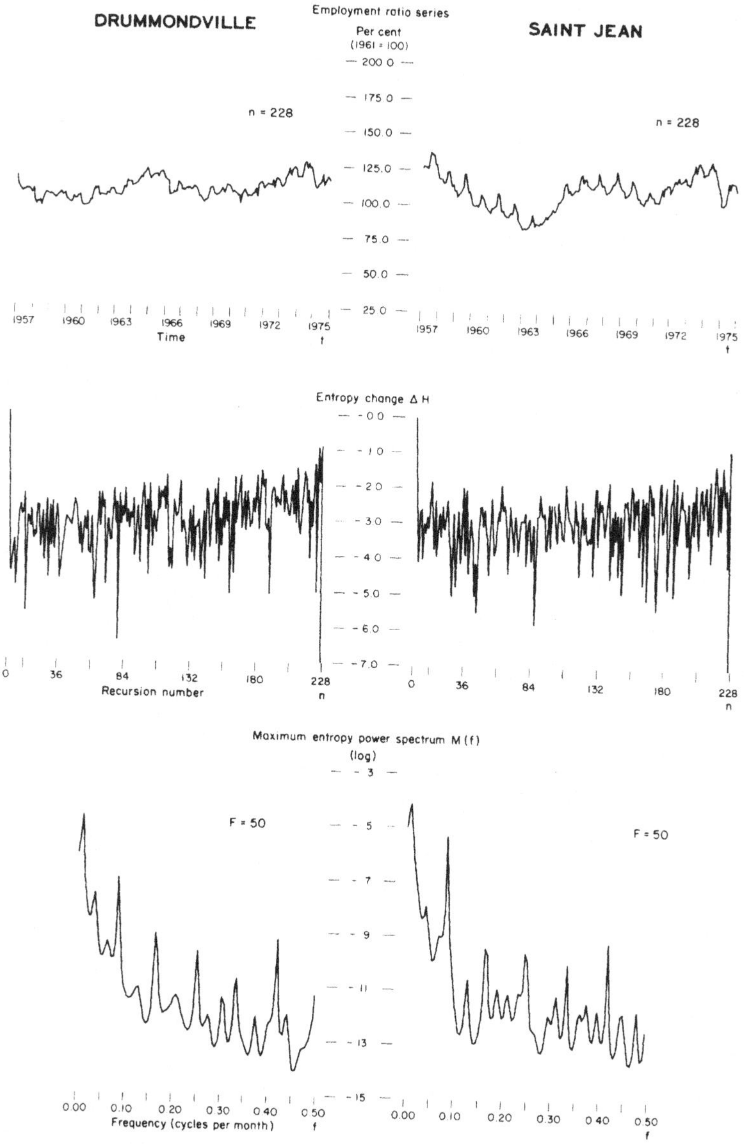

Fig. 2 Sample series of per capita employment, entropy change of the whitened series ε_j and spectra estimated with a whitening filter of length f = 50.

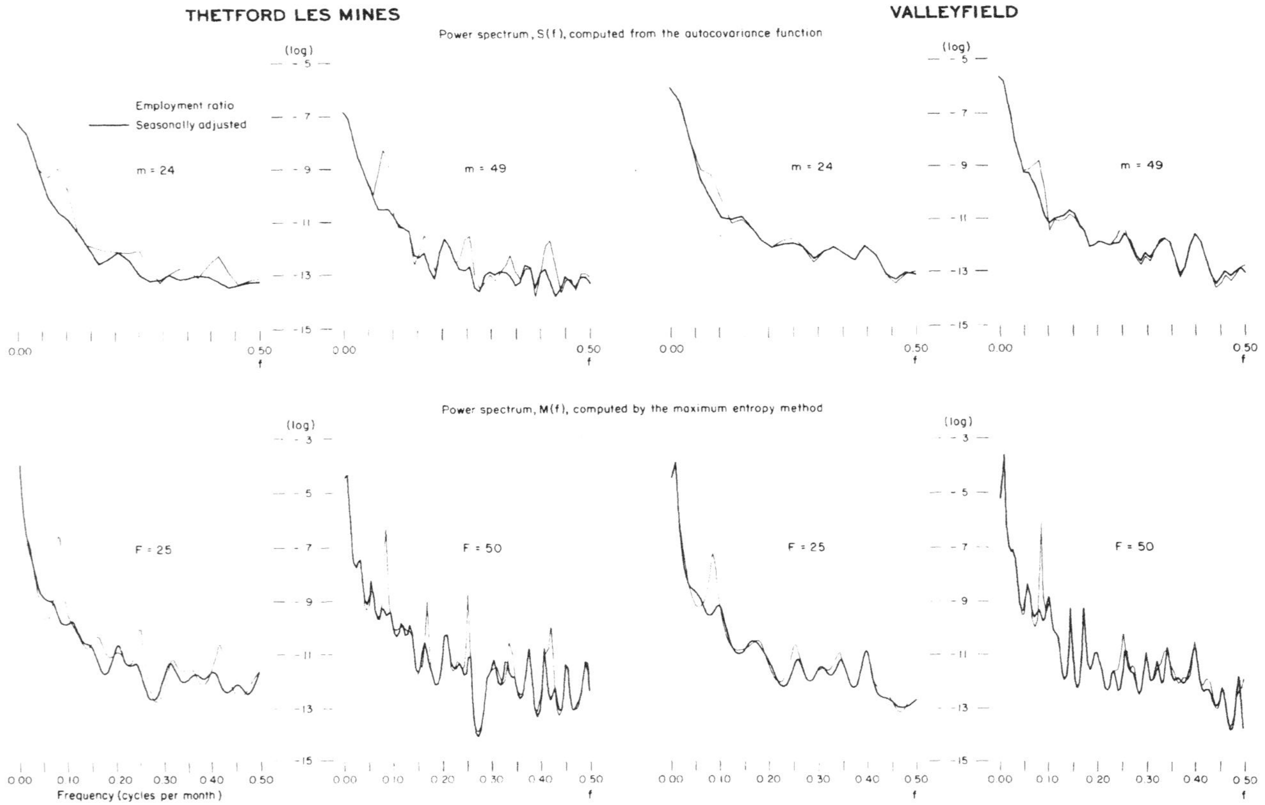

Fig. 3. Estimated spectra for per capita employment using (a) a Hamming window and (b) the maximum entropy method.

could be determined by visual inspection without confirmation by spectral analysis. The typical smooth shape of the spectrum at low frequency as described by Granger (1966) and reproduced by Haggett (1975), Hepple (1975), and Bartels, et al. (1977,1978) reflects the poor resolution afforded by conventional methods rather than an intrinsic property of the variable under study.

Moreover the application of the extremal approach to the analysis of regional systems verifies the theoretical assumptions of the model outlined in equation (6) and adapted from Brechling (1976). This is discussed at length in Marchand (1981). Suffice here to note that the ability to distinguish between cycles of different periodicity and different geographic incidence confirms the presence of two generating mechanisms of short term economic change corresponding to the $A_j(t)$ and $R_j(t)$ components. Employment change at the urban or regional level is affected not only by a systemic force but also by a second type of impulse whose amplitude in time and space is more limited as it is determined by local conditions. The specificity of such local cycles is well illustrated in Marchand (1981) as more than one minor cycle may affect an urban economy and the periodicity may vary from city to city.

Determination of individual importance of the major and minor cycles across the urban system also leads to a better understanding of the respective natures of aggregate and local demand conditions. Changes in aggregate demand conditions, as measured by the relative importance of the Juglar cycle, impinge only mildly on the employment levels of large cities. Seasonal fluctuations in the construction industry, and in some instances minor industrial cycles, determine the employment variations of such urban areas whose medium term behavior is well regulated by central controls. In contrast, the impact of national business fluctuations are strongest in small cities and particularly in those cities tied to the domestic market. Where foreign markets represent the major source of demand and/or investment, a commercial and financial zone is created that functions independently of domestic demand conditions, and leads to a considerable decrease in the influence of national business fluctuations. In effect, urban areas specializing in the production of staples have a quarter to a third of their employment variations accounted for by minor cycles.

Thus the aspect of urban performance which may be readily explained in terms of the export base is the local cycle. Spectral results refute a traditional approach whereby the export industries are singled out as the chief source of instability in a regional system and all other components of aggregate demand regarded as dependent rather than independent variables. Such a truncated view of the urban system presents serious limitations.

From a theoretical point of view it assumes away the network of information flows and controls that may be far more significant than import-export ties in diffusing destabilizing expectations across the entire system. In particular, it would be difficult to argue that changes in such diverse and patently unrelated industries as textiles and gold mining, for example, could constitute the single variable determining the simultaneous appearance of a Juglar cycle across the Quebec urban system. Rather, changes in the inventory levels of various industries create cyclical effects that are separate and additional to those effects generated by changes in aggregate demand conditions as determined by government policies and consumer expectations.

The determination of demand effects in both time and space is thus far more complex than traditionally posited, and the problem will be further examined in the following section where the performance of urban areas will be directly related to that of the metropolis.

CROSS-SPECTRAL ANALYSIS

Spectral analysis has enabled us to evaluate the amplitude, number and nature of employment fluctuations present in individual cities. Cross-spectral analysis of each city paired with the metropolis chosen as the norm, on the other hand, allows us to trace similarities in response to specific fluctuations. In particular, application of the cross-spectral method to economic time series generates two functions: the coherence spectrum shows the correlation between time series for cycles of various duration while the phase spectrum measures the leads or lags associated with such cycles. The frequency domain representation of cross covariances, like that of autocovariances, is thus more complete and precise than its time domain equivalent since it can distinguish between similarities of performance with respect to the industrial and Juglar cycles.

The cross-spectral approach is a direct generalization of the univariate case and will be discussed within the context of the cross covariance transform. Although a maximum entropy approach to the estimation of the cross-spectrum would improve resolution, such an approach has not yet been fully developed. However, the resolution of cross-spectra does not present the same problems as that of the spectra themselves, and the coherence function usually contains specific peaks at low frequencies. Because traditional procedures for spectral estimation provide a low level of resolution at low frequencies, Granger (1964) emphasizes that the coherence spectrum can be used as indirect evidence of the existence of business cycles.

This particular use of cross-spectral analysis has not been adopted in geographical research until Cho and McDougall (1978) interpreted the amplitude of the cross-spectra as measuring the significance of the interaction between two series as well as revealing the existence and importance of fluctuations in the original series. This practice, however, presents some difficulties when analyzing a diverse sample of time series. For instance, although the employment variations of cities tied to the domestic market is overwhelmingly dominated by the national cycle, not all urban centers cohere with Montreal at this frequency because of the differing regularity of the cycle among cities of the sample.

The coherence spectra can only be interpreted, therefore, in conjunction with the spectral lines resolved by MEM. Without this detailed knowledge, cross-spectral analysis is usually limited to an examination of phase patterns. Attempts to stretch the capabilities of the method beyond its limits does not provide greater insights than could have been generated by regression or correlation analysis. Cho and McDougall (1978), for example, have been forced to average the coherence squares for all low frequency bands, comparing urban economies that may vary in some instances only over a Juglar cycle and in other instances over several minor cycles as well.

Using the traditional estimation procedure, the covariance between X(t) and Y(t) is first computed from the data series x_j and y_j and subsequently Fourier transformed, like the respective autocovariances A_x and A_y and for identical purposes. The cross covariance suffers from the same disadvantages as the auto-covariance: neighboring values tend to be correlated. The cross spectrum on the other hand describes the interaction between X(t) and Y(t) in terms of the coherence and phase shifts of several orthogonal components. The same basic assumptions therefore apply for this method of estimation.

The bivariate process [X(t),Y(t)] must be stationary at least with respect to the first and second moments generated by the sample data series x_j and y_j. This implies that the means of both series, μ_x and μ_y, are constant and that the sample cross covariances C_{xy} and C_{yx}, like the sample autocovariances A_x and A_y, are independent of time and a function only of one variable, k, the integer interval between the observations of the series:

$$C_{xy}(k) = E(x_j \cdot y_{j+k}) = C_{yx}(-k) \quad , \tag{12}$$

$$C_{yx}(k) = E(y_j \cdot x_{j+k}) = C_{xy}(-k) \quad , \tag{13}$$

where $k = 0, \pm\Delta, \pm 2\Delta, \ldots, \pm m\Delta$ and $m < n$.

It is at once obvious that while the autocovariance function A(k) is even and centered at zero lag, the cross covariance is rarely symmetrical, because it is the summation of an even and odd part:

$$C_{xy}(k) = \Sigma\ l_{xy}(k) + q_{xy}(k) \quad . \tag{14}$$

As a result the cross-spectrum $S_{xy}(f)$ is generally a complex number such that

$$S_{xy}(f) = L_{xy}(f) - iQ_{xy}(f) \ , \tag{15}$$

where the real part is called the co-spectrum and the imaginary part $Q_{xy}(f)$ is called the quadrature spectrum.

Since the cross-spectrum is estimated from a series of finite length, it has a truncation error which can be minimized by smoothing the function with the fixed spectral window, W(f), here a Hamming window. The smoothed cross spectrum estimator becomes:

$$S'_{xy}(f) = \Delta \sum_{k=-\infty}^{\infty} w(k)\cdot C_{xy}(k)\cdot \exp(-i2\pi fk) \quad , \quad \frac{-1}{2\Delta} < f < \frac{1}{2\Delta} \ . \tag{16}$$

Writing in terms of the even and odd parts:

$$S'_{xy}(f) = \Delta \sum_{k=-\infty}^{\infty} w(k)\ l_{xy}(k) \cos 2\pi fk - i\Delta \sum_{k=-\infty}^{\infty} w(k) q_{xy}(k) \sin 2\pi fk \ . \tag{17}$$

The smoothed coherency and phase estimators are derived from the smoothed cross-spectrum in the following manner:

$$K'^{2}_{xy}(f) = \frac{L'^{2}_{xy}(f) + Q'^{2}_{xy}(f)}{S'_{xx}(f)\cdot S'_{yy}(f)} \quad , \quad 0 < K'^{2}_{xy}(f) < 1 \quad , \tag{18}$$

$$F'_{xy}(f) = \text{artan}\ [-Q'_{xy}(f)/L'_{xy}(f)] \quad . \tag{19}$$

Together these functions provide a complete description of a bivariate normal process. The coherency is a nondimensional measure of the correlation between two processes at each frequency. It is analogous to the R^2 derived for two variables, and is interpreted in a similar manner such that the larger the value of $K_{xy}^2(f)$ the more closely related the two frequency components are. The phase spectrum, on the other hand, measures the phase difference between two processes at each frequency.

In the present study, two truncated points have been chosen equivalent to m=24 and m=49. The spectra are generally stable with increased lag window for series which are already strongly coherent at m=24. The results for the broader lag window are discussed in cases where low frequencies are of interest and resolution improved.

The analysis produced four main types of coherence functions. Only the most significant ones and their corresponding phase estimates are displayed in Figures 4(a), 4(b) and 5. The first type of response [Figures 4(a), 4(b)] identifies coincident behavior with respect to the Juglar cycle as observed in nine cities. Figure 5 describes a second type of response dominated by the effects of coincident 2½-3 year cycles present in the metropolis and five other urban areas. The third type of profile, not shown here, portrays coherence only at certain seasonal harmonics (6, 4 and 3 months) that have remained in the seasonally adjusted spectra of Montreal and four other cities. Similarly, a last type of response describes the complete lack of coherence between Montreal and four small urban areas. About half the sample (14 cities) displays meaningful interaction with the metropolis.

The cities which cohere highly with Montreal at the lowest frequency (Juglar cycle) include the more diversified, usually northern, economies of the system, either because their manufacturing sector is composed of more than one industry or because they have added a significant service function to a narrow industrial base. Extreme specialization, on the other hand, characterizes urban economies which are not coherent with Montreal at low frequency. Thus Table 1 shows that the degree of coherence with Montreal is primarily explained by the presence of manufacturing, but is negatively related to a specialization in non-durable goods even though these two independent variables are otherwise positively and strongly related. A specialization in non-durable industries identifies those cities of the Quebec lowlands which are dependent on the domestic market and rank amongst the least diversified urban areas of Canada. Table 1 further shows that coherence values are only weakly associated with the importance of the Juglar cycle. The amplitude of the fluctuation in the original series is thus a poor predictor of coherence in bivariate series at the corresponding frequency. Since coherence, like correlation, is normalized for differences in amplitude, it is determined by the regularity of the cycle rather than its importance.

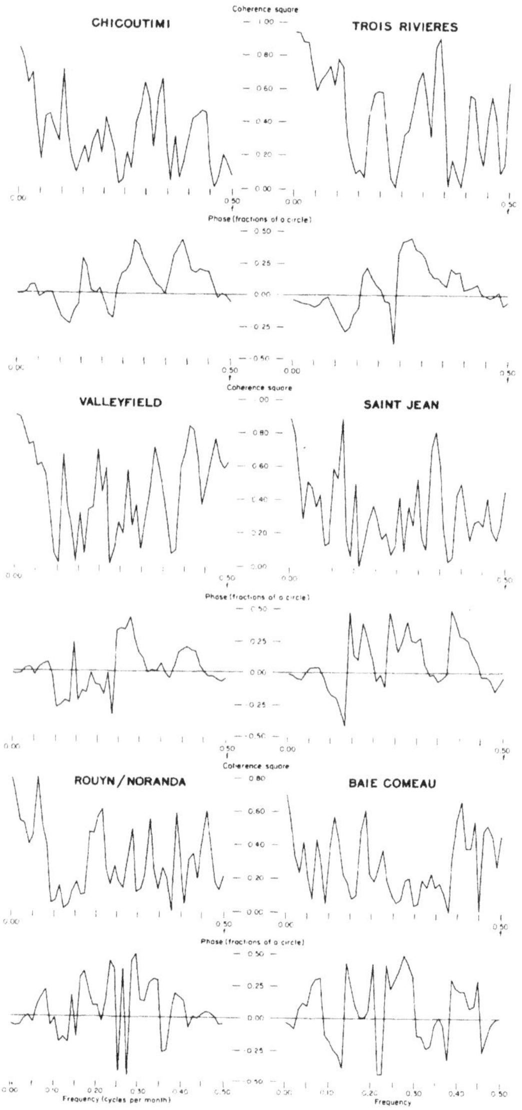

Fig. 4a. Sample coherence and phase angle of per capita employment (M = 49). Set of cities coherent with Montreal at the Juglar cycle.

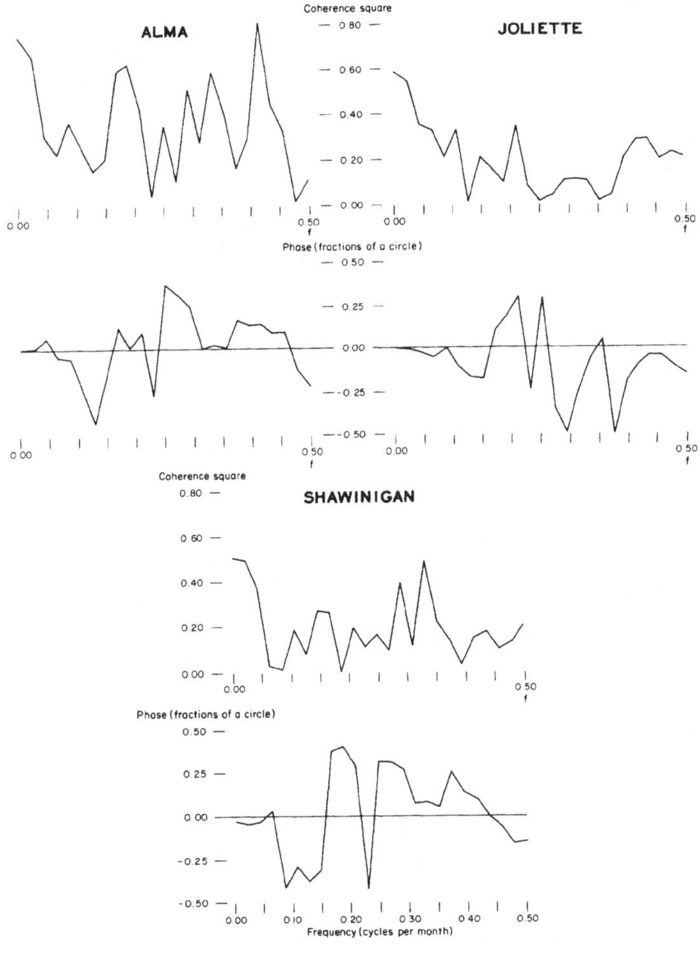

Fig. 4b. Sample coherence and phase angle of per capita employment (M = 24). Set of cities coherent with Montreal at the Juglar cycle.

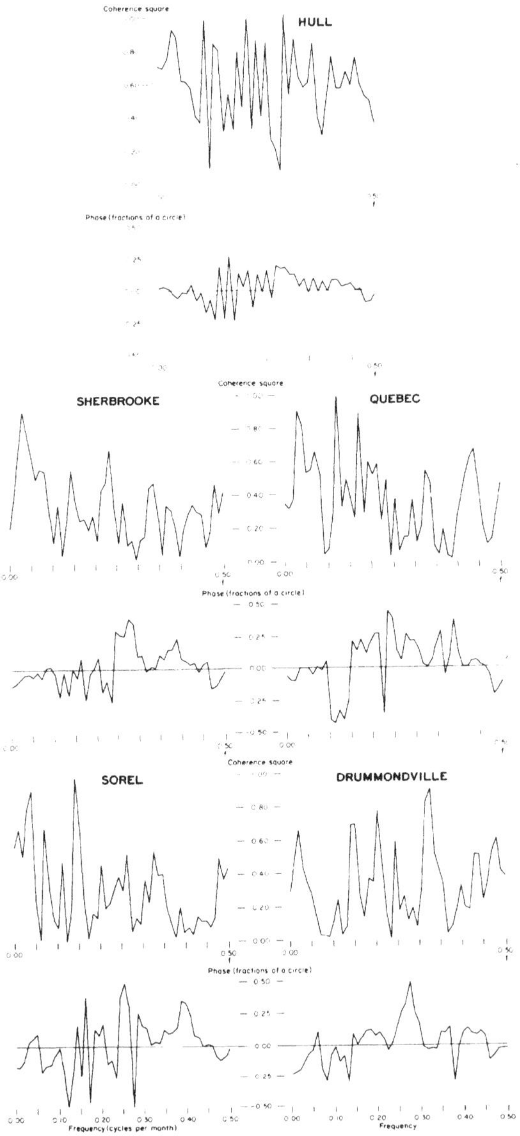

Fig. 5. Sample coherence and phase angle of per capita employment (M = 49). Set of cities coherent with Montreal at the 2-year cycle.

TABLE 1

Variables Explaining the Coherence in Employment with Montreal at the Juglar Cycle

CORRELATION MATRIX:

	K^2	JUG	JU1	LAG	TER	MFG	TEX
K^2		.24	.26	-.07	.05	.08	-.36*
JUG			.96	-.25	.36*	-.23	-.08
JU1				-.24	.25	-.10	-.01
LAG					.21	-.24	-.44*
TER						-.73*	-.61*
MFG							.68*
TEX							

REGRESSION EQUATION:

R^2	JUG	JU1	LAG	TER	MFG	TEX
.48	.66	-.12	-.03	-.27	1.51	-1.48
	(1.47)	(1.31)	(.03)	(.78)	(.71)	(.44)

JUG: Percental contribution of the Juglar cycle to total variance, 1957-1975.
JU1: Percental contribution of the Juglar cycle to total variance, 1957-1975. Non-seasonally-adjusted series.
LAG: Phase value at the Juglar cycle.
TER: Percentage of the labor force employed in the tertiary, 1961.
MFG: Percentage of the labor force employed in manufacturing, 1961.
TEX: Percentage of the labor force employed in textile, knitting and clothing, 1961.

In addition the phase information reproduced in Figures 4 and 5 implies a rejection of any hypothesis about a destabilizing metropolis propagating fluctuations throughout the system. Among the eleven cities which are coincident with Montreal at the Juglar cycle, one displays a small lag, three are in phase and the others lead by several months. Positive values in the phase function indicate that Montreal leads the given cities while negative values indicate a lag. At low frequency the phase angle generally varies around - .03 (fraction of a circle). This means that Montreal lags by approximately 3 months. Phase shifts on the order of two to seven months, however, do not represent even 10% of the cycle's length. The phase results for those cities approximate very nearly a zero lag situation. Longer phase shifts cannot be interpreted as they are associated with very low K^2.

The metropolis and, to a lesser degree, other centers which enjoy similar supply conditions therefore represent points in the system where cyclical fluctuations are dampened and/or spread over regular time intervals. Government and business and consumer spending are very important in metropolitan areas, as are the alternative opportunities for investment and consumption. Any contraction or expansion of production involves a small percentage of the labor force and is distributed more evenly over time. Urban ares which are smaller or more specialized display an immediate vulnerability to fluctuations in their employment base, since alternatives are scarce. Indeed for a smaller urban area proximity to metropolitan opportunities or a central place function constitutes the only alternative to setbacks in a narrow industrial base. While they may not dampen business fluctuations in small cities, such factors offset the more unpredictable behavior of the industrial base by promoting greater regularity in the occurrence of downturns and upturns.

Hence, in the economies of the northern communities, variations in aggregate demand may affect the various durable goods industries (pulp and paper, aluminium refinery, iron foundries, industrial chemicals) at different times and provoke contractions or expansions of the labor force over longer but more regular intervals than would otherwise occur if the economies were not somewhat diversified and had not acquired a service function. In contrast, specialized urban economies, the greater fraction of which are textile towns on the Quebec lowlands, display a more erratic behavior and are more prone to experience contractions and expansions of production as soon as the first shocks are introduced in the system.

Cross-spectral analysis thus brings periodicity into focus as an essential aspect of short term behavior. This factor, together with the amplitude and the phase shift of the oscillation, differentiates the performance of cities and reflects the role of structural characteristics. The presence of a large labor force dampens delays and regulates urban cyclic responses. Although a larger scale labor force guarantees diversification and a consequent flexibility in response to change, other supply conditions present in medium or small size cities may also promote regularity and delay in cyclical response. In particular, a specialization in durable goods industries and the presence of a large service function, or the satellization of cities by the metropolis, promotes a diversification of activity and the consequent greater predictability of behavior.

CONCLUSION

Spectral and cross-spectral results show that the short term behavior of an urban system cannot be explained simply as an extension of its export base. Thompson (1975) made the point when considering long term performance, arguing against an approach which would "cast local economies as mere bundles of industries in space." He emphasized the structural characteristics of the urban system rather than its industrial composition, suggesting that a sectoral approach could be adopted for the analysis of short term variations in urban income. In such a context, the urban system could be satisfactorily reduced to a matrix of inter-industry linkages.

The shortcomings of such an approach, however, prove to be just as great when considering short term behavior. Since it cannot explain how cities which exhibit the same export structure can respond quite differently to changes in aggregate demand in the metropolis--given different location, size and function in the service hierarchy. Only in the case of the minor cycle may one readily link the urban performance to the industrial base. The occurrence of such short term variations is extremely localized in space and the combination of minor cycles observed is often unique to a few cities, as is verified in Marchand (1981). Given its specificity it becomes reasonable to interpret coherence at that frequency from a narrow sectoral point of view. The products of the various hinterland economies constitute essential supplies for the major industries of the metropolis. Coincident behavior at a particular harmonic (2½-3 year cycle) may thus reflect the presence of production linkages and measure spatial interaction.

Cross-spectral results pertaining to coherences at the national cycle show, however, that the method cannot be used blindly as a measure of spatial interaction. The urban economies of the Montreal industrial complex do not cohere with the metropolis at the Juglar cycle. Similarities in response to changes in aggregate demand, rather than measuring interaction, mirrored a similarity in employment structure. In particular, a diverse mix of industries, typical of northern centers specialized in the production of staple and capital goods, promotes a greater regularity in the occurrence of downturns and upturns such as is found in the metropolis. The role of location is also important in an urban system where most of the large resource settlements have developed central place functions that provide a stabilizing element to the economies. The supply conditions that determine a short term behavior replicating that of the metropolis cannot be reduced to a matter of sectoral specialization. The scale and nodality of individual economies explains the propagation of aggregate demand conditions that obtains a regularity approaching that of the metropolis and the nation.

REFERENCES

Alonso, W., 1971. The Economies of Urban Size, Papers of the Regional Science Association, XXVI: 67-83.

Alonso, W., 1977. Industrial Location and Regional Policy in Economic Development, in Regional Policy. Readings in Policy and Applications, J. Friedmann and W. Alonso (eds.), M.I.T. Press, Cambridge, 64-96.

Bartels, C., 1977. The Structure of Regional Unemployment in the Netherlands, Regional Science and Urban Economics, 7: 103-105.

Bartels, C., Booleman, M. and Peters, W., 1978. A Statistical Analysis of Regional Unemployment Series, Environment and Planning A, 10: 937-954.

Brechling, F., 1976. Trends and Cycles in British Regional Unemployment, Oxford Economic Papers, 19: 1-21.

Burns, F. and Mitchell, W., 1947. Measuring Business Cycles, National Bureau of Economic Research, New York.

Burg, J., 1968. A New Analysis Technique for Time Series Data, Advanced Institute in Signal Processing, Netherlands.

Cho, D. and McDougall, D., 1978. Regional Cyclical Patterns and Structure, 1954-1975, Economic Geography, 54: 66-74.

Fishman, G., 1969. Spectral Methods in Econometrics, Harvard University Press, Cambridge.

Granger, C., 1966. The Typical Shape of an Economic Variable, Econometrica, XXIV: 150-161.

Haggett, P., 1975. Time Series Comparison: Unemployment, in Elements of Spatial Structure, A.D. Cliff, et al. (eds.), Cambridge University Press, Cambridge.

Harvey, P., 1956. Plein Emploi National et Plein Emploi Régional au Canada Dupuis la Guerre, L'Actualité Economique (Avril) 5-26.

Hepple, L., 1975. Spectral Techniques and the Study of Inter-regional Economic Cycles, in Processes in Physical and Human Geography, R. Peel, M. Chisholm and P. Haggett (eds.), Heineman, London, 392-408.

Jeffrey, D., 1974. Regional Fluctuations in Unemployment Within the Australian Urban Economic System: A Study of the Spatial Impact of Short Term Economic Change, Economic Geography 50: 111-123.

Jeffrey, D. and Webb, D., 1972. Economic Fluctuations in the Australian Regional System 1955-1970, Australian Geographical Studies 10: 141-160.

Jenkins, G and Watts, D., 1968. Spectral Analysis and its Applications, Holden-Day, Toronto-San Francisco.

Kanasewich, E., 1973. Time Series Analysis in Geophysics. University of Alberta Press, Edmonton.

King, L., Casetti, E. and Jeffrey, D., 1974. Cyclical Fluctuations in Unemployment Levels in the U.S. Metropolitan Areas, Tijdschrift Voor Economische en Sociale Geographie, 50: 345-352.

King, L. and Clark, G., 1978. Regional Unemployment Patterns and the Spatial Dimensions of Macro Economic Policy: The Canadian Experience, 1966-1975, Regional Studies, 12: 283-296.

Lacoss, R., 1971. Data Adaptative Spectral Analysis Methods, Geophysics XXXVI: 661-675.

Marchand, C., 1971. Maximum Entropy Spectra and the Spatial and Temporal Dimensions of Economic Fluctuations in an Urban System, Geographical Analysis, in press.

North, D., 1955. Location Theory and Regional Economic Growth, Journal of Political Economy LXIII: 243-258.

North, D., 1961. The Economic Growth of the United States, 1790-1860, Prentice Hall, Englewood Cliffs.

Schumpeter, J., 1935. The Analysis of Economic Change, Review of Economic Statistics 17: 1-10.

Thirlwall, A., 1966. Regional Employment as a Cyclical Phenomenon, The Scottish Journal of Political Economy 13: 204-219.

Thirlwall, A., 1975. Forecasting Regional Unemployment in Great Britain, Regional Science and Urban Economics, 5: 357-374.

Thompson, W., 1965. A Preface to Urban Economics, John Hopkins Press, Baltimore.

Thompson, W., 1975. Internal and External Factors in the Development of Urban Economics, in Regional Policy. Readings in Theory and Applications, J. Friedmann and W. Alonso (eds.), M.I.T. Press, Cambridge, 210-220.

VanDuijn, J., 1975. The Cyclical Sensitivity to Unemployment of Dutch Provinces, 1950-1972. Regional Science and Urban Economics, 107-132.

Watkins, M., 1963. A Staple Theory of Economic Growth, Canadian Journal of Economics and Political Science XXIX: 141-158.

SPATIAL ECONOMETRICS: PRINCIPLES AND RECENT RESULTS

J.-P. Ancot and J.H.P. Paelinck

Erasmus University &
Netherlands Economic Institute

INTRODUCTION

In general the specification, testing, and estimation of spatial econometric models has been poor and remains so. The best inducement to improve the state of affairs is to be obliged to cope with political problems, i.e., spatial model building that is intended to prepare policy decisions. An example of this is the so-called FLEUR-model, built for the EEC authorities in Brussels [Paelinck (1973)]. In this paper we will first review a number of results obtained in previous work [more details are to be found in Paelinck and Klaassen (1979) and Paelinck (1980)], and then go on to describe some recent attacks on policy-relevant spatial econometric problems.

SOME RESULTS

The first task to be performed was to elaborate on some principles of attack on the specification of the spatial dimension of econometric models. As a result of empirical work the following features could serve as a starting point:

(a) fundamental interdependence of spatial phenomenon;
(b) asymmetry of spatial economic mechanisms;
(c) allotopy [from the Greek "αλλος" (other) and "τοπος" (site)];
(d) ex-ante ex-post distinction; and,
(e) integration of topological data.

The first principle generalizes economic interdependence and requires the systematic set-up of spatial models in a spatially inter-dependent way. As an illustration one can take the following simple income-generating model

$$\underline{y} = \underline{Ay} + \underline{\varepsilon} \tag{2.1}$$

where $\underline{y}$ is a vector of regional incomes, $\underline{A}$ is a matrix based i.a. on contiguity matrices and $\underline{\varepsilon}$ is a vector of stochastic elements. [It is discussed at length in Paelinck and Klaassen (1979) and Paelinck (1981), Ch. 1.]

Introduction of spatially non-uniform elements leads to a first simple distinction between urban and non-urban regions among the elements of $\underline{y}$; that asymmetry, once introduced [for more details, see Chevailler and Paelinck (1978)], leads to more refined specifications and to a first attempt to integrate geographical analysis and spatial econometrics. Table 1 presents some results that could be attained through an appropriate specification of the parameter-component hypothesis, i.e., a hypothesis on the additivity, within one parameter, of a generic and a number of specific parts.

In fact, allotopy has already been introduced, as incomes in one site are explained by explanatory variables (here again incomes) in other sites. The same would apply to locational models, in which the locational evolution of activities would be explained by output and input factors available elsewhere in space. A case in point is the interregional intersectoral open-ended attraction model

$$\underline{q} = \underline{Bq} + \underline{C}\ \underline{X} + \underline{\eta} \tag{2.2}$$

in which $\underline{q}$ is a vector of regionalized sectoral production levels, $\underline{X}$ a vector of remaining explanatory factors, and $\underline{\eta}$ again a vector of stochastic variables.

Matrix $\underline{B}$ is composed of, among other things, friction functions which could be such that--for certain ranges of distance--friction could decline as distances increase.

Model (2.2) rests <u>ex post</u> on complex <u>ex-ante</u> locational decision processes, passing implicitly or explicitly through some optimizing mechanisms. Other distances can be introduced explicitly (integration of topological data). They would not rest <u>ex post</u> on the simple linear specification which would possibly have been introduced in the initial optimizing model (linear transport cost friction).

TABLE 1

One Appropriate Specification of the Parameter-Component Hypothesis

Parameters	Value*	Interpretaion
β_1	-.003	UU, contiguity 1
ε_1	-.016	UNU, contiguity 1
β_2	-.002	NUNU, contiguity 1
ε_2	-.006	NUU, contiguity 1
γ_1	-.019**	Urban, lag zero
γ_2	.010	Non-urban, lag zero
ξ_1	.066**	Urban, lag one
ξ_2	-.015	Non-urban, lag one
$\tilde{\beta}_1$	.405**	Urban, contiguity zero
$\tilde{\beta}_2$	.094**	Non-urban, contiguity zero
α	.270**	Direct generic (UU, NUNU)
δ	.215**	Cross-generic (UNU, NUU)
d_1	-.014**	Urban dummy
d_2	.007**	Non-urban dummy

*To be interpreted as elasticities; U = urban, NU = non-urban.

**Significant.

The models specified in this way will lead to some non-orthodox estimation procedures. In previous work a full-information maximum-likelihood approach was found to be a good reference and starting point for deriving estimators with some desirable properties. A central constraint that should be kept in mind is the fact that more often than not lengthy time series of regional observations are not available, and that estimation methods should be adapted to that state of affairs. Again parameter-component specifications have been found to be an efficient way for using up all the information contained in the poor statistical series available. Examples of demand functions for such situations appear in Ancot (1979).

The following results should be mentioned:

(1) For model (2.1) two estimation procedures have been investigated. One rests on a perturbation of the system matrix $(\underline{I}-\underline{A})$, of which the determinant has to be approximated in the log-likelihood function. This procedure leads to the solution of a polynomial, the degree of which increases with the size of the problem. Instrumental variables have been found to give a more practical solution, which could be easily combined with a parameter-component specification leading to a simple OLS estimation procedure.

(2) For model (2.2) a semi-separable maximum-likelihood procedure (ISSML: interregional semi-separable maximum likelihood) has been devised. It rests on deriving analytically $\ln|\det \underline{B}|$, correcting the OLS normal equation with appropriate terms from the resulting matrix. In a two-sector four-region test model, where one region has its sectors being exogenous, the procedure converged in one round at the one-percent level.

(3) A last aspect to be mentioned is the spatial autocorrelation of the stochastic terms. Again it was approached via maximum likelihood, and the following spatial autoregressive process was investigated:

$$\underline{\mu} = R\,\underline{\mu} + \underline{\nu} \qquad (2.3a)$$

$$R \triangleq \sum_{i=1}^{k} \rho^i C_i \qquad (2.3b)$$

$$\underline{\nu} \sim \{\underline{0}, \sigma^2 \underline{I}\} \; . \qquad (2.3c)$$

A scanning procedure with ensuing test [for testing, see Paelinck and Klaassen (1979), Ch. 4] led to the conclusion that for the

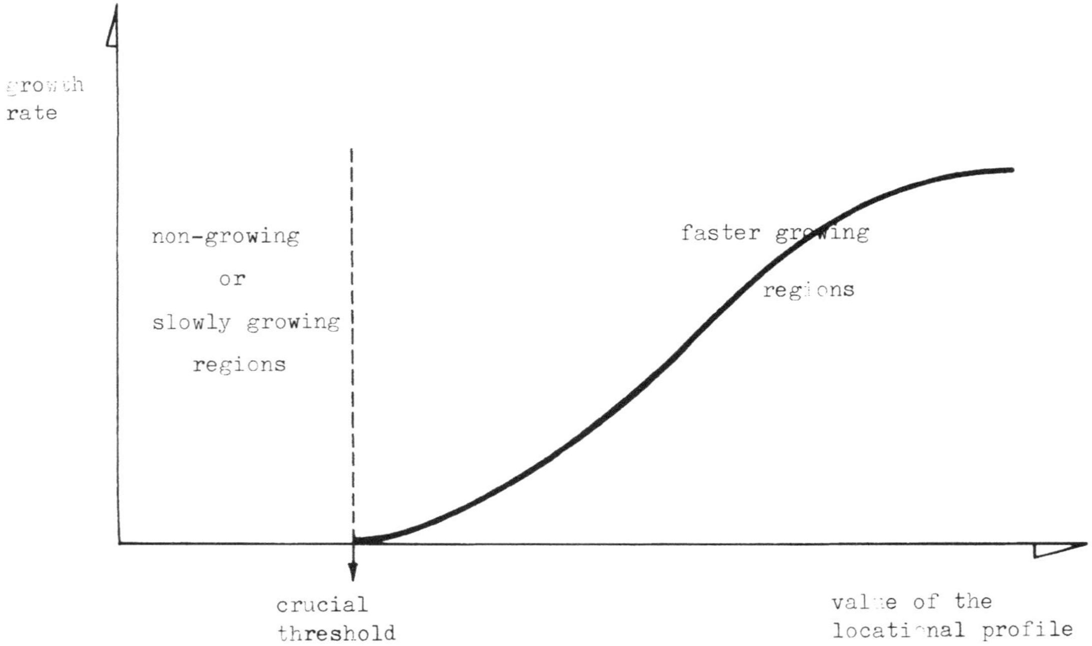

Fig. 1. Interregional Growth a la the Threshold Theory

data used a first-order contiguity process with $\rho \approx .15$ was the most plausible hypothesis.

No more could be done in this paper than give a quick overview of some approaches and results. It gives an introduction to some more recent experiences on which we will report now.

RECENT RESEARCH

This work has been directed essentially towards the construction or large interregional-intersectoral econometric models. The so-called European FLEUR-model has 70 sectors and 100 regions in it. [For its data bank inputs, see Ancot (1978) a.o. and Netherlands Economic Institute (1978) and (1980).] A threshold theory seems suited as a starting point for modelling the interregional growth phenomena [Paelinck (1977), (1978a), (1978b)], leading up to Figure 1.

Spanish data were used to test how workable the model could be. These data were collected for a study of the development of value added (at current prices) in seven sectors of the Spanish chemical industry in the period 1967-1971 [Ancot and Paelinck (1980a)]. The sectors are given in Table 2.

TABLE 2

Sectors of Chemical Industry

Code	Sector
40	Asbestos and rubber
41	Transformation of plastic material
42	Synthetic materials and artificial fibers
43	Basic chemicals and fertilizers
44	Soap, detergents, perfumes
45	Other chemical products
46	Oil products

The sample contains observations of the explanatory variables by Spanish province made in 1967 and observations of the sectoral growth figures to be explained by province dating from 1967 and 1971. The set of potential explanatory variables from which a location profile was to be selected by appropriate econometric methods can be split into five groups or profiles, which will not be commented upon here [see Ancot (1980), et al., pp. 8-9]. The results of parameter estimation on the basis of the data just described are reproduced in Tables 3, 4 and 5.

In a nutshell, the model consists of an equation for each sector, relating the critical threshold value for that sector to a subset of elements of the regional profile. This subset is selected by means of econometric methods and certain potentialized (in the case of the variables for which interregional dependencies are relevant, such as across to the labor market or to raw materials) and non-potentialized variables (whenever the corresponding element of the regional profile is relatively immobile, such as land availability or housing quality). To each of the selected variables, potentialized as well as non-potentialized ones, there corresponds a "propensity parameter," the value of which is an indicator of the relative importance of the corresponding element for the "attractiveness" of the regions. Estimated values of these parameters are given in Table 4 (potentialized variables) and 5 (non-potentialized variables). The potentialized variables are entered as space-discounted sums by means of a modified Tanner friction function, containing a sector- and a variable-specific friction parameter. The value of this friction parameter, the estimated values of which are given in Table 3, provides an indication of the spatial relation between the location of the sector and that of the corresponding element of the regional profile. If that value is larger than -1, the factor of attraction in one's own region will dominate, and increasingly so the higher the value. If, on the contrary, this friction parameter is smaller than minus unity, the region will be attractive because that element is present in another more or less distant region (for example, a region may be attractive because one neighboring region contains a market for the sector's products, another neighboring region has excess labor supply).

The quadratic-programming technique used here has led to a series of interesting results. However, it was not the only possibility, and recent thoughts have been devoted to alternative estimation methods with their relative merits and drawbacks. These thoughts will be briefly reproduced hereafter [see Ancot (1979), Ancot and Paelinck (1980a), Ancot and Paelinck (1980b), cfr. also Jansen (1975)].

Pre-Test Estimating

Consider the model

$$\underline{y} = \underline{X\beta} + \underline{\varepsilon} \tag{3.1}$$

TABLE 3

Estimation Results of Friction Coefficients ν

	Sector / Variable	40	41	42	43	44	45	46
(a)	Deliveries	--	--	-0,3429	--	--	5,5263	--
	Agriculture, forestry and fishery	--	--	--	--	--	-1,6000	--
	Alcoholic beverages (except wines)	--	--	--	--	--	0,8000	--
(b)	Paper	--	--	-4,8649	--	-5,2643	--	--
	Basic chemicals and fertilizers	-0,4700	--	-3,4848	--	--	--	--
	Electricity	--	--	-3,5333	-1,7030	--	--	--
	River transports	--	--	--	--	--	--	-0,6393
	Commerce	--	--	2,6000	--	--	--	--
	Other services	-1,1154	-1,4167	-3,3333	2,5950	--	-0,0606	7,8947
	Credit and insurance institutions	--	-0,6769	--	--	0,4320	--	--
	Active population	--	--	-1,9000	-1,5240	--	-2,5714	--
(c)	Agricultural labor	-1,0606	-0,8519	1,1212	-1,5000	--	-2,5128	--
	Executive staff (in %)	--	--	--	--	-1,2890	--	--
	Underemployment	-0,5796	0,6667	2,9091	--	-0,4020	-0,6174	-0,3846
	Net migration	4,0323	--	7,5500	18,3750	25,0000	2,6471	--

(a) Demand elements: regional (intermediate and final) deliveries.
(b) Supply elements: levels of production (value added) of sectors showing important input links with the sectors of chemical industry.
(c) Other locational factors.

TABLE 4

Estimation Results of "Propensities" β (Potentialized Variables)

	Variables \ Sector	40	41	42	43	44	45	46
(a)	Deliveries	--	--	0,0025	--	--	0,0955	--
	Agriculture, forestry, and fishery	--	--	--	--	--	0,00007	--
	Alcoholic beverages (except wines)	--	--	--	--	--	0,00033	--
(b)	Paper	--	--	0,00003	--	0,00002	--	--
	Basic chemicals and fertilizers	0,00063	--	0,00010	--	--	--	--
	Electricity	--	--	0,00004	0,00067	--	--	--
	River transports	--	--	--	--	--	--	0,0182
	Commerce	--	--	0,0337	--	--	--	--
	Other services	0,00085	0,00029	0,00003	0,0563	--	0.00062	0,2039
	Credit and insurance institutions	--	0,00033	--	--	0,0014	--	--
	Active population	--	--	0,00066	0,00092	--	0,00002	--
(c)	Agricultural labor	0,00011	0,00012	0,0101	0,00036	--	0,000003	--
	Executive staff (in %)	--	--	--	--	0,1707	--	--
	Underemployment	0,0088	0,0368	0,0202	--	0,0078	0,00062	0,0134
	Net migration	0,0350	--	3,8015	76,4280	43,2000	0,0096	--

(a) Demand elements: regional (intermediate and final) deliveries.
(b) Supply elements: production levels (value added) of the sectors showing important input links with the sectors of chemical industry.
(c) Other locational factors.

TABLE 5

Estimation Results of the Coefficients α (Non-Potentialized Variables)

	Variable / Sector	40	41	42	43	44	45	46
	Sites in coastal regions	0,0008	--	--	--	--	--	0,0058
	Mining	--	--	--	--	0,0134	--	--
(b)	Water, gas and electricity production	--	--	0,0139	0,0078	--	--	--
	Dwellings with shower (%)	5,0124	5,3878	43,2459	42,1174	4,8259	5,7830	0,3833
	Cultural density	1,9717	1,5725	1,7408	0,2249	1,6588	0,9624	0,1586
	Heads of households with university education	15,6928	7,9548	45,5130	38,0072	2,6070	7,7167	--
	Upward social mobility	0,3367	0,3331	0,0101	--	0,3628	--	0,7098
(c)	Cosmopolitan	32,8932	15,8838	236,8171	154,2538	18,5019	23,5717	--
	Population of towns of over 50,000 inh. (in %)	0,0088	0,0069	0,9704	1,2205	0,0379	0,0644	0,0113
	Capacity of higher education (in %)	0,1332	--	--	--	0,1150	--	0,4308
	Vacant housing	0,9181	0,2750	--	--	--	0,2183	0,4127
	Dynamism of provincial and local authorities	2,5375	2,0777	11,9672	7,5976	1,2368	2,9765	--
(d)	Dynamism of Madri lenian authorities	6,8038	4,8449	45,5589	43,1402	5,4184	5,5767	0,3466
	Public investments (per inhabitant)	0,0052	0,0053	0,0279	0,0274	--	0,0044	0,0042

(b) Supply elements: production levels (value added) of the sectors showing important input links with other sectors of chemical industry.

(c) Other locational factors.

(d) Regional policy factors.

where $\underline{y}$ is a T × 1 vector of observations on a dependent variable, $\underline{X}$ is a T × K matrix of T observations each on K non-stochastic variables, $\underline{\beta}$ is a K × 1 vector of parameters to be estimated, and $\underline{\varepsilon}$ is a T × 1 vector of random normal disturbances which are assumed to have zero mean and variance-covariance matrix $\sigma^2\underline{I}$. As is well known, ordinary least-square estimators of $\underline{\beta}$,

$$\underline{\beta} = (\underline{X}'\underline{X})^{-1}\underline{X}'\underline{y} \tag{3.2}$$

have the asymptotic distribution:

$$\underline{\beta} \sim N[\underline{\beta}, \sigma^2(\underline{X}'\underline{X})^{-1}] \quad . \tag{3.3}$$

Suppose one has structural information about $\underline{\beta}$ in the form of a set of linear inequality constraints

$$\underline{A\beta} \leq \underline{r} \tag{3.4}$$

where $\underline{A}$ is an S × K(S<K) matrix of known constants and $\underline{r}$ is an S × 1 known vector. The constrained least-squares problem then consists of finding the vector $\tilde{\underline{\beta}}$ which minimizes the quadratic function.

$$\underline{\varepsilon}'\underline{\varepsilon} = (\underline{y} - \underline{X\beta})'(\underline{y} - \underline{X\beta}) \tag{3.5}$$

subject to the constraints (3.4). This is the quadratic programming problem that has generated Tables 3 through 5. Apart from the non-negativity constraints $\underline{\beta} \geq \underline{0}$, the reduction of the above problem to a properly defined quadratic problem, when the constraints (3.4) do not directly imply $\underline{\beta} \geq \underline{0}$, does not affect the generality of the argument [see Judge and Takayama (1966)]. In the following we shall assume, for simplicity, that $\underline{\beta} \geq \underline{0}$.

The Lagrangian function corresponding to this problem is

$$L(\underline{\beta},\underline{\lambda}) = -(\underline{y} - \underline{X\beta})'(\underline{y} - \underline{X\beta}) - 2\underline{\lambda}'(\underline{r} - \underline{A\beta}) \tag{3.6}$$

where $\underline{\lambda}$ is an S × 1 vector of unknown Lagrange multipliers. If the optimum solution is represented by $\tilde{\underline{\beta}}$ and $\tilde{\underline{\lambda}}$, the first-order Kuhn-Tucker conditions imply

$$\left[\frac{\partial L}{\partial \tilde{\underline{\beta}}}\right]_{[\tilde{\underline{\beta}},\tilde{\underline{\lambda}}]} \equiv -2(\underline{X}'\underline{X}) + 2\underline{X}'\underline{y} + 2\underline{A}'\tilde{\underline{\lambda}} \leq \underline{0} \tag{3.7}$$

and

$$\left[\frac{\partial L}{\partial \underline{\lambda}}\right]_{[\tilde{\underline{\beta}},\tilde{\underline{\lambda}}]} \equiv -2(\underline{r} - \underline{A}\tilde{\underline{\beta}}) \leq \underline{0} \quad . \tag{3.8}$$

From (3.7), the vector $\tilde{\underline{\beta}}$ that maximizes the quadratic form (3.6) under constraints (3.4) is

$$\begin{aligned} \tilde{\underline{\beta}} &= (\underline{X}'\underline{X})^{-1}\underline{X}'\underline{y} + (\underline{X}'\underline{X})^{-1}\underline{A}'\tilde{\underline{\lambda}} \geq \underline{0} \\ &= \hat{\underline{\beta}} + (\underline{X}'\underline{X})^{-1}\underline{A}'\tilde{\underline{\lambda}} \geq 0 \quad . \end{aligned} \tag{3.9}$$

If the matrix $\underline{A}$ is partitioned into $\underline{A}_1$ and $\underline{A}_2$, of order $S_1 \times K$ and $S_2 \times K(S_1 + S_2 = S)$, respectively, and the vector $\underline{r}$ is partitioned accordingly into $\underline{r}_1$ and $\underline{r}_2$, such that $\underline{A}_1$ and $\underline{r}_1$ correspond to the binding restrictions in the optimum solution and $\underline{A}_2$ and $\underline{r}_2$ to the others, i.e.

$$\underline{A}_1\tilde{\underline{\beta}} = \underline{r}_1 \tag{3.10}$$

and

$$\underline{A}_2\tilde{\underline{\beta}} < \underline{r}_2 \quad , \tag{3.11}$$

the conditions (3.7) and (3.8) can be specialized to

$$\tilde{\underline{\lambda}}_1 > \underline{0} \tag{3.12}$$

and

$$\tilde{\underline{\lambda}}_2 = \underline{0} \quad . \tag{3.13}$$

The combination of (3.10) and (3.9) then yields

$$\begin{aligned} \underline{r}_1 = \underline{A}_1\tilde{\underline{\beta}} &= \underline{A}_1(\underline{X}'\underline{X})^{-1}\underline{X}'\underline{y} + \underline{A}_1(\underline{X}'\underline{X})^{-1}\underline{A}_1'\tilde{\underline{\lambda}}_1 \\ &= \underline{A}_1\hat{\underline{\beta}} + \underline{A}_1(\underline{X}'\underline{X})^{-1}\underline{A}_1'\tilde{\underline{\lambda}}_1 \end{aligned} \tag{3.14}$$

so that

$$\begin{aligned} \tilde{\underline{\lambda}}_1 &= [\underline{A}_1(\underline{X}'\underline{X})^{-1}\underline{A}_1']^{-1}[\underline{r}_1 - \underline{A}_1(\underline{X}'\underline{X})^{-1}\underline{X}'\underline{y}] \\ &= [\underline{A}_1(\underline{X}'\underline{X})^{-1}\underline{A}_1']^{-1}(\underline{r}_1 - \underline{A}_1\hat{\underline{\beta}}) \quad . \end{aligned} \tag{3.15}$$

Elimination of $\tilde{\lambda}$ between (3.9), (3.13) and (3.15) provides the solution

$$\tilde{\beta} = (X'X)^{-1}X'y + (X'X)^{-1}A_1'[A_1(X'X)^{-1}A_1']^{-1}[r_1 - A_1(X'X)^{-1}X'y]$$

$$= \hat{\beta} + (X'X)^{-1}A_1'[A_1(X'X)^{-1}A_1']^{-1}(r_1 - A_1\hat{\beta}) \quad . \qquad (3.16)$$

Finally, since

$$\hat{\beta} = \beta + (X'X)^{-1}X'\varepsilon \qquad (3.17)$$

one obtains

$$\tilde{\beta} = \beta + (X'X)^{-1}X'\varepsilon + (X'X)^{-1}A_1'[A_1(X'X)^{-1}A_1']^{-1} \times$$
$$[r_1 - A_1\beta - A_1(X'X)^{-1}X'\varepsilon] \quad . \qquad (3.18)$$

As noted in Judge and Takayama (1966), the partitioning of the matrix A into A_1 and A_2 and the vector r into r_1 and r_2 will in general vary from sample to sample, so that $\tilde{\beta}$ in (3.18) is no longer a linear function of a random vector (ε). Simulation procedures and numerical integration techniques have been advocated for examining the sampling properties and deriving the theoretical distribution of the restricted estimators (3.18). We, however, suggest testing the hypothesis that any particular constraint is binding in the optimum solution of (3.6), and determining the partitioning of A and r on the basis of such a test. Expression (3.18) can then be used in the usual way to derive sampling distributions of the estimators about which the same error-probability statements can be made as for the test on which they are based.

If all the constraints (3.4) are binding in the optimum solution, it follows from the Kuhn-Tucker first-order conditions that $A_1 = A$, $r_1 = r$, and $\tilde{\lambda} > 0$. Furthermore, specializing (3.15) yields

$$\tilde{\lambda} = [A(X'X)^{-1}A']^{-1}(r - A\hat{\beta}) > 0 \quad . \qquad (3.19)$$

Given the unconstrained least-squares estimators $\hat{\beta}$, the vector

$$l = [A(X'X)^{-1}A']^{-1}(r - A\hat{\beta}) > 0 \qquad (3.20)$$

can be computed, and for each element λ_i of $\underline{\lambda}$ the null hypothesis $\lambda_i > 0$ tested. Because under the null hypothesis l is a linear combination of the normally distributed variable (3.3), $\underline{l}$ is also normally distributed with mean

$$E(\underline{l}) = [\underline{A}(\underline{X}'\underline{X})^{-1}\underline{A}'] (\underline{r} - \underline{A\beta}) \tag{3.21}$$

and covariance matrix

$$E\{[\underline{l} - E(\underline{l})][\underline{l} - E(\underline{l})]'\} = \sigma^2[\underline{A}(\underline{X}'\underline{X})^{-1}\underline{A}']^{-1} \quad . \tag{3.22}$$

The test is a one-tailed t-test with T-K degrees of freedom. The test statistic is defined as follows:

$$t = \frac{l_i}{\hat{\sigma}\sqrt{m_{ii}}}$$

where σ^2 is an unbiased estimator of the regression variance in the restricted case and m_{ii} is the i-th diagonal element of the matrix $[\underline{A}(\underline{X}'\underline{X})^{-1}\underline{A}']^{-1}$.

Given the result of this test (3.16), the ordinary least-squares estimator of $\underline{\beta}$ and the partitioning of $\underline{A}$ can then be used to obtain the constrained least-squares estimator $\tilde{\underline{\beta}}$. The distribution of this estimator follows immediately from (3.18). It is a normal distribution with mean $\underline{\beta}$ (unbiasedness, if the constraints are correctly identified) and with covariance matrix (smaller variance than the ordinary least-squares estimator):

$$E[(\tilde{\underline{\beta}}-\underline{\beta})(\tilde{\underline{\beta}}-\underline{\beta})'] = \sigma^2\{(\underline{X}'\underline{X})^{-1}-(\underline{X}'\underline{X})^{-1}\underline{A}_1'[\underline{A}_1(\underline{X}'\underline{X})^{-1}\underline{A}_1']^{-1}\underline{A}_1(\underline{X}'\underline{X})^{-1}\}. \tag{3.23}$$

This sampling distribution, and hence the resuling t- and F-tests, <u>hold subject to the correctness of the above test</u>: The significance levels of t- and F-tests will thus be conditioned by that of the initial test. The consequences of this hierarchical dependence between the tests are a subject for further investigation. To our knowledge, indeed, this type of problem has been discussed only for the case where (3.4) is limited to one sign constraint on one coefficient [Lovell and Prescott (1970)]. The analysis suggests that a t-test based on the estimated equation when the constraint is present, is inappropriate when dealing with highly collinear economic time series, for it can lead to substantially exaggerated claims of significance. The authors conclude that "a valid test, which may be more powerful than the customary procedure, is to always report the t-statistic calculated from the initial regression regardless of the signs of the parameter estimates." It is the authors' opinion that multicollinearity should in principle pose no problem, being amenable to formal inclusion in the logic of the approach, which admits of three cases: there is no multicollinearity and the second stage gives the right results; multicollinearity is perfect and should be taken care of by exact constraints; or, there is near-multicollinearity, which should be taken care of by stochastic constraints. For further drawbacks in terms of mean square errors, the reader is referred to Judge and Bock (1978). It should be noted, however, that the Stein-rule estimators advocated there are not necessarily non-negative, and that the variance-covariance matrix is unknown.

A second line of thought, centering on the non-negativity problem, will be investigated hereafter.

PAREX-Estimators

Suppose in specification (3.1) we define

$$\underline{\beta} \stackrel{\Delta}{=} e^{\underline{\alpha}} \tag{3.24}$$

implying the strict positivity of all β_i's.

The first problem to be tackled is that of estimating the α_i's. One possible solution technique would be quadratic programming, but the results will be higher dependent on the numerical precision of the algorithm used. Maximum likelihood raises the problem of the type of distribution of the ε_t's in (3.1), least squares that of solving a system of equations for strictly positive values, a solution that rarely exists. Recent experience with canonical regression [see Paelinck (1979-1980) and (1980)] has suggested the solution to be presented here.

Redefine $\underline{y}^* = -\underline{y}$ and include $\underline{y}^*$ in a generalized matrix $\underline{X}^*$; then write

$$\underline{X}^*\underline{\beta}^* = \underline{\eta}^* \stackrel{\Delta}{=} -\underline{\varepsilon} \tag{3.25a}$$

$$\underline{\beta}^* > \underline{0} \quad . \tag{3.25b}$$

Canonical regression proceeds by minimizing $\underline{\eta}^{*\prime}\underline{\eta}^*$ subject to a Euclidean norm. This requires all $\underline{\beta}^*$'s to lie on a (unit) hypersphere centered around the origin. The following changes can be made:

(a) Center a hypersphere around the unit point, $\underline{i}^*$;

(b) Take a l_p-norm with p very large, to make the hypersphere extend towards the unit cube [see Paelinck (1981), Ch. 2];

(c) Reduce the radius to $1-\varepsilon$ so as to preclude zero values.

The full optimization problem then reads

$$\min \underline{\beta}^{*\prime}\underline{X}^{*\prime}\underline{X}^*\underline{\beta}^* \tag{3.26a}$$

$$\text{s.t. } (\underline{\beta}^* - \underline{i}^*)'(\hat{\underline{\beta}}^* - \underline{I}^*)^{p-2}(\underline{\beta}^* - \underline{i}^*) = 1 \quad . \tag{3.26b}$$

From the first-order conditions with dual parameter μ one obtains:

$$\underline{\beta}^{*o} = \mu p(\underline{X}^{*\prime}\underline{X}^*)^{-1}(\hat{\underline{\beta}}^{*o} - \underline{I}^*)^{p-2}(\underline{\beta}^{*o} - \underline{i}^*) \tag{3.27}$$

with the asymptotic property

$$\mu p \underline{\sim} -\underline{i}'(\underline{X}^{*\prime}\underline{X})\underline{\beta}^{*o} \tag{3.28}$$

given that for p very large,

$$\underline{i}'(\hat{\underline{\beta}}^* - \underline{I}^*)^{p-2}(\underline{\beta}^* - \underline{i}^*) \underline{\sim} -1 \quad . \tag{3.29}$$

A solution technique is presented in Appendix A.

The properties of $\underline{\beta}^*$ can be investigated by returning to specification (3.1). Paramter $\underline{\beta}_1$ belonging to variable $-\underline{y}$ has to be considered as fixed.

Property 1: partitioning $\underline{\beta}^*$ into $\underline{\beta}_1$ and $\underline{\beta}_2^*$, one gets

$$\begin{aligned} \Sigma\, \underline{\beta}_2^{*o} &= \underline{\beta}_1\underline{\beta}_2 + \underline{\beta}_1(\underline{X}'\underline{X})^{-1}\Sigma\, \underline{X}'\underline{\varepsilon} \\ &\quad + \Sigma\underline{i}'(\underline{X}^{*\prime}\underline{X}^*)\underline{\beta}^{*o}(\hat{\underline{\beta}}_2^{*o} - \underline{I})^{p-2}(\underline{\beta}_2^{*o} - \underline{i}) \quad . \end{aligned} \tag{3.30}$$

The second term of the right-hand side of (3.30) is zero under the usual OLS assumptions. As to the third term, it consists of mixed variance and covariance terms around 1, as it can be written

$$\begin{aligned} &\underline{i}'(\underline{X}^{*\prime}\underline{X}^*)\underline{\beta}^{*o}(\hat{\underline{\beta}}_2^{*o} - \underline{I})^{p-2}(\underline{\beta}_2^{*o} - \underline{i}) \\ &= \underline{i}'(\underline{X}^{*\prime}\underline{X}^*)[(\underline{\beta}_2^{*o}-\underline{i}^*)(\hat{\underline{\beta}}_2^{*o}-\underline{I})^{p-2}(\underline{\beta}_2^{*o}-\underline{i}) + \underline{i}(\hat{\underline{\beta}}_2^{*o}-\underline{I})^{p-2}(\underline{\beta}_2^{*o}-\underline{i})] \end{aligned} \tag{3.31}$$

which can now be made arbitrarily small by increasing p. Vector $\underline{\beta}_2^o$ can therefore be said to be p-unbiased.

Property 2: the elements of $\underline{\beta}_2$ are p-efficient in small samples within the class of p-unbiased estimators. Indeed, the variance-covariance matrix of $\underline{\beta}_2^o$ can be p-approximated by $\sigma^2(\underline{X}'\underline{X})^{-1}$, and to it the usual argument [see, e.g., Johnston (1963), pp. 110-112] for proving small-sample efficiency under common OLS assumptions can be applied.

A last point to be tackled in this paper is that of the distribution of $\underline{\beta}_2$. An introductory remark to be made is that the PAREX-formulation is compatible with distributions that have zero density around the zero-value of $\underline{\beta}_2$, e.g., a log-normal distribution. As already mentioned, the usual OLS variance-covariance matrix for $\underline{\beta}_2$, i.e., $\sigma^2(\underline{X}'\underline{X})^{-1}$, can be stated to rule the matter.

Our approach here will be of the inverse type, investigating whether there exists an unconstrained $\underline{\varepsilon}$-distribution comparible with a log-normal distribution of $\underline{\beta}_2^o$, meaning that these have expected values

$$E(\beta_{2i}^o) = e^{\mu_i + \frac{1}{2}\sigma_i^2} \tag{3.32a}$$

and variance expressions

$$\text{var}(\beta_{2i}^o) = e^{2\mu_i+\sigma_i^2}(e^{\sigma_i^2} - 1) \text{ [Aitchison and Brown (1957) p. 8].} \tag{3.32b}$$

Property 3: the computed PAREX parameters can follow a log-normal distribution for a given (compatible) distribution of $\underline{\varepsilon}$ along the extended real line.

Parameter $\underline{\beta}_2$ can indeed adopt all positive values over the open interval $(0,+\infty)$, for which values the log-normal probability densities are non-zero. From the equilibrium conditions (3.27) it can be seen that the same will be true for extreme values of $\underline{\varepsilon}$ which will reflect themselves in the $(\underline{X}^{*\prime}\underline{X}^*)^{-1}$ matrix via $\underline{y}^* \triangleq -\underline{y}$ and affect the computed, though still strictly positive values of $\underline{\beta}^{*o}$.

One-tailed log-normal tests can now be conducted for the $\underline{\beta}_2^o$ parameters, to test whether they are compatible (in probability) with extremely small positive values of $\underline{\beta}_2^o$ [expression (3.32a)].

Property 4: this Λp distribution of $\underline{\beta}_2^o$ is compatible to the first order with a normal distribution of the $\underline{\varepsilon}$.

To prove this start from (3.30) which gives

$$\ell n\ \underline{\beta}_2^{*o} \underset{\sim}{\sim} \ell n[\underline{\beta}_1\underline{\beta}_2 + \underline{f}(1o)] + [\underline{\beta}_1\underline{\beta}_2 + \underline{f}(\underline{o})]^{-1} \times \\ [\underline{\beta}_1(\underline{X}'\underline{X})^{-1} + J(o)]\underline{\varepsilon} \tag{3.33}$$

which implies first-order normality of the $\ell n\ \underline{\beta}_2^{*o}$ for normal $\underline{\varepsilon}$; $\underline{f}(\underline{o}) > \underline{0}$, the condition generated by (3.28)-(3.29), is the $\underline{0}$-value of the third right-hand side term of (3.30), $J(\underline{o})$ the $\underline{0}$-value of its Jacobian matrix.

CONCLUSIONS

Though results have been obtained for quadratic-programming estimates [see Ancot (1979)] and pre-test estimators, there is, as was reported, still some thought to be devoted to the spatial problem presented. In particular, research will be going on in the field of PAREX-estimators, and their generalization to other constraints.

A final possible approach to the estimation problem presented in section 2 is the Bayesian approach. According to this method the sample information is combined with the prior information expressed by the constraints (3.4), to obtain posterior distributions for the unknown coefficients, and location parameters of these posterior distributions could then be used as estimators for the coefficients. This is probably the most elegant approach, but it may prove analytically intractable, especially because of the presence of the specific spatial constraints. Analytical intractability can be circumvented by means of numerical integration techniques, but these too become quickly very expensive in terms of computer time and computer cost as the number of dimensions (i.e., the number of coefficients to be estimated) increases. However, it is also the intention of the authors to direct some research to the latter estimators.

APPENDIX A: SOLUTION TECHNIQUE FOR EQUATION 3.27

For solving (3.27) one could think of the following procedure. Start from an admissible initial $\underline{\beta}^*$-value near to zero, e.g.

$$\underline{\beta}^*_o = \{[1 - (K + 1)^{-p}]^{-1}\}\underline{i} \quad \text{(A1)}$$

and compute

$$(\mu p)_o \simeq -\underline{i}'(\underline{X}^{*\prime}\underline{X}^*)\underline{\beta}^*_o \quad . \quad \text{(A2)}$$

Compute $\underline{\beta}^*_1$ from (3.27) and $(\mu p)_1$ from (A2) with $\underline{\beta}^*_1$ substituted for $\underline{\beta}^*_o$. The procedure raises problems of convergence, however [Hughes-Hallett (1979)], towards the minimum solution [(3.27) has many solutions]. A constrained gradient method [Paelinck and Nijkamp (1975), Ch. 6] is probably preferable.

REFERENCES

Aitchison, J. and Brown, J., 1957. The Lognormal Distribution. University Press, Cambridge.

Ancot, J. and Paelinck, J., 1979. A Discriminant Analysis Approach to Regional Threshold Problems, with an Application to Spanish Data, Papers of the Regional Science Association, Vol. 42, 139-152.

Ancot, J., Kemp, P., Paelinck, J. and Smit, H., 1979. DATONEI, The Functional Databank of the Netherlands Economic Institute, Series: Foundations of Empirical Economic Research, 1978/8, Rotterdam, to appear in A. Kuklinski, Regional Dynamics of Socioeconomic Change, Volume II, Polarized Development and Regional Policies, Finnpublishers, Tampere.

Ancot, J., 1979. An Econometric Demand Model for Water with an Application to the Demand for Water in East Gelderland, Models for Water Management II, Provinciale Waterstaat-Gelderland, Arnhem.

Ancot, J., 1979. Une approche par analyse discriminante à des problèmes de seuils régionaux et d'analyse de localisation. Application à des données espagnoles, Series: Foundations of Empirical Economic Research, 1979/9, Rotterdam, to appear in Recherches Economiques de Lovain.

Ancot, J., Klaassen, L., Moole, W. and Paelinck, J., 1980. Main Problems of Regional Development Modelling (in Europe), Series: Foundations of Empirical Economic Research, 1980/4, Rotterdam, to be published by International Institute of Applied Systems Analysis, A. Andersson and F. Snickars (eds.).

Ancot, J. and Paelinck, J., 1980. Multiple Regression with Non-Stochastic Linear Inequality Constraint: A Simple Pre-Test and an Estimation Strategy, presented at the Colloque Structures Economiques et Econometriques, Lyon, May, to be published in its Proceedings.

Chevailler, J. and Paelinck, J., 1978. Parameter-Component Models in Spatial Econometrics, in The Econometrics of Panel Data, Annales de l'INSEE, Vol. 30-31: 83-98.

Hughes-Hallett, A., 1979. Efficient Methods of Model Solution: Some Comparisons of Gauss-Seidel and Jacobi Iterative Techniques, Erasmus University, Rotterdam (mimeographed).

Jansen, R., 1975. Effects of Linear Inequality Constraints on Distributions of Parameter Estimates in the Standard Linear Model, Erasmus University, Econometric Institute, Report 7509/E, Rotterdam (mimeographed).

Johnston, J., 1963. Econometric Methods, McGraw-Hill, New York.

Judge, G and Takayama, T., 1966. Inequality Restrictions in Regression Analysis, JASA, 61: 166-181.

Judge, G. and Bock, M., 1978. The Statistical Implications of Pre-Test and Stein-Rule Estimators in Econometrics, North Holland Publ. Co., Amsterdam.

Lovell, M. and Prescott, E., 1970. Multiple Regression with Inequality Constraints: Pretesting Bias, Hypothesis Testing and Efficiency, JASA, 65: 913-925.

Netherlands Economic Institute, DATONEI, de databank van het Nederlands Economisch Instituut, Fase 1, Rotterdam, 1978; fase 2 en 3, Rotterdam, 1980.

Paelinck, J., 1974. Modeles de politique économique régionale basés sur l'analyse d'attraction, L'Actualité Economique, 4: 559-564.

Paelinck, J. and Nijkamp, P., 1975. Operational Theory and Method in Spatial Economics, Saxon House and Lexington, Farnborough and Massachusetts.

Paelinck, J., 1977. Analyzing Regional Growth, in: The Strategy of Future Regional Economic Growth, IIASA, Laxenburg, 67-74.

Paelinck, J., 1978a. Estimation de systemes spatiaux complexes, in: Actes du Colloque Structures Econometriques et Econometrie, Universite de Lyon, Departement de Mathematique, Lyon.

Paelinck, J., 1978b. Une theorie des seuils de croissance regionaux, in: Seuils d'efficacité de la planification et de l'action régionale, IDEA, Mons., 101-107.

Paelinck, J., 1979. Deux modèles spatiaux. Institut National de Recherche Scientifique, Montréal, and Netherlands Economic Institute, Rotterdam (mimeographed).

Paelinck, J. and Klaassen, L., 1979. Spatial Econometrics, Saxon House, Teakfield.

Paelinck, J. a.o., 1979-80. Projet Tape, Etudes No. 1,2,3 (mimeographed).

Paelinck, J., 1980. An Axiomatic Reformulation of Regional Science, paper read at the First International Conference on Regional Science, Cambridge, Mass. (June 16).
Paelinck, J., 1981. Abstract Spatial Economic Analysis (to appear 1981).
Teekens, R. and Koerts, J., 1972. Some Statistical Implementation of the Log Transformation of Multiplicative Models, Econometrica, 40: 793-819 (No. 5).

---

*The authors gratefully acknowledge the help of L. Gérard-Varet, Peeters and Smith in the preparation of this paper.

APPLICATIONS OF OPTIMAL CONTROL THEORY APPROACHES TO SPACE TIME PROBLEMS

From the perspective of a planner, manager, or policy maker, system description is but a prelude to the ability to influence or control the trajectories of system state variables. The use of formal models of system control is perhaps problematic when we must acknowledge that neither the system dynamics nor the social objectives are known in sufficient detail or precision to implement an actual decision making strategy. Optimization models applied to poorly understood systems must be interpreted with caution. These models should be used either to give qualitative insights into the process under review or alternatively to serve as prototypes for future applied numerical control applications. The papers in this section illustrate both of these features--enhancement of qualitative understanding and demonstration and resolution of severe technical problems.

Bennett and Tan study the allocation of intergovernmental grants within an optimal control context. Their problem, involving a non-linear objective function, linear inequality constraints, and stochastic variables can be solved using either of two algorithms developed by the authors. The two algorithms presented will yield different solutions and neither may be the true optimum. However, the second algorithm does explicitly mirror the multi-stage nature of the problem and it is this solution method that is applied to the thirteen boroughs of Outer London using data from the 1974-79 period. Data limitations restrict the feasibility of using modern estimation methods, but the methodology developed appears to indicate that the spatial and temporal allocation of government grants may be susceptible to useful quantitative planning.

Barber and Ralston introduce another dynamic spatial planning problem, that of road investment, where the investment in transportation affects both supply of and the demand for an agricultural commodity. The optimal sequencing of road investment, even with very simple assumptions regarding the nature of the economic impacts of transportation, is more complicated than one might expect. Construction lags of different lengths and non-monotonic investment outlays emerge in the numerical example presented. Whether this problem can be imbedded in a more realistic spatial environment without abandoning its clear mathematical structure remains to be seen.

Quite a different use of optimization methods is illustrated by Fujita. He models the long run process of urban spatial development from both the planning and descriptive points of view.

It is well known that competitive systems are such that they often behave according to optimizing principles. Choice variables in Fujita's models include space and time specific construction and demolition rates and the number of activity units. Optimality conditions are derived from the maximum principle of optimal control theory. Variants of the basic model are developed to accomodate such features as urban renewal, mixtures of different buildings in the same zone, and even non-monotonic spatial variations of house size and rent. A variety of extensions and alternative models are identified.

SPACE-TIME MODELS OF FINANCIAL RESOURCE ALLOCATION*

R.J. Bennett
Department of Geography
University of Cambridge

K.C. Tan
Department of Comparative Social Sciences
Universiti Sains Malaysia

INTRODUCTION

A problem of resource allocation that occurs in most western countries is that of allocating grants from a central (or federal) government to a series of state or local governments. This intergovernmental grant allocation problem has been treated in economics as well as geography. This paper develops solutions to this allocation problem using the methods of optimum control theory, and then presents empirical results of applying this theory to allocation of the Rate Support Grant (hereafter to be referred to as RSG) in London.

The distribution of RSG based upon the methods of optimal control has been discussed earlier by the authors. Bennett and Tan (1979) for example give the control theory solution for RSG allocation with separate grant elements (of needs, resources and domestic elements). Tan (1979b), in contrast, discusses the manner in which control theory solutions can be applied to combined grant elements as a so-called "unitary" grant. To date, the only empirical analysis with these methods has been by Tan (1979a) for the twelve Inner London Boroughs and the City of London. The approach of the authors draws on the theory of optimal control derived in engineering by Kalman (1960). in economics by Chow (1975), Pindyck (1973) and Aoki (1976); in geography by MacKinnon (1975), Bennett (1979) and Bennett and Chorley (1978); and in regional science by Sakashita (1967) and Fujita (1981). The particular development of this paper is important in offering a solution to control problems when the control variables are subject to linear equality and inequality constraints. This problem also has been tackled by Nijkamp (1975), and Nijkamp and Verhage (1976), who offer alternative solutions that consider only constraints on the state variables. In addition, Sakashita (1967) offers constrained control solutions for deterministic continuous variable systems.

CRITERIA FOR ALLOCATING INTERGOVERNMENTAL GRANTS

The criteria that can be used for allocating intergovernmental grants concern four main variables: the local tax base, the local tax rate, the level of local financial need, and the local expenditure level. With these four variables six main criteria have been developed for allocating grants. Spatial criteria for grant allocation are frequently the most important in practice. They consist of two subsets:

(a) *Equalization.* Equalization concerns the attempt to place state or local governments in an equal position to provide a given service level at a given cost. Two criteria of equalization are usually employed, often simultaneously. First, it is important to compensate local governments for differences in levels of expenditure need. These arise because of different sizes of client groups, costs of servicing different groups, or differences in the economics of service delivery. A second criterion is resource equalization, whose aim is to compensate areas for differences in the size of their tax base. Tax base must be assessed against the revenue sources available, and hence may bear little relation to the distribution of personal wealth.

(b) *Spillover.* A second spatial criterion of grant allocation concerns attempts to use grants to compensate local areas for externality effects arising from the spillovers of costs and benefits from other local government units. Such spillovers frequently occur as a result of commuting, but also result from migration and tax exporting [see Tiebout (1956)]. Indeed Tiebout has erected an economic model of residential choice based on criteria of tax and benefit stimuli to migration. The use of intergovernmental grants to overcome such spillover effects was originally suggested by Pigou (1947) and is frequently referred to as the compensation principle.

Temporal criteria of allocation introduce comparisons between grant sums between years. Three factors are particularly important:

(a) *Stability*. This requires that grants maintain a stable pattern of distribution from one time period to another and that changes in distribution procedures result only from objective changes in circumstances.

(b) *Certainty*. This requires that once grants are allocated the total sum remains unchanged. In the allocation of the RSG, in particular, considerable uncertainty has been introduced through inconsistent use of Increase Orders.

(c) *Dynamics*. This is a very important criterion, and requires that past and future criteria are considered in present grant allocation. Past expenditure patterns are important in affecting future needs, and future needs should affect present allocations. Perhaps the best example is the analysis of population cohorts: knowledge of the present age distribution of a population gives good potential to forecast the future needs for schools of different categories, or services for the aged. Small present needs may therefore sometimes justify a large grant allocation in order to cater for large future needs.

Vertical allocation criteria concern fiscal balance. They refer to the redistribution of revenue between levels of government to meet hierarchical differences in expenditure requirements. Many public services can be sufficiently organized only at a local level, but most of the high yielding taxes are usually best reserved to central governments. Hence, resources must be redistributed between levels of government to achieve fiscal balance.

The aim of grants is to ensure a similar pattern of services in each local authority, and to achieve this goal with similar fiscal effort (ratio of tax rate to tax base). Few local services in Britain are redistributional in intent so that the main entry of social criteria is through revenue effort. However, in Britain, as in most countries, it usually has been deemed preferable to solve the income distribution issue of taxation at the national level [see Bennett (1980)] and this requires grants to produce similar fiscal efforts or tax burdens in each local authority.

Distinct from the social distribution problem is that of economic welfare. This is composed of two sub-elements:

(a) *Accountability*. This requires that clear lines of responsibility are visible to the voter. This in turn requires that local taxation represents a significant proportion of local revenue (as a perceived "cost"), and this requires grant levels to be kept in check.

(b) *Efficiency.* This also requires that grants do not affect too greatly the "cost" of services. Grants should have minimal impact on prices in the economy as a whole, and should not sudsidize profligate local spending and high labor/output ratios.

Clearly a very large number of other allocation criteria can be employed in grant determination. Two of the most important of these are feasibility and intelligibility. Feasibility requires that grant programs must be based on data that feasibly can be collected and the accuracy of which is beyond reasonable doubt. Intelligibility requires that the grant distribution programs are readily comprehensible both to the voters and to the elected representatives at each level of government concerned.

Although each of the six criteria discussed above should be important in determining methods of grant allocation, the control theory approach developed below emphasizes the first three, excluding (at this stage) treatment of spillover.

THE RATE SUPPORT GRANT (RSG) IN ENGLAND AND WALES

The rate support grant (RSG) is the major intergovernmental grant that is used in the United Kingdom, and is responsible for £7,000m of transfers of support from central to local government. As such, it is the largest single item of public expenditure after defense. It is a general or "block" grant that can be used relatively flexibly by local authorities. However, it is distributed by taking into account the expenditure needs and local tax capacity of different local governments. Initiated in its present form in 1974, various formulae have been applied to its allocation, many terms of which date back to 1929 [see Committee of Inquiry Into Local Government Finance (1976)].

Although transferred to local authorities as a single sum of money, the RSG is divided into three elements: needs, resources and domestic.

The need element (£3,712m in 1977/78 at 1976 prices) is aimed at allocating to local authorities an amount equal to their expenditure requirements. It is assessed according to a number of needs criteria that are based on three factors. One is the variation in the proportion of people who consume each local authority's services (e.g., the number of old people, the number of people in various levels of education). A second is the variation in the quantity of inputs required to provide services in each local authority (e.g., variations in pupil-teacher ratios between urban and rural areas). Another is the variations in costs of inputs in different local authorities (e.g., salary weightings in London and the South East).

The method of needs allocation that has been in use since the fiscal year 1974/75 is based on a so-called regression analysis. It attempts to "explain" total local authority expenditure in each need area in terms of "needs" indicators by use of multiple regression. The dependent variable in this procedure is the total expenditure per head of each needs authority (the County and District sums). The independent variables vary from year to year, but are consistent in including measures of population size, population characteristics, and size and density variation of local authorities. Hence, for any fiscal year the resulting regression equation is given by:

$$Y_{ti} = \underline{\alpha}_t' \underline{X}_{ti} \quad , \qquad (1)$$

where Y_{ti} is a scalar, $\underline{X}_{ti}$ is a vector of dimension $(n \times 1)$ that represents the various n needs indicators in area i, and $\underline{\alpha}_t$ an $(n \times i)$ vector of coefficients, or regression parameters. The prime denotes the transpose of this vector. There are N needs authorities $(i = 1,2,...,N)$, and there are n needs indicators in each of the N local authorities. From this multiple regression the values of the regression coefficients are determined, their estimates being denoted by $\underline{\alpha}_t$. The final regression formula used in allocation then is modified in three ways. First, any variable included in the original regression that yields a negative coefficient $(-\alpha_j)$ is omitted and the regression recalculated. This represents a *positivity constraint.* This does not apply to one needs indicator, namely, the number of low income households in authority i, that is deemed too important to omit. This variable is included in the final regression calculation, but omitted from the formula used for the distribution of the needs element. A second modification is to incorporate the effect of the total financial resources available for allocation under the needs element of RSG. This is decided exogenously to the RSG procedure by central government, and represents a *global expenditure constraint.* A third modification is imposed to ensure that the needs allocation to each authority does not fluctuate too wildly from year to year. In 1975/76 this modification restricted the use of the regression formula to allocation of only 28.7% of the needs element. The remaining 71.3% was allocated solely on the basis of the needs element entitlement in the previous year (1974/75). The use of this procedure imposes a strong damping factor on needs allocation, and represents a *holding harmless constraint.*

The effect of the three modifications noted above can be expressed as three constraints on the calculation of the needs formula (1). These are given as follows:

Needs positivity constraint

$$Y_{t,i} \geq 0 \; ; \qquad (2)$$

Global needs expenditure constraint

$$\sum_{i=1}^{N} Y_{t,i} = f_t^N \; ; \text{ and,} \qquad (3)$$

Needs continuity constraint

$$y^a_{t+1,i} = Y_{t+1,i} + K\, Y^a_{t,i} \; . \qquad (4)$$

The first constraint (2) requires the needs expenditure allocated to authority i to be positive. In other words, there can be no loss of finance from local to central government under the needs formula. The second constraint (3) requires the total needs allocation to all authorities N to equal the sum f_t^N set aside by central government for this purpose. The third constraint (4) requires the part of next year's allocation is equal to a proportion k of the present year's allocation. This gives the proportion determined by the needs formula as only the first term on the right hand side of equation (4), and the total allocation each year is given as $Y^a_{t,i}$. Since only the first term $u_{t,i}$ can be determined freely, only this element is discussed further below.

The resources element (£1,788m in 1977/78 at 1976 prices) is designed to compensate differences between local authority rateable values across the country. Since these control the ability to raise revenue locally by rates, a subsidy is paid to each local authority that falls below a pre-determined per capita rateable value standard for England and Wales. Moreover,

$$R_{t,i} = (\sigma P_{t,i} - V_{t,i}) r_{t,i} \; , \qquad (5)$$

where $R_{t,i}$ is the finance given to local authority i under the resources element, $P_{t,i}$ is the population in authority i, $V_{t,i}$ is the total rateable value of authority i, and $r_{t,i}$ is the tax rate (or rate poundage) exacted in the authority, all at time t. The constant σ is defined by the central government prior to allocation of the resources element, and is usually termed the "standard rateable value per head."

The amount of resources element received by an authority is governed by the setting chosen for the standard rateable value per head σ. The value of σ multiplied by the population $P_{t,i}$ gives a measure of the total tax capacity of any authority based on its population, if the authority has the standard rateable value. The difference between the standard and the actual rateable value of the authority $V_{t,i}$ determines two conditions for allocation of the resources element. First, when the standard rateable value exceeds the local rate base, then a resources element is paid to that authority. The magnitude of this payment is equal to the difference in standard and actual rate bases times the rate of local tax exacted $r_{t,i}$. The second condition applies when the standard rateable value is less than or equal to the local rate base. In this situation, no resource payment is made to the authority since it is assumed that the authority has sufficient resources of its own to cover its required expenditure. Hence, a positivity constraint is imposed also on the resources element to ensure that, like the needs element, an allocation by central government does not result in an actual loss of finance to a local authority. This is written as follows:

Resources positivity constraint

$$R_{t,i} = \begin{cases} (\sigma P_{t,i} - V_{t,i})r_{t,i} & , \sigma P_{t,i} \geq V_{t,i} \\ 0 & , \sigma P_{t,i} \leq V_{t,i} \end{cases} \quad . \tag{6}$$

The allocation of the resources element is also subject to a second constraint that arises from the central government's need to limit the total funds payable under this element of RSG:

Global resources expenditure constraint

$$\sum_{i=1}^{N} R_{t,i} = f_t^R \quad . \tag{7}$$

This restriction states that the sum of all resource element payments over the N local authorities must equal, and not exceed, a predetermined amount f_t^R set aside by central government for this purpose.

The domestic element (£657m in 1977/78 at 1976 prices) is a subsidy to domestic ratepayers by which the full poundage of their rateable values is reduced by 18½p in England and 36p in Wales; non-domestic (commercial) property owners receive no subsidy, i.e.,

$$D_{t,i} = \delta \, V^{d}_{t,i} \quad , \tag{8}$$

where $D_{t,i}$ is the RSG allocation under the domestic element, $V^{d}_{t,i}$ is the total domestic rateable value of authority i, and δ is the tax rate (rate poundage) by which domestic rates are to be subsidized.

The domestic element is hypothecated such that the local authorities are obliged to use this element to reduce the tax rate to domestic ratepayers. Hence, it differs from both the needs in resources elements that are block grants, and is an incongruity in the RSG mechanism. As a result, it has often been suggested that it should be moved to become part of the specific grants category, and accordingly is not discussed further in this paper.

RATE SUPPORT GRANT AS A CONTROL PROBLEM

The control theory approach incorporates the dynamics of dependence of expenditure and revenue levels on previous levels of expenditure, grant levels, resource base and tax rates by the use of a dynamic model. Although this could be specified in a number of forms, that usually chosen as the most convenient for subsequent mathematical developments is the following equation:

$$\underline{X}_t = \underset{\sim}{A} \, \underline{X}_{t-1} + \underset{\sim}{B} \, \underline{U}_t + \underline{b}_t + \underline{e}_t \quad , \tag{9}$$

where

$\underline{X}_t$ is a vector of need and grant variable (the *state* vector);

$\underline{U}_t$ is a vector of grant allocations (the *control* vector);

$\underline{b}_t$ is a vector of exogenous variables (uncontrolled variables);

$\underline{e}_t$ is a vector of independently distributed error terms;

$\underset{\sim}{A}$ is a matrix relating present to previous years' needs resources, and the such (the *state* matrix); and,

$\underset{\sim}{B}$ is a matrix relating present grant allocation to the level of need, satisfied (the *control* matrix).

The precise structure of the dynamic model (9) depends upon the form of grant allocation procedure adopted, i.e., whether a unitary, separate element, or some other grant structure is used. The two most important approaches are given below.

Case A: Control theory unitary grant

$$\begin{pmatrix} E_{ti} \\ R_{ti} \end{pmatrix} = \begin{pmatrix} \alpha_{1i} & 0 \\ 0 & 1 \end{pmatrix} \begin{pmatrix} E_{t-1i} \\ R_{t-1i} \end{pmatrix} + \begin{pmatrix} \alpha_{3i} & \alpha_{2i} RV_{ti} \\ 0 & RV_{ti} \end{pmatrix} \begin{pmatrix} T_{ti} \\ \Delta r_{ti} \end{pmatrix} + \begin{pmatrix} \alpha_{0i} \\ 0 \end{pmatrix} , \quad (10)$$

where

E_{ti} = total expenditure;

R_{ti} = total rate yield;

RV_{ti} = rateable value;

T_{ti} = unitary grant allocation

Δr_{ti} = change in rate poundage from t-1 to t; and,

α_{ji} = parameters to be determined.

Each term is defined for local authority i at time t.

Case B: Control theory separate grant elements

Needs

$$\begin{pmatrix} \underline{x}_{ti} \\ T^N_{ti} \end{pmatrix} = \begin{pmatrix} \underset{\sim}{\alpha}_1 & \underset{\sim}{\alpha}_2 \\ 0 & 0 \end{pmatrix} \begin{pmatrix} \underline{x}_{t-1} \\ T^N_{t-1} \end{pmatrix} + \begin{pmatrix} \underline{\alpha}_3 \\ 1 \end{pmatrix} \quad T^N_{ti} , \quad (11)$$

Resources

$$\begin{pmatrix} RV_{ti} \\ T^R_{ti} \end{pmatrix} = \begin{pmatrix} 1 & \frac{1}{r_{t-1i}} \\ 0 & 0 \end{pmatrix} \begin{pmatrix} RV_{t-1i} \\ T^R_{t-1i} \end{pmatrix} - \begin{pmatrix} \frac{1}{r_{ti}} \\ -1 \end{pmatrix} T^R_{t-1} + \begin{pmatrix} 1 & -1 \\ 0 & 0 \end{pmatrix} \begin{pmatrix} \sigma_t \\ \sigma_{t-1} \end{pmatrix} , \quad (12)$$

where

$\underline{x}_{ti}$ = vector of need indicators;

T^N_{ti} = needs grant allocation;

T^R_{ti} = resources grant allocation;

σ_t = standard rateable value per head; and,

$\underset{\sim}{\alpha}_j$ = parameters to be determined.

Other terms are as defined in the unitary grant structure, and each term is defined for local authority i at time t.

Two points should be noted with respect to these equations. First, each equation (10) to (12) is symmetrical in structure to the control equation (9), and terms can be identified with each other. Second, each equation captures the desired form of dependence of grant, need and rate levels on previous years. This is achieved chiefly through the definition of the state matrix $\underset{\sim}{A}$ in each case.

Using either the unitary or separate grant structures the control theory approach proceeds by the same route. The state variables in equation (9) are used to define target levels that are sought, within the RSG allocations procedure, to match. Depending upon the grant structure used, these targets will be levels of need matching, expenditure matching or tax rate matching. Having defined target levels, as $\underline{a}_t$, the control approach seeks to achieve that level of grant allocation that renders as a minimum the difference between targets and the actual level of needs, expenditure, and so forth; i.e., it makes the term $(\underline{x}_t - \underline{a}_t)$ as small as possible. It is usual to weight these differences differentially for each target, and to sum the differences over the planning time horizon adopted. A frequent choice of weighting is the quadratic:

$$\underset{\sim}{W} = \sum_{t=1}^{T} (\underline{X}_t - \underline{a}_t)'\underset{\sim}{M}(\underline{X}_t - \underline{a}_t) \quad , \tag{13}$$

where $\underset{\sim}{W}$ is the total (weighted) deviations of actual levels from targets, $\underset{\sim}{M}$ is the weighting matrix chosen, and T is the final time for which it is sought to make grant allocations. The quadratic weighting criterion (13) yields simple mathematical solutions and is the one used below, but it does have a number of conceptual problems, such as it gives equal weight to both positive and negative deviations from target.

Having defined target levels and chosen a criterion to measure deviations from targets, the control approach then yields an optimal grant allocation in the following form:

$$\underline{U}_t = f(\underline{X}_t) \quad . \tag{14}$$

These grant allocations (the control variable $\underline{U}_t$) are defined as some function f(•) of the need, expenditure and resource indicators (the state variable $\underline{X}_t$). However, the specific character of the RSG problem requires that the criterion (13) is minimized to yield grant settings (14) subject to a series of special constraints summarized in equations (2), (3), (4), (6) and (7). The solution to the problem of determining that level of grant allocation to each local authority, that yields a minimum to the deviation of need and expenditure allocation from target [i.e., renders (13) a minimum], while simultaneously satisfying the four constraints, is not simple. Two possible solution strategies will be presented now.

CONTROL THEORY SOLUTION

There has been very little development of research for determining optimal control solutions in which the control variables are subject to constraints. The following discussion draws from the constrained solutions to optimal control problems given by Tan (1979a,b). Specifically, the control problem may be stated as:

given the linear control model (9) dropping the b_t term,

$$\underline{X}_t = \underset{\sim}{A}\,\underline{X}_{t-1} + \underset{\sim}{B}\,\underline{U}_t + \underline{e}_t \quad . \tag{15}$$

With initial conditions $\underline{X}$, and the parameters $\underset{\sim}{A}$ and $\underset{\sim}{B}$ unknown but constant, the control problem is to find the set of control variables that minimizes the quadratic function

$$V = E_o(W) = E_o \left[\sum_{t=1}^{T} (\underline{X}_t - \underline{a}_t)' \underset{\sim}{M}_t (\underline{X}_t - \underline{a}_t) \right] , \tag{16}$$

when the control variables are subject to a set of q linear constraints and the non-negativity restrictions

$$\left.\begin{array}{ll} \underset{\sim}{F}_t \underline{U}_t = \underline{f}_t & (17a) \\ \underline{U}_t \geq 0 & (17b) \end{array}\right\} \quad t = 1,\ldots,T \quad ,$$

where $\underset{\sim}{F}_t$ is a (q-by-m) diagonal identity matrix, $\underline{f}_t$ is a (q-by-1) vector of government spending limits, and q<m is the number of linear equality restrictions on the m control variables.

The random error term is assumed to have zero mean, with

$$\text{Cov}(e_t, e_s) = \begin{cases} \underset{\sim}{R} \ (t=s) & , \\ 0 \ (\text{otherwise}) & , \end{cases} \tag{18}$$

and to be statistically independent of A and B. Moreover, it is assumed that the probability distribution of A and B is given at the initial conditions, and then remains constant. In (16), E_o denotes the expectation conditional on the initial conditions $\underline{X}_o$.

The strategy proposed by Tan is composed of two parts. First, the control problem is solved with only the equality constraints imposed. Second, if the resulting solution is not feasible in the sense that there exists at least one local authority for which the inequality constraint is not satisfied, then the grant allocations need to be adjusted for feasibility. Two methods are proposed, one of which is based on the dual simplex method [Hadley (1962)], the other is referred to as the method of imposing constraints. This two-step procedure may be summarized as follows:

Step 1

With just the constraint (17a) present, the optimal control solution is given by

$$\underline{U}_t^* = \hat{\underset{\sim}{G}}_t \underline{X}_{t-1} + \hat{\underline{g}}_t \ , \tag{19}$$

where the covariance equations are given by

$$\hat{\underset{\sim}{G}}_t = \underset{\sim}{G}_t - \underset{\sim}{D}_t^{-1} \underset{\sim}{F}_t' \underset{\sim}{P}_t^{-1} \underset{\sim}{F}_t \underset{\sim}{G}_t \ , \text{ and} \tag{20}$$

$$\hat{\underline{g}}_t = \underline{g}_t - \underset{\sim}{D}_t^{-1} \underset{\sim}{F}_t \underset{\sim}{P}_t^{-1} \underset{\sim}{F}_t' \underline{g}_t + \underset{\sim}{D}_t^{-1} \underset{\sim}{F}_t' \underset{\sim}{P}_t^{-1} \underline{f}_t \ , \tag{21}$$

and where dynamic programming soltuions define

$$\left.\begin{aligned} \underset{\sim}{D}_t &= E(\underset{\sim}{B}' \underset{\sim}{H}_t \underset{\sim}{B}) \\ \underset{\sim}{G}_t &= -\underset{\sim}{D}_t^{-1} E(\underset{\sim}{B}' \underset{\sim}{H}_t \underset{\sim}{A}) \\ g_t &= \underset{\sim}{D}_t^{-1} E(\underset{\sim}{B}') \underline{h}_t \\ \underset{\sim}{P}_t &= \underset{\sim}{F}_t \underset{\sim}{D}_t^{-1} \underset{\sim}{F}_t' \ . \end{aligned}\right\} \tag{22}$$

In addition, the terms $\underset{\sim}{H}_t$ and $\underline{h}_t$ satisfy the Ricatti equations

$$\left.\begin{aligned} \underset{\sim}{H}_T &= \underset{\sim}{M}_T \\ \underset{\sim}{H}_t &= \underset{\sim}{M}_{t-1} + E(\underset{\sim}{A}'\underset{\sim}{H}_t\underset{\sim}{A}) + E(\underset{\sim}{A}'\underset{\sim}{H}_t\underset{\sim}{B})\hat{\underset{\sim}{G}}_t \\ \underline{h}_t &= M_T\underline{a}_T \\ \underline{h}_{t-1} &= \underset{\sim}{M}_{t-1}\underline{a}_{t-1} + E(\underset{\sim}{A}')\underline{h}_t - E(\underset{\sim}{A}'\underset{\sim}{H}_b\underset{\sim}{B})\hat{\underline{g}}_t \end{aligned}\right\} \quad (23)$$

By assuming certainty equivalence, the random term $\underline{e}_t$ can be set to its expected value of zero. The above equations (19) to (23) then allow us to obtain the solution path $\underline{U}_1^*,\ldots,\underline{U}_T^*$, subject to only linear equality control constraints (17a). If this solution path satisfies the non-negativity constraints (17b), then it is also the optimal solution to the control problem stated above. However, if this solution path does not satisfy the inequality constraints (19b), then the second step of the solution must be followed.

Step 2

Method A (based on dual simplex algorithm)

This method requires rewriting equations (19) to (23) as a single-stage solution strategy. The one-stage control solution is then calculated, and a check is made to see if the constraint $U_t \geq 0$ is satisfied. If the constraint is not satisfied, then the dual simplex solution strategy is followed. First, re-write the multi-stage problem as a step-by-step single stage optimization problem. Re-write the objective function (16) as a one-stage criterion:

$$V(1,t) = E(\underline{X}_t - \underline{a}_t)'\underset{\sim}{M}_t(\underline{X}_t - \underline{a}_t) \quad . \qquad (24)$$

Then the single-stage control problem at time t given all information until t is given by:

$$\text{MIN } V(1,t)$$

$$\text{s.t.:} \quad \underset{\sim}{F}_t\underline{U}_t = \underline{f}_t \quad , \text{ and} \qquad (25a)$$

$$\underline{U}_t > 0 \quad . \qquad (25b)$$

The optimal control solution with just (25a) present is easily found to be given by

$$\underline{U}_t^* = \hat{G}_t^* \underline{X}_{t-1} + \hat{\underline{g}}_t^* \quad , \tag{26}$$

where

$$\hat{G}_t^* = G_t^* - \hat{D}_t^{-1} F_t' \hat{P}_t^{-1} F_t G_t^* \quad , \text{ and} \tag{27}$$

$$\hat{\underline{g}}_t^* = \underline{g}_t^* - \hat{D}_t^{-1} F_t' \hat{P}_t F_t \underline{g}_t^* + \hat{D}_t^{-1} F_t' \hat{P}_t^{-1} \underline{f}_t \quad , \tag{28}$$

and where

$$\left.\begin{aligned} \hat{D}_t &= E(B'M_tB) \\ G_t^* &= -\hat{D}_t^{-1} E(B'M_tA) \\ \underline{g}_t^* &= \hat{D}_t^{-1}[E(B')M_t\underline{a}_t] \\ P_t &= F_t \hat{D}_t^{-1} F_t' \quad . \end{aligned}\right\} \tag{29}$$

This is a single stage version of equations (19) to (23). However, note that since this is a single-stage solution, the Ricatti equations (23) are no longer required. If the solution $\underline{U}_t^*$ given to (26) does not satisfy (25b), then it is adjusted using the dual simplex method. This is based on formation of a new programming problem:

$$\text{MAX } z = \underline{1}_v'(F_t\underline{U}_t - \underline{f}_t) \tag{30}$$

$$\left.\begin{aligned} \text{s.t.:} \quad \underline{U}_t - \bar{B}_t \underline{\beta}_t &= \underline{U}_t^* \\ \underline{U}_t &\geq 0 \\ \underline{\beta}_t &\geq 0 \end{aligned}\right\} \tag{31}$$

$$\underline{\beta}_t'\underline{U}_t = 0 \quad ,$$

where $\underline{\beta}_t$ is a vector of Lagrange multipliers, and $\bar{B}_t = \hat{D}_t^{-1} - \hat{D}_t^{-1} F_t' \hat{P}_t^{-1} F_t \hat{D}_t^{-1}$, and $\underline{1}_v$ is a (q-by-1) vector whose elements are all unity. Except for the restriction $\underline{\beta}_t'\underline{U}_t = 0$ this problem is linear. An initial basic solution to (3) is $\underline{U}_t = \underline{U}_t^*$. Of course $\underline{U}_t^* \geq 0$, otherwise the control problem with non-negativity constraints is solved. Thus, the initial basic solution to the primal problem is infeasible.

However, it is dual feasible. Hence, the dual simplex algorithm can be used to solve this problem (30). However, because of the constraint $\underline{\beta}'\underline{U}_t = 0$, a variation to this algorithm is needed in which U_{it} is not allowed to enter the basis when $\beta_{it} > 0$, and vice versa. This variation is identical to that found in Wolfe's (1959) algorithm for solving quadratic programming problems. For well-defined problems the algorithm will terminate to an optimal solution in a finite number of steps, yielding to the adjusted solution $\underline{U}_t^*$ that satisfies the constraint (25b) as well.

A criticism of this method is that in finding the sequence of single-stage control solutions, the future outcomes are not taken into consideration. Moreover, it will not be optimal in the stochastic case. As a result a second method of imposing constraints merits attention.

Method B (method of imposing contraints)

This method is stimulated by the property that at each iteration in the dual simplex algorithm the basic variable with the largest negative value is made non-basic (i.e., it is set to the value of zero). When this basic variable is a control variable, then making it non-basic amounts to adding another control constraint, namely that this control variable is equal to zero.

Using this stimulus the method of imposing constraints consists of a three-step strategy:

(i) Set the largest negative control variable to zero (its non-basic value).

(ii) Resolve the T-stage control problem using equations (19) to (23).

(iii) Check if all control variables are non-negative. If not, set the largest negative control variable to zero as in step (i) and repeat until all control variables are non-negative. Hence constraint (25b) will be satisfied.

It is not necessarily true that the control solution obtained in this way is the true optimum. This is because of two reasons. First, the set of added control constraints at the zero level may not be corrected. Second, once a control variable is set to zero, it remains so throughout the analysis, whereas in the dual simplex algorithm this is not necessarily true. However, the solution obtained by this method of imposing constraints takes into account the multi-stage nature of the problem, and will be different from the solution obtained by the sequence of single-stage solutions of the dual simplex method.

EMPIRICAL RESULTS

Empirical results of applying the optimal control algorithm discussed above are developed in this section for thirteen of the Outer London boroughs for the five years 1974/75 to 1978/79. This is a relatively small scale investigation, but evidences the general properties of the control results. Since it is a preliminary analysis, only the results of the unitary grant structure (10) are calculated.

Three assumptions are required prior to the start of the analysis. First the target values $\underline{a}_t$ have to be specified. These are not available from the present RSG allocation methods, and in the present analysis the target values of the state variables are set equal to their estimated values deriving from estimates of the model (10). A second assumption relates to the choice of weights $\underset{\sim}{M}_t$ to enter the objective function. Again these are not available from the present RSG allocation procedure. Those used below are derived by taking into account the varying magnitude of each state variable, i.e.

$$m_{iit} = \left\{\frac{\text{value of largest target variable}}{\text{target value of variable of i the authority}}\right\}^2 . \qquad (32)$$

Using this method the targets will vary as a function of both time and location, and will each be adjusted to the same scale at any location.

A third assumption related to the initial conditions. These are assumed to derive from the historic past for the fiscal year 1973/74.

Using these assumptions the resulting control theory solution follows three stages: estimation of the state equation (10), derivation of the equality constrained control solution, and derivation of the inequality constrained solution.

(i) *Estimation of the state equation.* Ideally this problem should be tackled with relatively sophisticated techniques such as the recursive filters suggested by Kalman (1960). However, with the present five-stage control problem, the shortness of the available data preclude such an approach. Instead, the estimates of the regression parameters are obtained below using OLS. In this context OLS can be interpreted only as an averaging tool, and certainly statistical significance tests cannot be employed. The resulting estimates of the parameter values are given in Table 1. These show in general a predictable pattern. The response of local expenditure to grants (α_{3i}) gives a generally high and positive elasticity. Because of the generally increasing proportion of grants within the local authority budgets, the parameter relating to changing expenditure levels (α_{1i}) is mostly negative. Hence grants have either had a supply expansion pressure on local services, or have substituted for local revenues.

(ii) *Equality constrained solution.* A second stage of analysis is the solution to the control problem with the constraint (16a) imposed, but ignoring (17b). The results of this calculation are given in Table 2. There are two control variables: the central grants, and the local rate poundage. Table 2 demonstrates that, without inequality constraints imposed to prevent removal of grants from local areas, one area Hillingdon pays a considerable contribution to the grants of other areas. This results from the high positive expenditure elasticity estimated for this borough, as shown in Table 1 (i.e., 1.19). However, too much should not be read into this result, given the wide sampling error

TABLE 1

Estimates of the Parameters of Equation (10)

	α_{0_i}	α_{1_i}	α_{2_i}	α_{3_i}
Barnet	7.84	0.26	1.42	1.26
Brent	23.05	-0.17	0.64	1.58
Croydon	85.75	-3.12	1.23	5.03
Ealing	20.54	0.07	0.32	1.40
Enfield	26.90	-0.62	0.76	2.08
Haringey	5.66	0.17	3.30	1.17
Harrow	12.64	-0.06	0.56	1.49
Hillingdon	8.63	1.19	-0.89	-0.67
Hounslow	16.23	-0.02	0.81	1.56
Kingston-on-Thames	11.71	-0.29	0.73	1.93
Merton	16.53	-0.46	1.26	1.82
Richmond-on-Thames	14.02	-0.29	0.33	1.81
Sutton	7.78	0.59	0.06	0.43

TABLE 2

Optimal Grants and Rate Poundages in Equality Constrained Case (17a)

	Grants (£m) 1974/75 - 1978/79					Rate Poundage (p in £) 1975/75 - 1978/79				
Barnet	20.0	23.9	29.8	35.9	41.9	35.5	37.1	37.5	38.6	39.5
Brent	47.3	57.9	78.2	97.9	111.6	38.5	40.5	40.5	41.2	42.6
Croydon	4.7	26.1	44.7	32.4	19.4	38.0	39.0	39.0	40.3	42.9
Ealing	105.3	115.1	146.9	175.8	192.8	46.1	48.1	51.0	52.4	55.2
Enfield	33.1	35.7	51.3	59.4	66.1	39.0	41.0	41.2	41.9	43.0
Haringey	9.5	12.3	14.0	17.9	21.1	51.0	52.6	53.2	54.0	54.6
Harrow	31.5	41.9	54.7	63.1	64.3	39.8	41.7	41.9	42.8	44.8
Hillingdon	-494.8	-587.3	-697.6	-788.9	-836.5	41.6	43.0	43.3	45.0	50.0
Hounslow	30.9	37.4	50.5	55.5	59.9	42.5	44.7	44.7	45.4	46.2
Kinston-on-Thames	20.7	23.5	29.8	35.2	37.5	47.3	49.2	49.8	50.7	51.9
Merton	14.7	17.1	22.2	25.4	29.9	47.7	49.2	49.4	50.6	51.2
Richmond-on-Thames	45.5	50.5	70.6	80.4	80.6	41.1	42.7	43.1	44.4	46.8
Sutton	240.7	318.3	348.4	413.6	476.9	44.4	45.5	45.7	47.1	49.9

in estimation of the coefficients in Table 1. The lack of inequality constraint also means that Hillingdon pays the bulk of the historic rise in local expenditures due to its negative value of α_{2i} and α_{3i}. This results in the rate of increase in local rate poundages being very much smaller than actually occurred, and Hillingdon has the only significant overall increase resembling the historical pattern.

(iii) *Inequality constrained solution.* A final stage of analysis is the solution to the control problem with both constraints (17a) and (17b) imposed. The results of this calculation are given in Table 3. In comparison with the equality constrained case of Table 2 this shows a much more even level of grant distribution between the local boroughs, and a much more general rise in rate poundage overall. Hillingdon changes from being a large net contributor to a position in which it receives no grants and has to raise its rate poundage fairly rapidly. However, it is joined in receiving no grants by Croydon and Sutton, and also by Ealing for 3 out of 5 years. These are boroughs that also have positive expenditure elasticity, as shown in Table 1.

Historically, each local borough receives some grant support, and that for Croydon and Hillingdon has been subject to a slower rate of increase than most other boroughs. Hence, the results of the control solution given in Table 3 exhibit the special features of the unitary grant approach; that is, by combining the separate elements of needs and resources it is possible to achieve a much higher degree of variation in grant aid and also potentially have a much higher degree of equalization. Analysis of the control solutions with separate grant elements [using equation (11) and (12) instead of equation (10)] would give a closer correspondence to the historical record.

CONCLUSION

This paper has presented the preliminary results of applying optimal control theory to allocation of the Rate Support Grant. The results are interim in three respects. First, it is necessary to carry out analysis with the separate grants equations (11) and (12) to determine how successful the control approach can be in more closely replicating existing practice. Second, it is necessary to extend the scale of the optimization problem to larger spatial units. Third, it is necessary to improve the method of estimating the state equation, which proves to be the most sensitive component of the present analysis. This will be accomplished using historical records of central government decisions on grant allocations. These three developments are being pursued by the authors, and will be reported in future papers.

TABLE 3

Solution to Control Problem with Equality and Inequality Constraints (17a) and (17b)

	Grant (£m) 1974/75 - 1978/79					Rate Poundage (p in £) 1974/75 - 1978/79				
Barnet	14.1	18.6	23.6	28.5	33.7	36.0	37.7	38.6	40.7	43.3
Brent	11.8	25.5	41.6	58.4	73.8	40.0	42.4	43.5	46.0	50.4
Croydon	0.0	0.0	0.0	0.0	0.0	37.9	40.4	43.1	46.4	50.2
Ealing	0.0	0.0	0.0	52.3	48.3	46.4	48.7	51.8	58.3	67.4
Enfield	13.3	18.3	29.9	36.3	40.3	39.8	42.3	43.7	46.2	50.0
Haringey	8.5	11.2	12.9	16.8	20.1	51.1	52.8	53.6	54.8	56.0
Harrow	12.1	23.8	34.0	40.9	35.8	40.5	42.7	44.0	46.6	52.3
Hillingdon	0.0	0.0	0.0	0.0	0.0	38.3	40.4	42.5	48.3	53.6
Hounslow	18.1	24.6	36.4	42.8	48.6	41.7	44.2	44.8	46.7	49.4
Kingston-on-Thames	14.2	17.4	22.2	26.7	26.5	42.3	44.4	45.8	47.9	51.5
Merton	11.0	14.4	18.6	21.5	25.3	48.0	49.7	50.4	51.8	53.8
Richmond-on-Thames	6.2	18.8	24.2	26.6	13.1	42.1	44.2	46.6	50.9	58.6
Sutton	0.0	0.0	0.0	0.0	0.0	43.2	44.7	44.1	47.1	59.8

REFERENCES

Aoki, M., 1976. Optimal Control and System Theory in Dynamic Economic Analysis, North Holland, Amsterdam.

Bennett, R., 1979. Spatial Time Series: Analysis - Forecasting - Control, Pion, London.

Bennett, R., 1980. The Geography of Public Finance: Welfare Under Fiscal Federalism and Local Government, Methuen, London.

Bennett, R. and Chorley, R., 1978. Environmental Systems; Analysis, Forecasting and Control, Methuen, London.

Bennett, R. and Tan, K., 1979. Allocation of the UK Rate Support Grant by Use of the Methods of Optimal Control, Environment and Planning A, 11: 1011-1027.

Chow, G., 1975. Analysis and Control of Dynamic Economic Systems, Wiley, New York.

Committee of Inquiry Into Local Government Finance, Local Government Finance, Cmnd. 6453, London, 1976.

Fujita, M., 1981. Spatial Dynamics of Urban Land Use, in: Dynamic Spatial Models, D. Griffith and R.D. MacKinnon (eds.), Plenum, New York, 404-439.

Hadley, G., 1962. Linear Programming, Addison-Wesley, New York.

Garbade, K., 1976. On the Existence and Uniqueness of Solutions to Multi-Period Linear/Quadratic Optimal Control Problems, International Economic Review, 17: 719-731.

Lee, R., 1964. Optimal Estimation, Identification and Control, M.I.T. Press, Cambridge, Mass.

Kalman, R., 1960. A New Approach to Linear Filtering and Prediction Problems, Trans. ASME, D, J. of Basic Engineering, 82: 35-45.

MacKinnon, R., 1975. Controlling Interregional Migration Processes of a Markovian Type, Environment and Planning A, 7: 781-792.

Musgrave, R. and Musgrave, P., 1976. Public Finance in Theory and Practice, McGraw-Hill, New York.

Nijkamp, P., 1975. Spatial Interdependencies and Environmental Effects, in: Dynamic Allocation of Urban Space, A. Karlqvist, L. Lundqvist and F. Snickars (eds.), Saxon House, Farnsborough.

Nijkamp, P. and Verhage, C., 1976. Cost Benefit Analysis and Optimal Control Theory for Environmental Decisions: A Case Study of the Dollard Estuary, in: Environment, Regional Science and Interregional Modelling, M. Chatterji and P. van Rompuy (eds.), Springer, Berlin.

Oates, W., 1972. Fiscal Federalism. Harcourt Brace Jovanovich, New York.

Pigou, A., 1947. A Study in Public Finance, 3rd Ed., Macmillan, London.

Pindyck, R., 1973. Optimal Planning for Economic Stabilisation, North Holland, Amsterdam, 1973.

Sakashita, M., 1967. Regional Allocation of Public Investment, Papers Regional Science Association, 19: 161-182.

Tan, K., 1979a. Optimal Control of Linear Econometric Systems with Linear Equality Constraints on the Control Variables, _International Economic Review_, 20: 253-258.
Tan, K., 1979b. Constrained Control of Spatial Systems: An Application to Allocation of the Rate Support Grant, unpublished Ph.D. Thesis, University of London.
Tiebout, C., 1956. A Pure Theory of Local Expenditures, _Journal of Policital Economy_, 64: 416-424.
Wolfe, P., 1959. The Simplex Method for Quadratic Programming _Econometrica_, 27: 382-398.

DYNAMIC ROAD INVESTMENT POLICY: A RECURSIVE VERSION OF THE ELLET-WALTERS MODEL*

Gerald M. Barber

Department of Geography
University of Victoria

Bruce A. Ralston

Department of Geography
The University of Tennessee

In this paper a recursive version of a dynamic road investment problem is developed by extending the Ellet-Walters developmental model of transportation so as to account explicitly for the effects of road investment on the supply and demand of the port hinterland's product. An explicit lagged supply function dependent upon commodity price and road penetration is related to a given linear demand function assuming annual market clearance. A recursive formulation of a cobweb-like dynamic market is thus obtained. Secondly, the optimal road penetration policy of a road planning authority faced with a given cost function and this form of market dynamics is derived. The time paths of both the market and road penetration policies can be traced by forward recursion through a set of solution equations. A numerical example is used to illustrate this problem.

Approaches to the optimization of transport network investments generally fall into one of two categories. On the one hand are mathematical programming formulations of discrete space network planning problems; and on the other a variety of continuous space geometric problems. The former determine the set of link additions, modifications, or deletions which optimizes some criterion function, usually the minimization of internodal flow costs [See for example Bergendahl (1969), Ochoa-Rosso (1968), Quandt (1960) and Steenbrink (1974)]. In the second class of problems

* The authors wish to acknowledge the assistance of Ole Heggen and Ken Josephson in the preparation of the figures appearing in this paper.

the goal is to determine the specific geometric structure of a single route or network with known topology which optimizes some measure of efficiency such as minimal length [See for example Fawaz and Newell (1976), Sen (1971), and Werner (1968a,b; 1969)]. Attempts to link these two approaches are rare though they are quite complementary, at least in principle.

However, it is characteristic of both sorts of problems to assume at the outset that the economic environment within which the system operates is unaltered by the changes in the transport network. In the first case, the typical assumption is that the nodal supply and/or demand of the commodities (or trip production and attraction in urban cases) is invariant with network link additions. In the second class of problems the characteristic assumption is either that trip density and orientation are uniform or else trip production is known and all trips are oriented to a single destination. In either instance, it is possible if not probable that any derived network is inconsistent with the pattern of transport demand from which it has been generated.

There are relatively few models which attempt to account explicitly for the interdependencies that relate the transport network to the larger economic system. Most models have attempted to effect a rather complex and at times curious combination of input-output, transportation, and econometric sub-models [See for example Amano and Fujita (1970) and Bronzini et.al. (1974)]. To date, the most ambitious attempt has been the Brookings Transport Research Program simulation model which attempted to trace the effect of transport system improvements to their ultimate effect on gross regional and national product [See Kresge and Roberts (1971)]. Though soundly criticized, the major shortcoming of the model appears, in hindsight, to have been that its proponents were overly ambitious [See Holland (1972)]. Transport investment sets in motion any number of complex interrelated processes many of which involve behavioral decisions and time lags. There is simply insufficient knowledge of these processes to include them in an empirical investment model at the present time.

In this paper, a model is developed that effects a compromise between the demands for simplicity and analytical tractability of a geometric network synthesis problem and the requirement that the model account, at least in the simplest sense, for the interactions between the economic system and the transport network. It has at its core a developmental model of transport initially put forward by Charles Ellet in 1836 and subsequently extended in great detail by Walters (1968a,b). To this framework has been added a recursive marketing structure with a cobweb-like solution, summarizing and extending the analysis in Barber and Ralston (1980).

First, the structure of the Ellet-Walters model is presented;

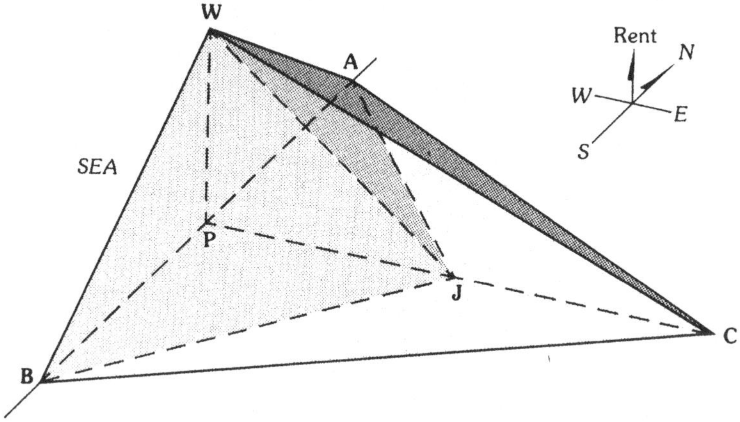

Fig. 1. Rent Pattern Before and After Road Construction

then the manner in which a comparative static network synthesis solution can be obtained is explained. Next, the model is extended by incorporating a lagged supply function and market clearing requirement in order to produce various types of dynamic behavior depending upon the relationship of supply and demand. Finally, the problem of specifying an optimal road penetration policy under these conditions is addressed. Several numerical examples are then used to illustrate the nature of the solution procedure and the characteristics of the structure of the optimal road penetration policy.

THE BASIC ELLET-WALTERS MODEL

Consider an isolated port P located on the coast of a country where land is homogeneous in all respects. This port serves as the sole market for the single agricultural commodity produced on the arable land within its hinterland. For purposes of simplification it is also assumed that the productivity of land is constant and, without loss of generality, yields one ton of the commodity per cultivated acre. For the moment, further suppose that the price of the commodity is given as a constant price of \$k per ton no matter what amount of the commodity is actually supplied. This assumption of a perfectly elastic demand will be relaxed as the analysis proceeds.

There are two modes of transportation. The existing domestic technology--headloading--has a unit cost of \$b per ton-mile and takes place only in directions parallel to or at right angles to the coast. If we imagine the coast running north-south then headloading is limited to the north-south and east-west directions (See Figure 1). A new technology in the form of a uniform quality road is to be introduced to the country. It will run due east from port P in the agricultural interior. Agricultural produce is carried by motor vehicles along this road at a rate of \$a per ton-mile. By assumption $a < b$ and the road is the superior technology.

Any farm or plot of land can be denoted by its coordinates on a map created by selecting the port P as the origin, the east-west road as the x-axis, and the sea coast as the y-axis. First, consider the pre-road situation. The area of cultivation around the port is determined by the condition

$$b(x + |y|) \leq k \qquad x \geq 0, \tag{1}$$

and encompasses the triangular area ABJ. This condition simply requires the total cost of transportation of the commodity to be less than the prevailing market price of \$k per ton. All points on the frontier are k/b Manhattan miles from the port P. Given the assumption of unit yield total production and thus supply at the port is k^2/b^2.

Suppose now that a road is extended eastward from the port to the limit of agricultural production. The new area of cultivation is determined by the condition

$$a x + b |y| \leq k \qquad x \geq 0, \tag{2}$$

and encompasses the triangle ABC. As in a simple von Thünen environment the area of cultivation is extended along the artery of cheaper transport. Total supply is thus increased to k^2/ab. Implicit in both the pre-road and post-road situations is a system of location rents in which the rent at any location is determined by the difference in the price at the port P and the transport cost from that location to P. In the pre-road situation the pattern of location rents appears as the half-pyramid ABJW with the summit at P equal to \$k. Total rent is given by the volume of this pyramid and is simply $\$k^3/3b^2$. For the road extended to the limit of agricultural production the rent pyramid is defined by ABCW and total rent is $k^3/3ab$.

Of course it would almost be certainly foolish to build a road eastward from P to the limit of agricultural production, a distance of k/a miles. For a road of length ℓ, $0 \leq \ell \leq k/a$, the situation is illustrated in Figure 2. The area of cultivation ABDFE can be shown to be $k^2/ab + [(k - a\ell)^2/b](1/b - 1/a)$. The total rent generated by the road is equal to the sum of the truncated half-pyramid WABDEG and the half-pyramid DEGF. This can be shown to be

$$R (\ell) = k^3/3ab + [(k - a \ell)^3 /3b] (1/b - 1/a) \tag{3}$$

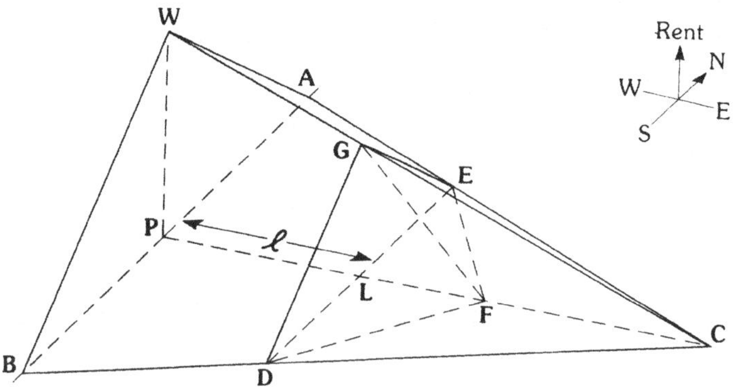

Fig. 2. Rent Pattern of a Truncated Road of Length ℓ

which is maximized when ℓ = k/a and when ℓ = 0, R (0)= $k^3/3b^2$, the pre-road rent.

OPTIMUM ROAD EXTENSION POLICY: COMPARATIVE STATICS

It is useful to examine the method utilized to determine the optimum road investment policy in a comparative statics framework. Assuming there is free entry to a perfectly competitive road haulage industry, all reductions in the cost of transportation are passed on in their entirety to users. Secondly, the total benefits from road construction are limited to the increases in location rent, there being no externalities of consequence. The authority controlling highway investment utilizes a self-financing rule which requires the benefits (and in this case the rents) generated by road investment to cover all construction and maintenance costs. It is therefore assumed that the authority is capable of levying taxes that extract the increment in rent due to road construction.

If the total annual cost of construction and maintenance of an infinite life road of length ℓ is G (ℓ), then the total net benefit of a road of length ℓ is $R(\ell) - G(\ell) - k^3/3b^2$. This is simply the difference between the total rent generated by the road and the sum of construction costs and pre-road rent. Since pre-road rent is constant at $k^3/3b^2$ the maximum benefit is obtained by the equation of the marginal rent and marginal cost functions. The optimum road penetration in the comparative static case is

$$\ell^* = k/a - b/a\ [G'\ (\ell)/b-a]^{\frac{1}{2}}\ . \qquad (4)$$

From (4) it is apparent that the optimal length of road penetration approaches a maximum of k/a miles as the marginal annual cost of road construction and maintenance approaches zero. And, as would be expected, the existence of economies of scale in road construction necessarily leads to deeper road penetration.

A RECURSIVE FORMULATION

Consider now several dynamic aspects of this investment problem. Let us first relax the key assumption in the Ellet model of a perfectly elastic demand for the commodity at the port. It is this assumption which precludes the sensitivity of the commodity market to changes in the transportation network. This problem will be addressed in the following manner. First, it is shown how the supply of the commodity in any one time period is related to both the price of the commodity and the length of the penetration road. Secondly, a marketing framework is specified in which a new price is reached in any period based on a market clearing principle. These interactions can be shown to lead to a simple recursive cobweb-like system. In the subsequent section of this paper this framework is extended and the problem of specifying an optimal investment policy within this recursive system is addressed.

The Supply Function

At each time interval t farmers in the hinterland of the port must make their production decisions for the subsequent time period t+1. Knowing the current price of the commodity k(t) and the length of the penetration road into the hinterland from the port $\ell(t)$, they decide to produce the commodity so long as they can ship the product to the port without incurring a loss. Again assuming unit yields, the total supply in the subsequent period t+1 is therefore the total area in production. Letting q(t+1) be the supply of the commodity at the port in time period t+1, it is easily shown that

$$q(t+1) = \frac{k^2\ (t)}{ab} + \frac{[\,k(t)-a\ell(t)\,]^2}{b}\,(1/b - 1/a)\ . \qquad (5)$$

Note that when there is no road and $\ell=0$ then total supply is equal to k^2/b^2 as in the pre-road situation of the static problem. Similarly, when ℓ approaches the maximum length of k/a miles total supply equals k^2/ab as in the post road situation of the static problem. The principal difference here is that both k and ℓ will change as road development proceeds.

The Demand Function and Market Clearance

Unknown to the farmers, demand at the port is actually

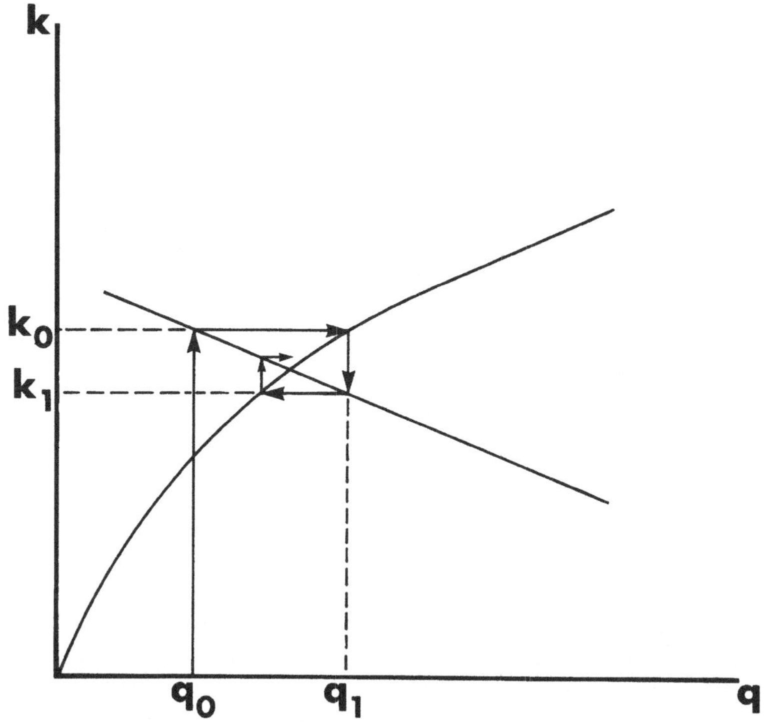

Fig. 3. Conventional Cobweb Situation Characterized by Equation 7

governed by a simple function of commodity price. Assuming that in each time period the market price is always set at a level which clears the market, and, for simplicity, assuming price is a linear downward sloping function of commodity supply, then the demand and market clearing conditions can be expressed as

$$k(t+1) = \alpha - \beta\,[q(t+1)] \; . \tag{6}$$

The dynamic structure of the market is thus completely specified by the lagged supply function (5) and the unlagged demand function (6).

Suppose for the moment that the length of road penetration is a constant. By substituting (5) into (6) and then simplifying the dynamic structure is specified by the single non-linear difference equation

$$k(t+1) = -\frac{\beta}{b^2}\,k^2(t) + \frac{2\beta\ell(t)(a-b)}{b^2}\,k(t) + \alpha + \frac{a(b-a)\beta\ell^2(t)}{b^2} \tag{7}$$

which can be shown to be a rather conventional cobweb situation (See Figure 3). The non-linear difference equation (7) is of the

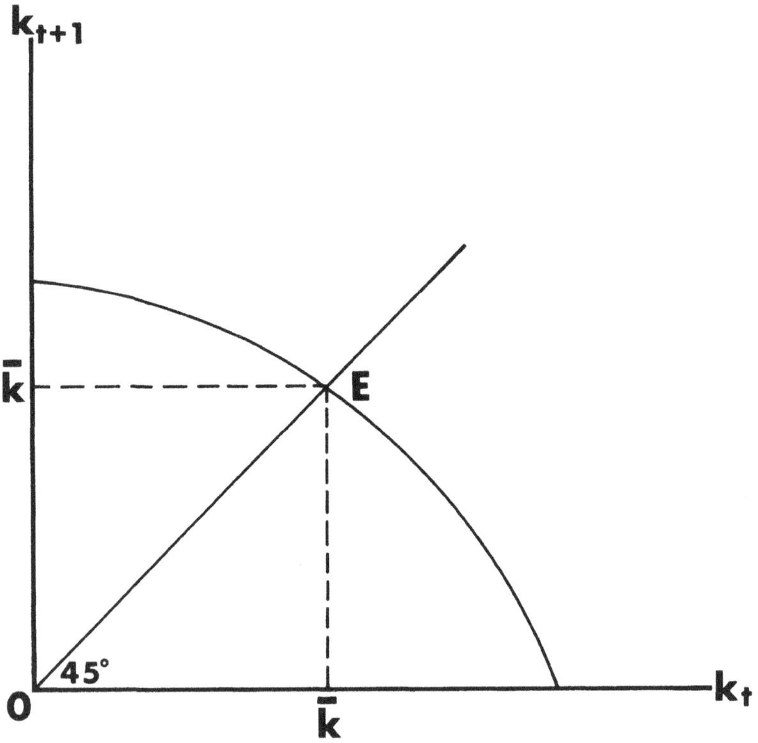

Fig. 4. Phase Diagram Associated With the General Form of Equation 7

general form $\psi[k(t)] = A\ k^2(t) + B\ k(t) + C$. Since $\psi'[k(t)]$ is clearly negative the phase diagram takes the form illustrated in Figure 4. It is possible to determine the point of intertemporal equilibrium to which the system tends and examine its stability, but that is of only passing interest since any actions taken by the highway authority can affect both the stability and existence of this equilibrium.

OPTIMUM ROAD EXTENSION POLICY: RECURSIVE DYNAMICS

Superimposed on this recursive structure are the actions taken by the road planning authority. Under similar operating conditions as those specified for the static problem, we can see that the road authority faces the problem of selecting a road penetration and investment policy which maximizes net benefits at the same time anticipating its implications on commodity supply and hence the course of market prices. Let us define $\ell(t)$ as the length of road penetration at time t and u(t) as the additional road length constructed in time interval t. Of course these two decision variables are linked by the identity

$$\ell(t+1) = \ell(t) + u(t), \tag{8}$$

but each gives rise to different costs. First, a maintenance cost is incurred in each time period which depends solely on the length of the road at that time. If c(t) is the unit cost of road maintenance and is assumed to be invariant with time then total road expenditure amounts to $c(t)\cdot\ell(t)$ in each time period. Secondly, the additional road development undertaken in time period t incurs a construction cost. If d(t) is the construction cost per mile then total construction costs amount to $d(t)\cdot u(t)$ and total expenditure in time period t equals $c(t)\ell(t) + d(t)u(t)$. Faced with these costs and the revenue to be extracted from any additional land rent which accrues to landowners because of authority actions, the authority must select the road investment policy that maximizes net benefits. Their objective is thus to maximize the difference between the rent generated by the road over the planning horizon and their expenditures on construction and maintenance; that is,

$$\text{MAX} \sum_{t=0}^{\infty} \{k^3(t)/3ab-[k(t)-a\ell(t)^3/3b]\ (1/b - 1/a)\} -$$

$$c(t)\ell(t)-d(t)u(t)\ , \tag{9}$$

where the first part of the maximand is a modified form of the dynamic version of the rent function (3) and the second part is the total authority expenditure. Since supply depends on the price of the commodity in the previous period, the dynamic version of the rent function contains a term in $k(t+1)/k(t)$. It has been omitted here (thus assumed to be one) and has the effect of insuring price stability into the model. Besides being an attractive feature in any dynamic model, this stability in commodity price can be seen to be an especially desirable feature of a dynamic road development problem. Quite clearly, it is in the interests of the authority to reduce unnecessary uncertainty in expected price for farmers.

This objective function is constrained by the recursive supply-demand behavior implicit in equation (7), the identity relating the two decision variables $\ell(t)$ and u(t), equation (8), and the usual non-negativity requirements on the decision variables, that is $u(t) \geq 0$. Though it is not possible to solve this problem directly for u(t) and $\ell(t)$, the solution to this multistage synthesis problem yields a set of equations from which it is possible to derive the optimal road penetration policy by forward recursion.

These are

$$\lambda(t) = \frac{k^2(t)}{ab} - \frac{[k(t) - a\ell(t)]^2}{b} \frac{1}{b} - \frac{1}{a} +$$

$$\lambda(t+1) \left(\beta\left\{ \frac{2k(t)}{ab} + \frac{2[k(t) - a\ell(t)]}{b} \right\} \frac{1}{b} - \frac{1}{a}\right) \tag{10}$$

$$0 = \frac{a[k(t) - a\ell(t)]^2}{b} \left(\frac{1}{b} - \frac{1}{b}\right) - c(t) + \gamma(t+1) +$$

$$\lambda(t+1) \left(\beta\left\{ \frac{2a[k(t) - a\ell(t)]}{b} \right\} \frac{1}{b} - \frac{1}{a}\right) \tag{11}$$

$$0 = -d(t) + \gamma(t+1) - \mu(t) \tag{12}$$

where the variables $\lambda(t)$, $\gamma(t)$ and $\mu(t)$ correspond to constraints (7), (8), and the non-negativity constraints on $u(t)$ respectively.

First, if in any period construction takes place and $u > 0$, then it follows that $\mu = 0$. Otherwise $\mu > 0$. If $\mu = 0$ then from equation (12) it is possible to solve for $\gamma(t+1)$, since $\gamma(t+1) = d(t)$. Together with the initial conditions (or previous period values) on all parameters, it is now possible to determine $\lambda(t+1)$ from equation (11), and then $\lambda(t)$ using (10). Finally, using equation (8) the decision variables are updated and the process carried forward to the next time period.

AN ILLUSTRATIVE NUMERICAL EXAMPLE

In order to illustrate the nature of the solutions generated by these recursive equations, consider the following numerical example. The optimal schedule of road penetration is to be determined for the case in which $\alpha = 12$, $\beta = .08$, $a = 1$, $b = 2$, $c = 10$, $d = .1$, and the initial conditions are $k(0) = 10$ and $q(0) = 25$. The optimal schedule of road penetration reveals five periods of road construction, two construction lags of different lengths, and a non-monotonic structure (See Figure 5). The massive investment in the initial period leads to an increase in supply, a reduction in price, and a smaller increment in road construction in the following time period. Two lags in construction are required to ensure stability in supply and demand and once a critical level of stability is achieved construction is resumed. Ultimately, construction ceases and both price and supply settle down to their equilibrium values (See Figures 6 and 7).

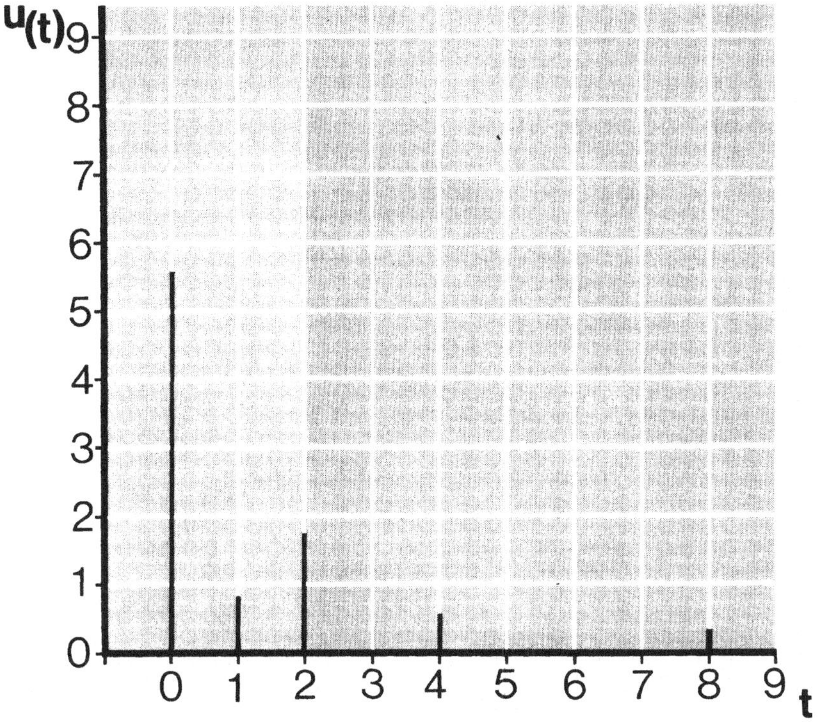

Fig. 5. Optimal Schedule of Road Construction

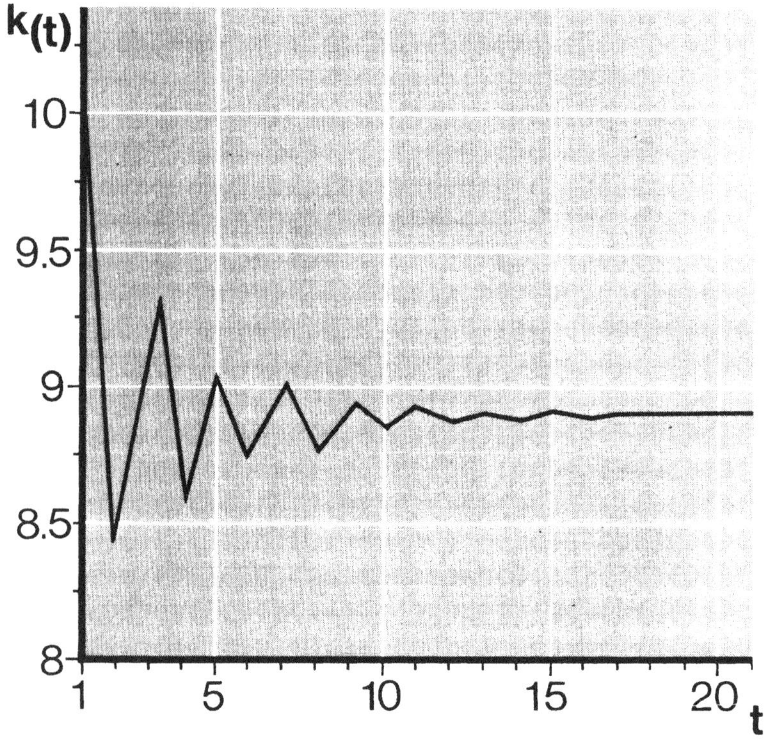

Fig. 6. Time Series of the Price of the Commodity

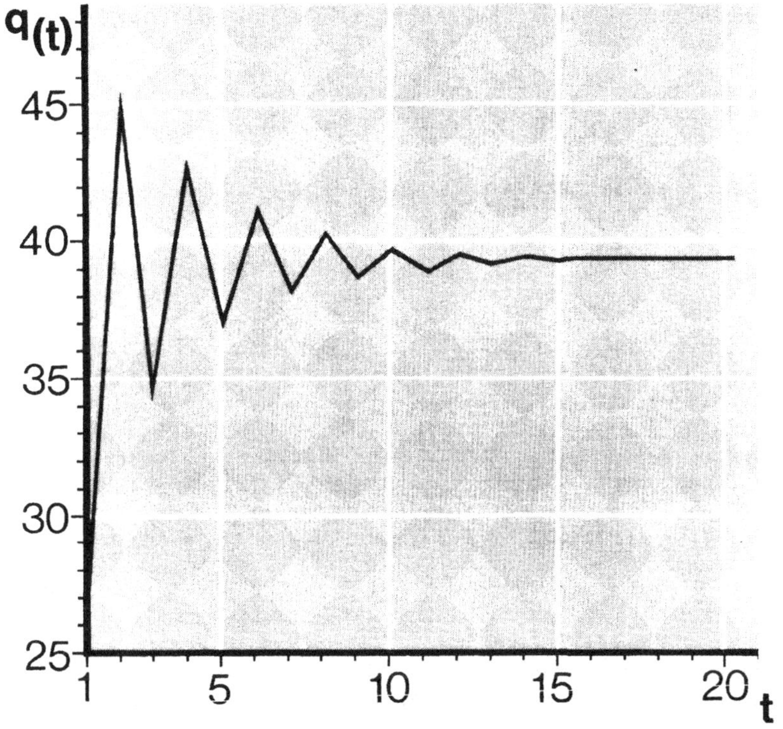

Fig. 7. Time Series of the Supply of the Commodity

CONCLUSIONS

In this paper it has been shown how a fully dynamic road investment model can be developed by coupling a simple cobweb model of commodity supply and demand with the Ellet-Walters developmental model of transportation. Moreover, the model explicitly accounts for the interrelationships between the road penetration policy followed by a highway planning authority and the marketing conditions of the commodity. It is essentially this feature which sets this model apart from most existing network development models. It is hoped that the model presented in this paper will extend in two principal directions. The first problem to be addressed is the specification of road penetration policies under different financing arrangements. The most obvious case is funding through a ton-mile or vehicle-mile tax. Secondly, the dynamics of more complex network situations must be analyzed. Two problems present themselves. First, the conditions underlying the emergence of feeder roads off the main east-west road must be specified. Next, a range of alternative topologies which can be used to link a node to its hinterland must also be examined. In any event, the framework developed in this paper can be used to address all these problems.

REFERENCES

Amano, K. and Fujita, M., 1970. A Long Run Economic Effect Analysis of Alternative Transportation Facility Plans--Regional and National, Journal of Regional Science, 10: 297-324.

Barber, G. and Ralston, B., 1980. "The Elementary Dynamics of Road Development, Geographical Analysis, 12: 258-263.

Bergendahl, G., 1969. Models for Investments in a Road Network, University of Stockholm, Department of Business Administration, Monograph No. 1.

Bronzini, M.S., et al., 1974. A Transportation Sensitive Model of a Regional Economy, Transportation Research, 8: 45-63.

Fawaz, M.Y. and Newell, G.F., 1976. Optimal Spacings for a Rectangular Grid Transportation Network--I: A Hierarchy Structure. II: Revised Assignments. Transportation Research, 10: 111-130.

Holland, E.P., 1972. Economic System Models for Transport Sector Studies, Internal Bank for Reconstruction and Development, Staff Working Paper, No. 131, Washington, D.C.

Kresge, D.T. and Roberts, P.O., 1971. Systems Analysis and Simulation Models. The Brookings Institution, Washington, D.C.

Ochoa-Rosso, F., 1968. Applications to Discrete Optimization Techniques to Capital Investment and Network Synthesis Problems. Ph.D. Dissertation, Dept. of Civil Engineering, Massachusetts Institute of Technology.

Quandt, R.E., 1960. Models of Transportation and Optimal Network Construction, Journal of Regional Sciences, 2: 27-45.

Sen, L., 1971. The Geometric Structure of an Optimal Transport Network in a Limited City-Hinterland Case, Geographical Analysis, 3: 1-14.

Steenbrink, P., 1974. Optimization of Transport Networks. Wiley, London.

Walters, A.A., 1968a. A Developmental Model of Transport, American Economic Review, Papers and Proceedings, 58: 360-377.

Walters, A.A., 1968b. The Economics of Road User Charges. Washington, D.C.: International Bank for Reconstruction and Development, World Bank Staff Occasional Paper No. 5.

Werner, C., 1968a. The Role of Topology and Geometry in Optimal Network Design, Papers of the Regional Science Association, 21: 173-190.

Werner, C., 1968b. The Law of Refraction in Transportation Geography: Its Multivariate Extension, Canadian Geographer 12: 28-40.

Werner, C., 1969. Network of Minimum Length, Canadian Geographer, 13: 47-69.

SPATIAL DYNAMICS OF URBAN LAND USE*

Masahisa Fujita

Regional Science Department
University of Pennsylvania

INTRODUCTION

The Alonso (1964)-Muth (1969) type model of urban land use describes, in effect, the equilibrium land use under the assumption that the entire metropolis is built at one point in time (the instant metropolis) or that land use adjustment costs are zero. In reality, though, most cities develop over a longer period of time, and the adjustment of land use is very costly and slow. Hence, satisfactory explanation of land use patterns in actual cities requires the use of dynamic models. As pointed out by Anas (1976a), a consideration central to all dynamic analyses of urban land use is the durability of the costly replacement of urban infrastructure (housing, non-residential buildings and transport facilities). This consideration is essential in distinguishing a dynamic analysis from a comparative static treatment of urban spatial growth. But this is also the main cause of the slow progress being made in the development of dynamic theory since it requires that we treat space and time simultaneously, an analytically and conceptually complex undertaking. Given this difficulty, any single study is obliged to focus on only a part of the dynamic aspects of urban land use. Therefore, many alternative approaches are conceivable for modeling dynamics of urban land use.

The purpose of this paper is to present a summary exposition of our research on the development of a dynamic theory of urban land use, based on the model developed in Fujita (1976a). Our model is located at an extreme position (i.e., perfect expectation with no uncertainty with respect to all conceivable models, and has the following major characteristics. First, the model aims to investigate the long-run efficient process of urban spatial development (under the condition of no uncertainty). Although it is possible to construct a variety of dynamic models of urban land use, this model provides a good point of departure. That is, since the relative efficiency of a solution path for any model becomes clear when it is compared to the long-run efficient path (i.e., the most efficient path), the information obtained from the analysis of the long-run model would provide a basis of comparison for other dynamic urban land use models. Second, our model investigates the long-run efficient process of urban spatial development from both the normative (i.e., planning) and positive (i.e., descriptive) points of view. Hence, the information obtained from the analysis of this model could be used, for example, to conduct a systematic comparison of the similarities and differences between urban land use patterns in capitalist and socialist economies. Third, the model can generate a variety of special problems of interest by systematic addition of further assumptions. Hence, by solving each special problem separately and unifying them afterwards we would be able to analyze thoroughly the complex phenomena of urban land use dynamics. In particular, in this paper we study in detail three special problems described below. Finally, the model is constructed so as to enable us to use the maximum principle in optimal control theory which has proved to be a very powerful tool in economic dynamics.

THE BASIC MODEL

In this section we formulate the *basic model* for efficient urban land use development. This formulation presents the model as an optimal planning problem. However, Fujita (1976a) gives the reformulation of the model as a market equilibrium problem, and shows that the two formulations of the model are mathematically equivalent. Hence, in the following discussion, we must keep in mind that for each optimal problem there exists a corresponding market equilibrium problem which has the same solution.

Suppose that a city is to be developed on an isolated featureless plain. We divide the area into g districts, and let

d_ℓ = the distance of district ℓ from the predetermined center of the city,

s_ℓ = the size of district ℓ,

and assume that

$$d_1 < d_2 < \ldots < d_\ell < \ldots < d_g \text{ , and}$$

$$s_\ell > 0; \quad \ell = 1,2,\ldots,g \text{ .}$$

Since g can be arbitrarily large, locations within each district are considered to be homogeneous; and, the total area $\sum_{\ell=1}^{g} s_\ell$ is sufficiently large so that there will be no spatial constraint on the size of the city in the future.

The city is to consist of m types of buildings, each of which is identified not only by its structure but also by its lot size. We define

$x_{i\ell}(t)$ = the number of buildings of type i existing in district ℓ at time t,

$u_{i\ell}(t)$ = the number of buildings of type i constructed in district ℓ at time t,

$v_{i\ell}(t)$ = the number of buildings of type i demolished in district ℓ at time t, and

k_i = the constant lot size of a building of type i.

Then, at each time t we have

$$\dot{x}_{i\ell}(t) = u_{i\ell}(t) - v_{i\ell}(t), \; u_{i\ell}(t) \geqq 0, \; v_{i\ell}(t) \geqq 0 \text{ ,}$$

where "$\cdot$" denotes the time derivative, and the following spatial restriction must be satisfied:

$$\sum_{i=1}^{n} k_i x_i (t) \leqq s_\ell \; ; \quad \ell = 1,2,\ldots,g \text{ .}$$

Next, we divide all the building-using activities (such as business activities, commercial activities and households) into n types, depending on their characteristics (j). We assume that the total number of behaving units belonging to each activity type j is given exogenously at each time, and define

$N_j(t)$ = the number of behaving units belonging to activity type j at time t, and

$y^j_{i\ell}(t)$ = the number of behaving units of activity type j allocated to buildings of type i in district ℓ at time t.

Then, since each behaving unit must use some building in some district in the city, the following restriction must be satisfied at each t:

$$\sum_i \sum_\ell y^j_{i\ell}(t) = N_j(t), \quad y^j_{i\ell}(t) \geqq 0 \;; \quad j = 1,2,\ldots,n \quad .$$

We assume that each behaving unit uses one and only one unit of buildings of some type at each time. Hence the following restriction should hold:

$$\sum_j y^j_{i\ell}(t) \leqq x_{i\ell}(t); \quad i = 1,2,\ldots,m; \quad \ell = 1,2,\ldots,g \; .$$

Finally, we define the objective of a dynamic planning problem of urban land use as the maximization of the present value of net revenues from the development of the city over time:

$$\int_0^\infty \exp(-\gamma t) \; [\sum_i \sum_j \sum_\ell \Psi^j_{i\ell}(t) y^j_{i\ell}(t) - \sum_i \sum_\ell B_i(t) u_{i\ell}(t)$$

$$- \sum_i \sum_\ell C_i(t) v_{i\ell}(t)]dt \quad ,$$

where

$B_i(t)$ = the construction cost per unit of buildings of type i at time t;

$C_i(t)$ = the demolition cost per unit of buildings of type i at time t;

$\Psi^j_{i\ell}(t)$ = the momentary value attached to a unit of building services of type i when a building of type i in district ℓ is occupied by activity type j at time t; and,

λ = the time-discount rate of the net revenue.

$B_i(t)$, $C_i(t)$, $\Psi^j_{i\ell}(t)$ are exogenously given functions of time t, and γ is a positive constant.

When we consider $\Psi^j_{i\ell}(t)$ to be a subjective monetary value attached by the planning authority to each building service, our planning problem can be considered to represent an optimal plan of the city by the central authority. On the other hand, when we compare our planning problem with the corresponding market equilibrium problem, we must consider that function $\Psi^j_{i\ell}(t)$ defined by

$$\Psi^j_{i\ell}(t) \equiv \Psi^j_{i\ell}[\overline{\xi}_j(t),t] \quad , \tag{2}$$

where $\Psi^j_{i\ell}[\overline{\xi}_j(t),t]$ is the *bid rent* for a building of type i in district ℓ at time t corresponding to a given utility (or profit) level, $\overline{\xi}_j(t)$, for activity type j at time t. In this case, if function $\overline{\xi}_j(t)$, $0 \leq t < \infty$, is the one which is realized on the market equilibrium path, then the solution of our planning problem here coincides with that market equilibrium solution [see Fujita (1976a)].

Optimum Problem

Choose values of construction speed $u_{i\ell}(t)$, demolition speed $v_{i\ell}(t)$, and the number of activity units $y^j_{i\ell}(t)$ so as to maximize

$$\int_0^\infty \exp(-\gamma t) \Big[\sum_i \sum_j \sum_\ell \Psi^j_{i\ell}(t) y^j_i(t) - \sum_i \sum_\ell B_i(t) u_{i\ell}(t)$$

$$- \sum_i \sum_\ell C_i(t) v_{i\ell}(t) \Big] dt \quad ,$$

subject to the following restrictions:

a) variation of building stock,

$$\dot{x}_{i\ell}(t) = u_{i\ell}(t) - v_{i\ell}(t), \; u_{i\ell}(t) \geqq 0 \; , \; v_{i\ell}(t) \geqq 0 \; ;$$

b) building stock constraint,

$$\sum_j y^j_{i\ell}(t) \leqq x_{i\ell}(t) \; , \; y^j_{i\ell}(t) \geqq 0 \; ;$$

c) total behaving-unit number constraint,

$$\sum_j \sum_\ell y^j_{i\ell}(t) = N_j(t) \; ;$$

d) land constraint,

$$\sum_i k_i x_{i\ell}(t) \leqq s_\ell \ ;$$

e) non-negativity constraint,

$$x_{i\ell}(t) \geqq 0 \ ; \text{ and,}$$

f) initial condition

$$x_{i\ell}(0) = \bar{x}_{i\ell}(0) \quad .$$

In the above problem we assume that:

Assumption OP-1. $B_i(t) > 0$, $C_i(t) > 0$ and $N_i(t) \geqq 0$ for all $t \geqq 0$; $B_i(t)$, $C_i(t)$ and $N_i(t)$ are continuously differentiable with respect to t; $d \exp(-\gamma t)B_i(t)/dt < 0$ and $d \exp(-\gamma t)C_i(t)/dt < 0$ for all $t \geqq 0$; $\max\{N_j(t) \mid t \in [0,\infty)\} < \infty$ and

$\sum_i k_i \max\{N_j(t) \mid t \in [0,\infty)\} < \sum_\ell s_\ell$; $\sum_{\ell=1}^{g} \bar{x}_{i\ell}(0) \geqq N_j(0)$; each $\Psi^j_{i\ell}(t)$ is piecewise continuously differentiable with respect to t, where $i = 1,2,\ldots,m$; $j = 1,2,\ldots,n$; and $\ell = 1,2,\ldots,g$.

Given Assumption 1, the optimality conditions for the above problem can be obtained by applying the maximum principle in optimal control theory as follows:

Optimality Conditions. For all $u_{i\ell}(t)$, $v_{i\ell}(t)$ and $y^j_{i\ell}(t)$ to be optimal construction, demolition and allocation processes for the optimum problem above, it is necessary and sufficient that there exist $x_{i\ell}(t)$ and a set of multipliers $R_{i\ell}(t)$, $Q_j(t)$, $P_{i\ell}(t)$, $P_\ell(t)$ $(i = 1,2,\ldots,m;\ j = 1,2,\ldots,n;\ \ell = 1,2,\ldots,g;\ 0 \leqq t < \infty)$ such that

(i) the rental market equilibrium condition is

$$R_{i\ell}(t) = \max\{\max_j [\Psi^j_{i\ell}(t) + Q_j(t)],\ 0\} \quad ,$$

$$R_{i\ell}(t) = \Psi^j_{i\ell}(t) + Q_j(t) \text{ if } y^j_{i\ell}(t) > 0 \quad ,$$

$$\sum_j y^j_{i\ell}(t) \leqq x_{i\ell}(t) \ ,$$

$$\sum_j y^j_{i\ell}(t) = x_{i\ell}(t) \text{ if } R_{i\ell}(t) > 0 \ , \text{ and}$$

$$\sum_j \sum_\ell y^j_{i\ell}(t) = N_j(t) \ , \ y^j_{i\ell}(t) \geqq 0 \quad ;$$

(ii) the construction and demolition market equilibrium condition is

(ii-1) construction:

$$P_{i\ell}(t) \leqq B_i(t) + k_i P_\ell(t) \ ,$$

$$P_{i\ell}(t) = B_i(t) + k_i P_\ell(t) \text{ if } u_{i\ell}(t) \geqq 0 \ , \text{ and}$$

(ii-2) demolition:

$$k_i P_\ell(t) \leqq P_{i\ell}(t) + C_i(t) \ ,$$

$$k_i P_\ell(t) = P_{i\ell}(t) + C_i(t) \text{ if } v_{i\ell}(t) > 0 \ ;$$

(iii) the asset market equilibrium condition is:

(iii-1) building:

$$\dot{P}_{i\ell}(t) \leqq \gamma P_{i\ell}(t) - R_{i\ell}(t) \ ,$$

$$\dot{P}_{i\ell}(t) = \gamma P_{i\ell}(t) - R_{i\ell}(t) \text{ if } x_{i\ell}(t) > 0, \text{ and}$$

(iii-2) land:

$$\dot{P}_\ell(t) \leqq \gamma P_\ell(t) \ ,$$

$$\dot{P}_\ell(t) = \gamma P_\ell(t) \text{ if } \sum_i k_i x_{i\ell}(t) < s_\ell \ ;$$

(iv) the variation of building stock and land constraint is

$$\dot{x}_{i\ell}(t) = u_{i\ell}(t) - v_{i\ell}(t) \ , \ u_{i\ell}(t) \geqq 0, \ v_{i\ell}(t) \geqq 0 \ , \text{ and}$$

$$x_{i\ell}(t) = 0 \ , \ \sum_i x_{i\ell}(t) \leqq s \ ; \quad \text{and,}$$

(v) the transversality condition is

(v-1) initial:

$$x_{i\ell}(0) = \bar{x}_{1\ell}(0) \text{ , and}$$

(v-2) terminal:

$$\lim_{t\to\infty} \exp(-\gamma t)P_{i\ell}(t) = 0 \text{ , } \lim_{t\to\infty} \exp(-\gamma t)P_{\ell}(t) = 0 \text{ .}$$

The economic meaning of these optimality conditions becomes clear when we compare them with the equilibrium conditions for the competitive market problem associated with the optimum problem. It is shown in Fujita (1976a) that these optimality conditions can be viewed as the equilibrium conditions in a competitive market through which the public authority tries to realize the optimum path by using appropriate subsidy (or tax) policy in the building rental market. Under this market interpretation of the above optimality conditions, the economic meanings of auxillary variables are:

$R_{i\ell}(t)$ = the rent for a building of type i in district ℓ at time t;

$Q_j(t)$ = the amount of building rent subsidy (or tax) for a renter of type j at time t;

$P_{i\ell}(t)$ = the price of a building of type i in district ℓ at time t; and,

$P_{\ell}(t)$ = the price of a unit of land in district ℓ at time t.

Thus the first equation of Optimality Condition (i) means that at each time t the market building rent $R_{i\ell}(t)$ is equal to the maximum of all the *bid rents* (with subsidies), $\Psi^j_{i\ell}(t) + Q_j(t)$, for buildings of type i in district ℓ by all the renter types, j = 1,2,...,n. The second equation says that only the highest bidders can occupy buildings of each type in each district. Optimality Condition (ii) says that no construction or demolition activity can obtain positive profit, and that if a construction or demolition activity takes place at some point in time, then that activity must get zero profit at that time. Optimality Condition (iii) represents the non-profitability for speculators in the asset market. For example, the first equation of (iii-1) says that the sum of the capital gain $\{\dot{P}_{i\ell}(t)\}$ and the income gain $\{R_{i\ell}(t)\}$ from holding a building of type i in district ℓ for a unit of time at t cannot exceed the corresponding opportunity costs $\{P_{i\ell}(t)\}$ that would be obtained if that building were sold at price $P_{i\ell}(t)$ and the money earned from this sale were deposited at interest rate γ. Condition (iii-2) represents the similar condition in the land market. Since vacant land gains no rental revenue (by assumption), we do not have land rent in (iii-2). Condition (iii-2) says that the rate of land price appreciation cannot exceed the interest rate, and the rate of land price appreciation in a district is equal to the interest rate where there remains vacant land in that district.

The optimum problem, as outlined above, is too complex for an explicit solution to be obtained. Therefore, in each of the following sections, we study the solution for a special problem which is derived by adding some special assumptions to the original problem.

PROBLEM A: SPATIAL PROCESS OF URBAN GROWTH

First, let us consider the problem that is obtained by adding the following two assumptions to the optimum problem put forth in the previous section:

(i) There is a one-to-one correspondence between activity types and building types. That is, activities of type i use only buildings of type i.

(ii) The possibility of the demolition of any building can be neglected throughout the planning horizon. That is, we can assume that $v_{i\ell}(t) = 0$ for all $t \geqq 0$ and for all i and ℓ.

The first assumption enables us to remove index j, allowing index i to represent building type i as well as activity type i, with the result that m = n. Hence, the new optimum problem is simplified as follows.

Optimum Problem A (OPA). Choose values of construction speed $u_{i\ell}(t)$ and numbers of activity units $y_{i\ell}(t)$ ($i = 1,2,\ldots,m$; $\ell = 1,2,\ldots,g$) at each time t so as to maximize

$$\int_0^\infty \exp(-\gamma t)\left[\sum_i \sum_\ell \Psi_{i\ell}(t)y_{i\ell}(t) - \sum_i \sum_\ell B_i(t)u_{i\ell}(t)\right]dt \quad ,$$

subject to the following restrictions:

a) variation of building stock

$$\dot{x}_{i\ell}(t) = u_{i\ell}(t) \ , \ u_{i\ell}(t) \geqq 0 \ ;$$

b) a building stock constraint,

$$0 \leqq y_{i\ell}(t) \leqq x_{i\ell}(t) \ ;$$

c) a total behaving-unit number constraint,

$$\sum_{\ell} y_{i\ell}(t) = N_i(t) \ ;$$

d) a land constraint

$$\sum_{i} k_i x_{i\ell}(t) \leqq s_\ell \ ; \text{ and,}$$

e) an initial condition,

$$x_{i\ell}(0) = \bar{x}_{i\ell}(0) \ .$$

It must be noted that in the above problem index i now represents building type as well as activity type i. Therefore, $y_{i\ell}(t)$, for example, represents the number of behaving units of activity type i allocated to buildings of type i in district ℓ at time t.

To obtain the explicit solution for OPA, hereafter we will make Assumption OPA-1.

Assumption OPA-1. For all i = 1,2,...,g, we have

$$\Psi_{i\ell}(t) = \Psi_i(t) - a_i(t)d_\ell \quad ,$$

where $a_i(t)$ is a continuous function of time t, and $a_i(t) > 0$ for all $t \geqq 0$. That is, we assume that the slope of the bid rent curve for each building type is linear with respect to the distance from the city center.

In the following we explain the solution of OPA assuming that

$$\dot{N}_i(t) \begin{cases} > 0 & \text{for} \quad t \in (0,T), \\ = 0 & \text{for} \quad t \geqq T, \end{cases} \qquad i = 1,2,\ldots,m, \text{ and} \qquad (3)$$

$$\bar{x}_{i\ell}(0) = 0 \quad \text{for } i = 1,2,\ldots,g \quad \text{and} \quad \ell = 1,2,\ldots,g \ . \qquad (4)$$

We can prove under the two assumptions given above that no building in the city is vacant at any time [see Fujita (1976b, 1979)].

Next, to obtain the optimal process of spatial growth we define the *bid land price function*, $p^i_\ell[t|Q_i(\tau)]$, for each building type i as follows:

$$p^i_\ell[t|Q_i(\tau)] = \frac{\exp(-\gamma t)[\int_t^\infty \exp\{-\gamma(\tau-t)\}\{\Psi_{i\ell}(\tau)+Q_i(\tau)\}d\tau - B_i(t)]}{k_i} \ . \quad (5$$

By definition, $p^i_\ell[t|Q_i(\tau)]$ represents the total net revenue (*discounted back to time* 0 by interest rate γ) obtainable from a unit of land in district ℓ if developers construct buildings of type i in district ℓ at time t, and if the future building rent will be $\Psi_{i\ell}(\tau) + Q_i(\tau)$ at each time $\tau \geqq t$. As such it can be shown under assumption (3) that

$$p_\ell(0) = \max\{\max_{i,t} p^i_\ell[t|Q_i(\tau)], 0\} \ , \text{ and} \qquad (6)$$

$$p_\ell(0) = p^i_\ell[t|Q_i(\tau)] \quad \text{if} \quad u_{i\ell}(t) > 0 \ , \qquad (7)$$

where the functions $Q_i(\tau)$, $0 \leqq \tau < \infty$, $P_\ell(0)$ and $u_{i\ell}(t)$ are those in the solution of OPA. Relation (3.4) says that the land price curve $P_\ell(0)$ at time 0 is the upper envelope of all bid rent curves $P^i_\ell[t|Q_i(\tau)]$ for all i and t; and, (7) says that if construction of buildings of type i occurs in district ℓ at time t, then the corresponding bid land price $p^i_\ell[t|Q_i(\tau)]$ must have the same value as the initial land price $P_\ell(0)$ in that district. Hence, the relation between the initial land price curve $P_\ell(0)$ and bid land price curves $p^i_\ell[t|Q_i(\tau)]$ for two building types, i = i' and i", at time t ($0 < t < T$) can be depicted as in Figure 1. The construction of buildings of type i occurs in district $\ell_i(t)$, where the initial land price curve $P_\ell(0)$ and bid land price curve $p^i_\ell[t|Q_i(\tau)]$ are tangent at distance $d_{\ell_i}(t)$. Note that under Assumption OPA-1

$$\frac{\partial P^i_\ell[t|Q_i(\tau)]}{\partial d_\ell} = -\frac{A_i(t)}{k_i} \ , \qquad (8)$$

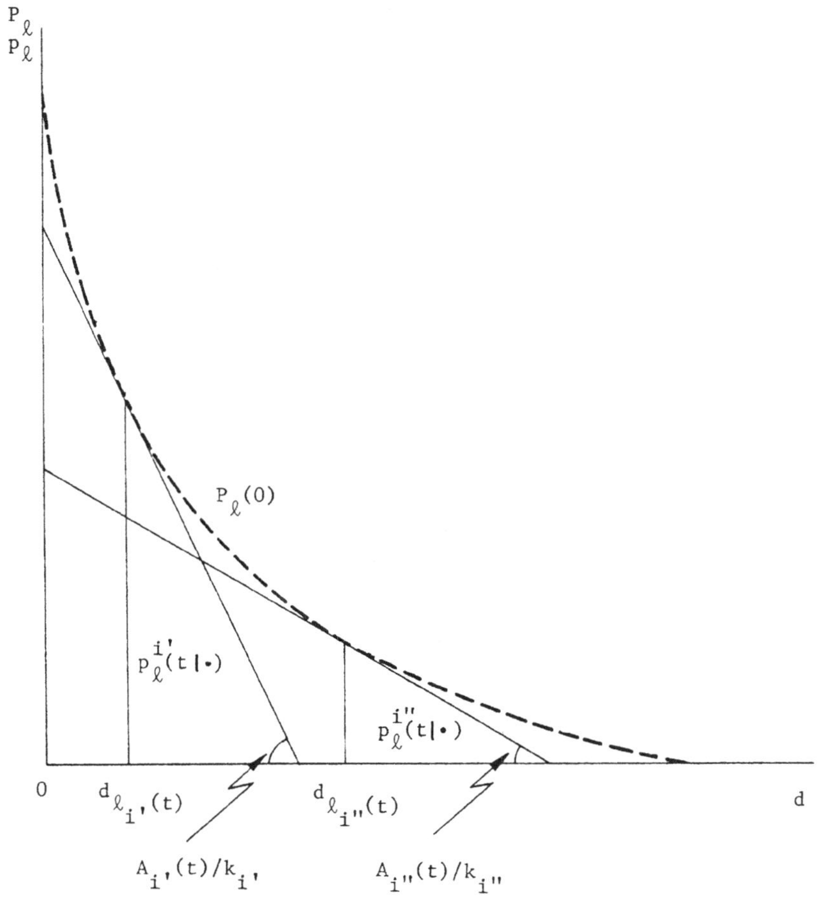

Fig. 1. Relation Between Locations of Construction Sites and Slopes of Bid Land Price Curves

where

$$A_i(t) = \int_t^{\infty} \exp(-\gamma\tau) a_i(\tau) d\tau \quad . \tag{9}$$

Hence, we conclude that at each t construction districts $\ell_i(t)$ $(i = 1,2,...,m)$ are ranked by distance from the city center in the same order in which they are ranked by the steepness of their bid rent curves, $A_i(t)/k_i$ $(i = 1,2,...,m)$. The construction district of the building type with the steepest bid land price curve is closest to the city center, followed by the building type with the next steepest bid land price curve, and so on. We also see from Figure 1 that since slope $A_i(t)/k_i$ of the bid land price curve of each building type i decreases with time [by definition (9)], construction district $\ell_i(t)$ of each building type i moves outward with time.

Next, assuming that the city is to be composed of only two types of buildings $(i = 1,2)$, the optimal process of urban spatial growth can be studied in detail. Consider the following examples:

Example 1: i = 1; business offices; i = 2; residential houses; and,

Example 2: i = 1, high rise residential houses; i = 2 flat residential houses.

For convenience, we assume

$$A_1(t)/k_1 > A_2(t)/k_2 \quad \text{for all} \quad t \geqq 0 \ . \tag{10}$$

Then, from the discussion above we see that

$$\ell_1(t) < \ell_2(t) \quad \text{for all } t \geqq 0 \ . \tag{11}$$

That is, a construction district of building type 1 always locates closer to the city center than that of building type 2. Assuming $T < \infty$, there are essentially two different cases for the shape of urban growth.

$$\left.\begin{array}{ll} \text{Case (a).} & A_1(T)/k_1 < A_2(0)/k_2 \ , \\ \text{Case (b).} & A_1(T)/k_1 \geqq A_2(0)/k_2 \ . \end{array}\right\} \tag{12}$$

Given values of parameters k_i and $a_i(t)$ $(t = 1,2, t \geqq 0)$, Case (a) occurs when growth period T is relatively long, and Case (b) when T is relative short [recall definition (9)].

Let us first examine Case (a). Considering (11), we can summarize the optimal process of the urban spatial growth in this case as follows (refer to Figure 2).

(1) Construction of buildings of type 2 begins in some district $\ell_2(0)$ far from the city center. Then, while some vacant land remains in that district, construction moves to the next district $\ell_2(0) + 1$. In this manner it gradually moves outward, leaving some vacant land in each district.

(2) Construction of buildings of type 1 begins in the district nearest to the city center (i.e., $\ell = 1$). Then, occupying the whole area in each district, construction moves toward district $\ell_2(0) - 1$. After leaving that district it gradually moves outward, occupying all the vacant land remaining in each district after construction of buildings of type 2.

(3) Construction of buildings of type 2 ends in the fringe district $\bar{\ell}$ at time T, and construction of buildings of type 1 ends at the same time in some district $\ell_1(T)$ that is inside the urban fringe.

Figure 2 depicts this process of urban spatial growth. The left (right) vertical axis represents *land use ratio* θ_ℓ^1 (θ_ℓ^2) in each district ℓ, where $\theta_\ell^i = k_i x_{i\ell}/s_\ell$, $i = 1,2$, and the horizontal axis shows the distance from the city center. One of the outstanding characteristics of this growth process is that the construction of buildings of type 2 moves toward the suburbs while leaving a large area vacant in each district. Hence, this growth process of urban space has the property of urban sprawl, and results in a mixture of different buildings in many districts. To prevent the mixture of different buildings, we may adopt appropriate zoning ordinances. Figure 5(a) depicts an example of zoning restriction for achieving the optimal spatial growth for Case (a), where we assume that these external diseconomies among different building types sharply decrease with distance.

Next let us examine Case (b). The optimal process of urban spatial growth in this case can be depicted in Figure 3. Construction of buildings of type 1 begins in district 1, and gradually moves outward, occupying the whole area in each district. On the other hand, construction of buildings of type 2 begins in some district $\ell_2(0)$ far from the city center, and gradually moves outward occupying the whole area in each district. Hence, buildings of type 1 eventually occupy the whole area between district 1 and district $\ell_2(0)$ $[= \ell_1(T)]$, and buildings of type 2 occupy the whole

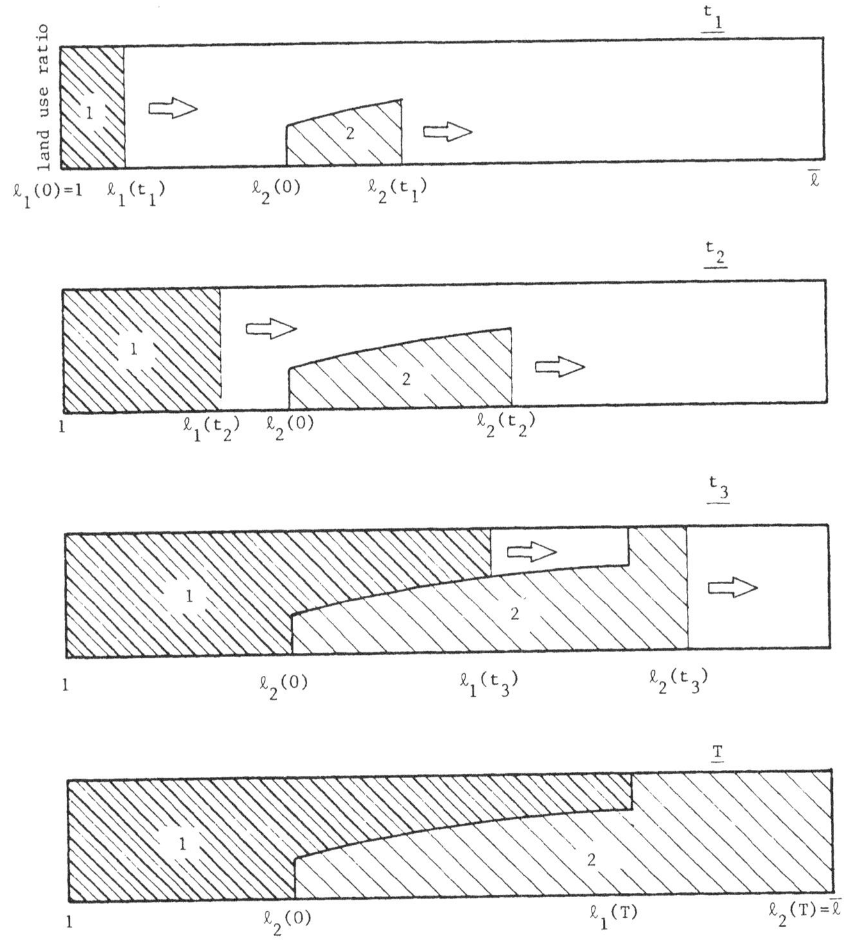

Fig. 2. Spatial Process of the Urban Growth, Case (a), $t_1 < t_2 < T$.

area between district $\ell_2(0)$ and fringe district $\bar{\ell}$. There is no mixture of buildings of different types in any district [except possibly in district $\ell_2(0)$]. Thus, the so-called *Thünen rings* are completed at the end of city growth. Figure 5(b) gives a two-dimensional picture of this process of urban spatial growth.

So far, we assumed $T < \infty$. If $T = \infty$ [Case (c)], that is if the city grows forever, then the corresponding optimal process of urban spatial growth can be depicted as in Figure 4. The growth process for Case (c) is essentially the same as that for Case (a). The only difference is that in Case (a) construction in buildings of type 2 occupy all of the land beyond district $\ell_1(t)$ (see Figure 2), while in Case (c) construction of buildings of type 2 moves outward, leaving some vacant land at every distance. Figure 5(c) shows the effects of applying zoning restrictions for achieving the optimal spatial growth for Case (c).

Finally, let us consider the case where the city is to be composed of buildings of three different types ($m = 3$). For example, $i = 1$, business offices; $i = 2$, high rise residential houses; and, $i = 3$, flat residential houses. Let us assume that, at each time t (>0),

$$A_1(t)/k_1 > A_2(t)/k_2 > A_3(t)/k_3 \quad . \tag{13}$$

Then assuming $T = \infty$, the optimal process of urban spatial growth in this case can be depicted as in Figure 6. At each time t, construction districts, $\ell_1(t)$, $\ell_2(t)$ and $\ell_3(t)$, for buildings of these three types have the following relation:

$$\ell_1(t) < \ell_2(t) < \ell_3(t) \quad . \tag{14}$$

Hence, again, the growth process of urban space has the property of urban sprawl; and, without zoning restrictions we eventually have a mixture of different buildings in many districts. Figure 7 depicts the effects of applying zoning restrictions for achieving optimal spatial growth without mixture of different buildings.

PROBLEM B: SPATIAL PROCESS OF URBAN RENEWAL

For the second special problem consider the problem that is obtained when we add the following assumption to the optimum problem in section 2:

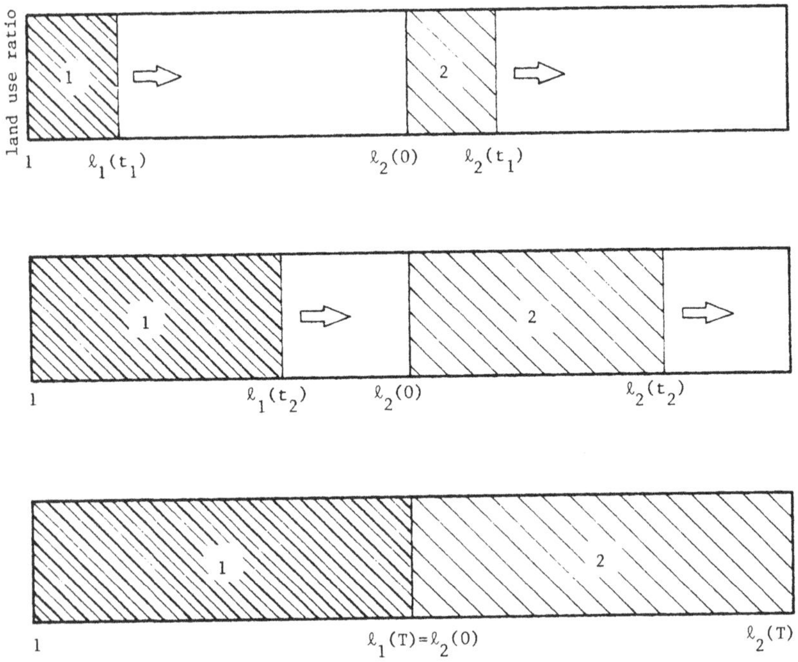

Fig. 3. Spatial Process of Urban Growth, Case (b), $t_1 < t_2 < T$.

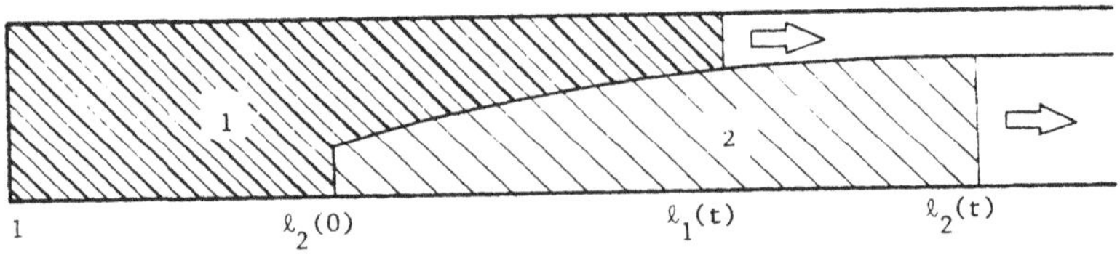

Fig. 4. Spatial Process of Urban Growth, Case (c).

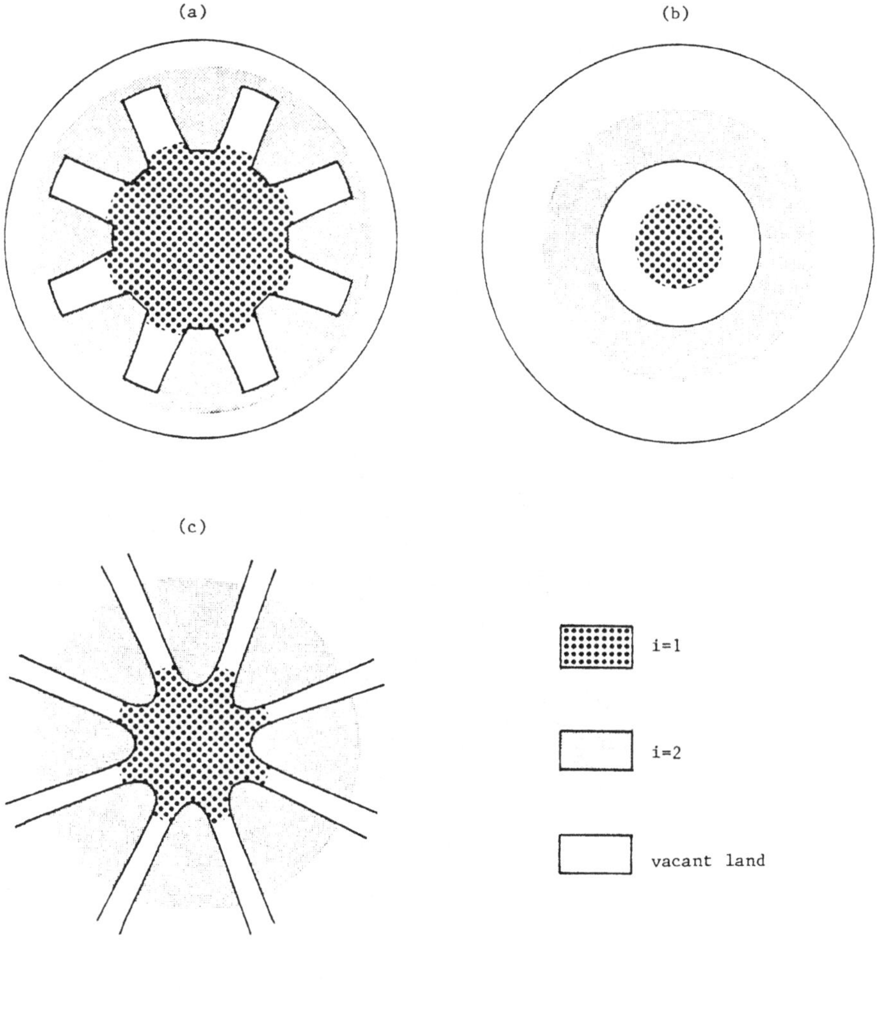

Fig. 5. Efficient Zoning Restriction

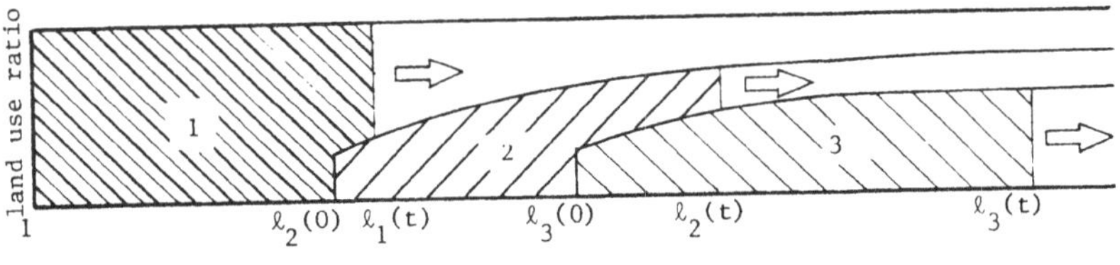

Fig. 6. Spatial Process of Urban Growth (m = 3).

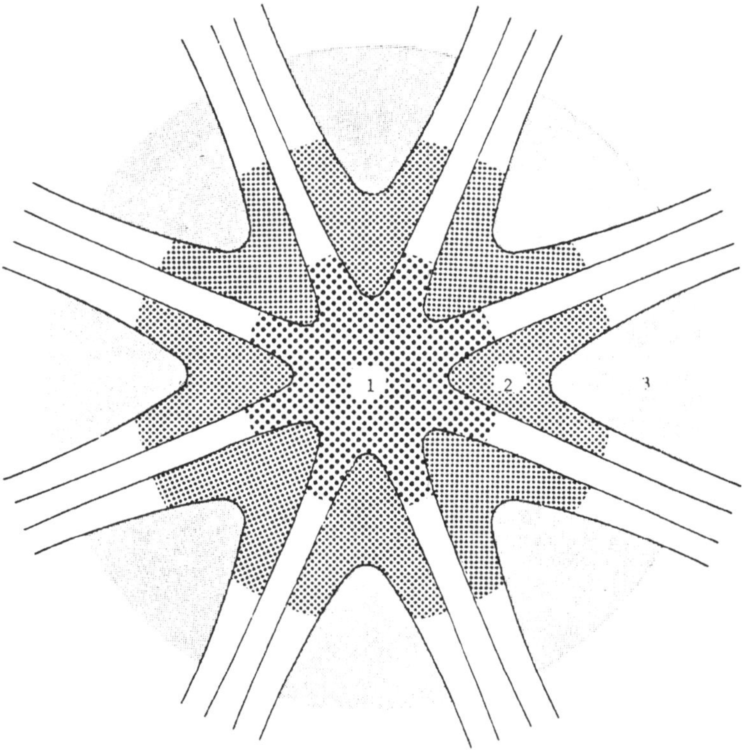

Fig. 7. Efficient Zoning Restriction (m = 3).

(i) There is a one-to-one correspondence between activity types and building types. That is, activities of type i use only buildings of type i.

We see that this second problem is obtained by removing assumption (ii) from Problem A in the previous section. The problem now becomes the following optimum problem.

Optimum Problem B (OPB). Choose values of construction speed $u_{i\ell}(t)$, demolition speed $v_{i\ell}(t)$, and the number of activity units $y_{i\ell}(t)$ so as to maximize

$$\int_0^\infty \exp(-\gamma t) \sum_{i=1}^{n} \sum_{\ell=1}^{g} [\Psi_{i\ell}(t)y_{i\ell}(t)-B_i(t)u_{i\ell}(t)-C_i(t)v_{i\ell}(t)]dt,$$

subject to the following restrictions:

a) a variation of building stock,

$$\dot{x}_{i\ell}(t) = u_{i\ell}(t) - v_{i\ell}(t),\ u_{i\ell}(t) \geqq 0,\ v_{i\ell}(t) \geqq 0;\ \text{and,}$$

b)
c)
d)
e) } the same as b), c), d) and e) of OPA, respectively.

In the following we examine the solution of OPB under the following set of assumptions (in addition to Assumption 1 of section 2):

$$\Psi_{i\ell}(t) = \Psi_i(t) - a_i d_\ell,\ a_i > 0,\ i = 1,2,\ldots,m; \tag{15}$$

$$\dot{N}_i(t) \begin{cases} > 0 \text{ for } t \in [0,T) \\ = 0 \text{ for } t \geqq T;\ \text{and} \end{cases} \tag{16}$$

$$\bar{x}_{i\ell}(0) = 0 \quad \text{for } i = 1,2,\ldots,m,\ \ell = 1,2,\ldots,g, \tag{17}$$

where in (16) T is a finite constant.

Although the solution procedure is essentially the same for the two problems OPA and OPB, it is much more complicated in OPB. The reason is that in problem OPA the type of buildings on a piece of land never changes (since demolition of buildings is not considered), but in problem OPB the types of buildings on a piece of land will change over time according to the renewal process. Hence, in solving OPB *bid land prices* must be defined as functions of subsidy level $Q_i(\tau)$ and *land use sequences*. A land use sequence designates how a piece of land is utilized over the planning horizon; that is, on a piece of land, which type of building is first constructed and when, at which time are these buildings demolished, which type of building is next constructed and when, and so on. By using these defined bid land price functions, we can obtain the solution for OPB via a procedure similar to that used in solving OPA [see Akita and Fujita (1979)].

We first explain the solution of OPB assuming that the city is to be composed of only two types of buildings (i = 1,2). For convenience we assume

$$a_1/k_1 > a_2/k_2 \quad ; \tag{18}$$

that is, the slope of the bid rent curve for the type 1 building per unit area is greater than that for the type 2 building [recall assumption (15)]. Under assumptions (15) to (18) there are essentially four different solution patterns for OPB.

One of the spatial growth patterns is described as follows (refer to Figure 8).

(a) Construction of buildings of type 2 starts from a district f far from the city center. Buildings of type 2 occupy a portion of this district with some land being left vacant, and construction of buildings of type 2 moves toward the suburbs while leaving some vacant land in each district. However, after reaching a district h construction of buildings of type 2 moves toward the suburbs without leaving any vacant land, and finally reaches a district $\bar{\ell}$ at time T.

(b) Construction of buildings of type 1 starts in the district nearest to the city center, namely, district 1. Buildings of type 1 occupy all of the land in district 1 and construction of buildings of type 1 moves to the next district, 2. The construction district moves step-by-step toward district f-1 in a similar way (i.e., buildings of type 1 occupy all the land in each district). After construction passes district f-1, buildings of type 1 occupy all of the remaining vacant land (left after the construction of buildings of type 2) in each district step-by-step from district f to district h-1.

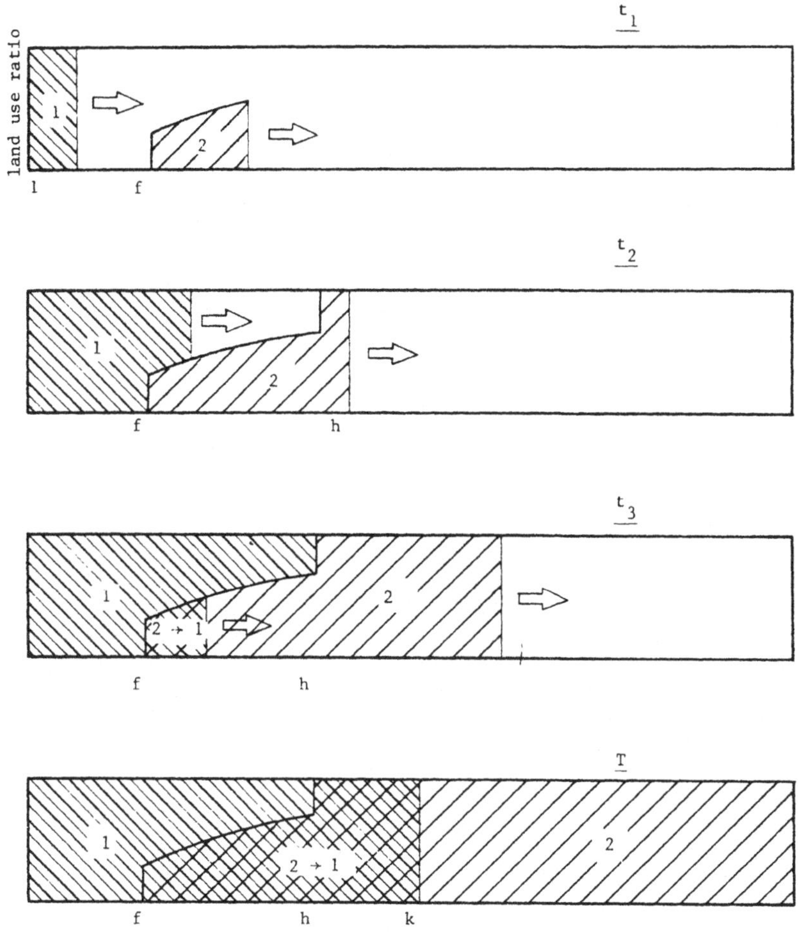

Fig. 8. Spatial Process of Urban Growth with Renewal, Pattern 1 ($t_1 < t_2 < t_3 < T$).

(c) After the construction district for buildings of type 1 readies district h-1 and buildings of type 1 occupy all of the remaining vacant land in this district, the construction district for buildings of type 1 moves back to district f. In district f, construction of buildings of type 1 is carried out after the demolition of the existing type 2 buildings. After the type 1 buildings renew all the existing type 2 buildings in district f, construction of buildings of type 1 moves to the next district, f+1. The construction district for buildings of type 1 moves toward the suburbs in a similar way (i.e., type 1 buildings renew all existing type 2 buildings in each district step-by-step from district $\bar{f}$), and finally reaches a district k at time T (where $h \leq k < \bar{\ell}$).

In this spatial pattern called pattern 1 we assume $k \geq h$. Figure 8 depicts this process of urban spatial growth. The vertical axis represents the land use ratio between the two types of buildings in each district, and the horizontal axis shows the distance from the city center. At the end of the growth period, T, the city consists of two von Thünen rings. This pattern of urban spatial growth process occurs when cost $B_2(t) + C_2(t)$ for renewal of type 2 buildings to type 1 buildings is relatively small and/or growth period T of the city is very long.

The next pattern of urban spatial growth process, called pattern 2, is the same as pattern 1 except that we assume, in this case, $k < h$ (see Figure 9). This pattern will be realized when the renewal cost is relatively small (but not too small) and the growth period is relatively long. Without appropriate zoning ordinances we have a mixture of different buildings at the end of city growth, as well as during the process of city growth.

The last two patterns, pattern 3 and pattern 4, of the optimal spatial growth process for problem OPB are the same as those of Cases (a) and (b), respectively, in section 3 (see Figures 2, 3 and 9). In both patterns no renewal occurs in the city at any time. That is, if renewal costs are very large, renewals of buildings never occur in the growth process; and, hence, the two solutions, OPA and OPB, given the same locations.

PROBLEM C: SPATIAL PROCESS OF RESIDENTIAL DEVELOPMENT

The common assumption in the previous two problems, A and B, is that there is a one-to-one correspondence between activity types and building types. This assumption excludes the possibility of substitutions among different building types for each activity type. However, the substitutability of different building types is an important aspect of some types of locational problems.

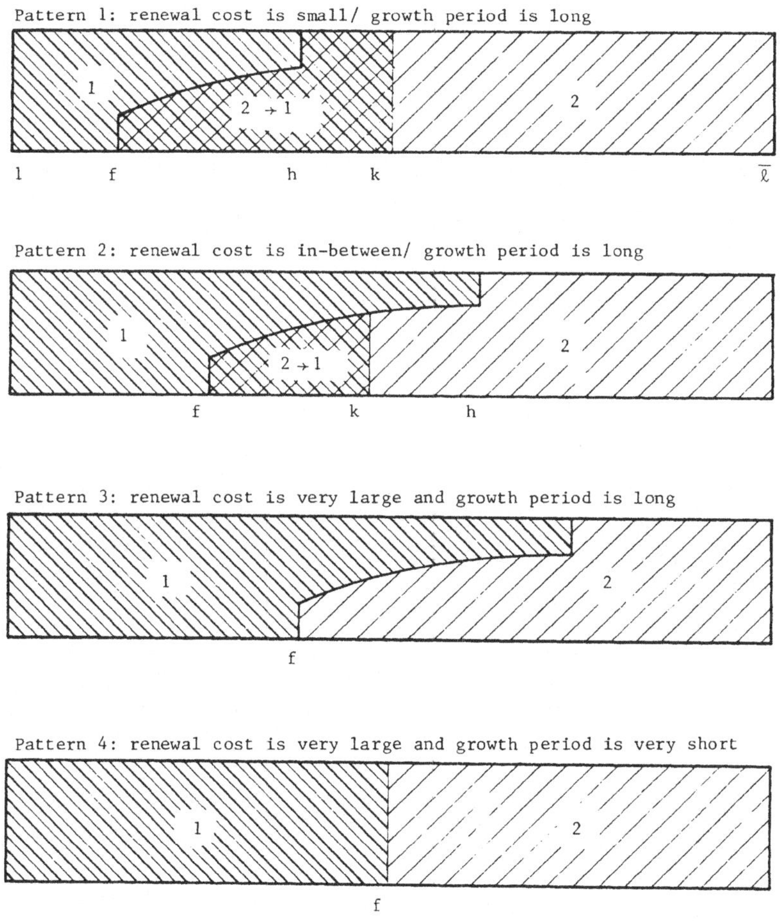

Fig. 9. Spatial Patterns of the City at the End of the Growth Period

For simplicity we assume that all the urban firms are going to locate at the center (i.e., in district $\ell = 0$) whose size is neglected (or predetermined); therefore we need to pay attention only to the residential development in the city. We also assume for simplicity that all the households in the city are identical to their locational characteristics; that is, there is only one type of household in the city [for the analysis in the case of multiple household types, see Ando and Fujita (1979)].

Formally, the problem to be studied in this section is obtained by adding the following assumption to the original problem in section 2 (i) $n = 1$. That is, there is only one type of activity that represents the identical households in the city. The formulation of the problem and the corresponding optimality conditions are obtained, respectively, if we eliminate index j from the optimal problem and the optimality conditions in section 2. However, instead of directly studying the planning formulations of this problem, we study in this section the corresponding market problem that is described immediately below. In the following discussions, for convenience we take continuous representations of space and building types, respectively.

Suppose we have a monocentric city in which the residential area expands as the number of households increases. Each household rents one and only one residential house of some type at some location in each period of time. Households can move among rental houses without cost at any time. Each household is myopic in the sense that in each time period the entire income is expended for the composite good, a house and transport costs. Based on these assumptions the residential choice behavior of each household is formulated as follows: at time $\tau (0 \leqq \tau < \infty)$,

$$\left.\begin{aligned} &\max_{z,s,d} \; U(z,s) \, , \\ &\text{subject to } z + R(d,s,\tau) + D(d,\tau) = Y(\tau) \, , \end{aligned}\right\} \qquad (19)$$

where

$U(\cdot)$ = the utility function of the household;

z = the amount of the composite good whose price is the numeraire;

s = the lot size of the house;

d = the distance from the city center;

$R(d,s,\tau)$ = the rent of a house with lot size s at distance d at time τ;

$D(d,\tau)$ = the transport cost per unit of time for a household at distance d at time τ; and,

$Y(\tau)$ = the income of each household at time τ.

We assume in the above that functions $D(d,\tau)$ and $Y(\tau)$ are exogenously given for each time τ $(0 \leq \tau < \infty)$. Though it is difficult to express the character of residential houses by a single index, we assume for simplicity that the character of each house can be designated by its lot size, s.

Next, we assume that all the residential houses are to be built and owned by outside developers. These developers must determine, for each unit of land at each distance d, the time t at which the land is converted from agricultural use to residential use and the type of houses (i.e., the lot size s) to be constructed on the land. It is assumed that developers have perfect foresight about the future time path of housing rent $R(d,s,\tau)$ and the time path of agricultural land rent $R^A(t)$. The agricultural land rent is assumed to be exogenously given, and to be independent of distance d. Under these assumptions we define the developer's problem to be to choose his optimal *development strategy* (s,t), for each unit of land at each distance d so as to maximize

$$P(d,s,t) \equiv \int_0^t \exp(-\gamma\tau)R^A(\tau)d\tau + \exp(-\gamma\tau)[\int_\tau^\infty \exp(-\gamma(\tau-t)) \times R(d,s,\tau)d\tau - B(s)]/s \quad , \qquad (20)$$

where

t = the time to convert the land at distance d from agricultural use to a house of lot size s;

$B(s)$ = the construction cost (excluding the land cost) for a house with lot size s, which is exogenously given and is constant over time; and,

γ = the discount rate for future revenues and costs.

We call function $P(d,s,t)$ the *bid land price* at distance d under development strategy (s,t) which represents the present value of the future net revenues from a unit of land at distance d when development strategy (s,t) is adopted.

Next, since all the households in the city are the same, on the equilibrium path all the households should have the same utility level, $U(\tau)$, at each time τ ($0 \leqq \tau < \infty$). From (19) this utility level is given by

$$u(\tau) = \max_{z,s,d} \{u(z,s) \mid z + R(d,s,\tau) + D(d,\tau) = Y(\tau)\} .$$

Hence, the equilibrium rent $R(d,s,\tau)$ for a house (s,d) at time τ (assuming that it is not vacant at that time) is given by

$$R(d,s,\tau) = Y(\tau) - D(d,\tau) - z[s,u(\tau)] , \tag{21}$$

where $z[s,u(\tau)]$ is the derived demand function from composite good z that is obtained by solving equation $u(\tau) = U(s,z)$ for z.

Therefore, from (20) and (21) we obtain the conditions for the optimal development strategy as follows: at each $t > 0$,

first order conditions

$$Y(t) - D(d,t) - Z[s,u(t)] = sR^A(t) , \tag{22}$$

$$\overline{Y}(t) - \overline{D}(d,t) - \overline{Z}(s,t) = - s\overline{Z}_s(s,t) , \text{ and} \tag{23}$$

second order conditions

$$(\dot{Y} - D_t - z_u\dot{u} - s\dot{R}^A)\overline{Z}_{ss} > \gamma (R^A + Z_s)^2 , \tag{24}$$

where

$$Z(s,u) = z(s,u) + \gamma B(s)$$

$$\overline{Z}(s,t) \equiv \overline{Z}[s,t|u(\tau)] = \gamma\int_t^\infty \exp(-\gamma(\tau-t))Z[s,u(\tau)]d\tau ,$$

$$\overline{Y}(t) = \gamma\int_t^\infty \exp(-\gamma(\tau-t))Y(\tau)d\tau ,$$

$$\overline{D}(d,t) = \gamma\int_t^\infty \exp(-\gamma(\tau-t))D(d,\tau)d\tau ,$$

"$\cdot$" denotes the total derivative of a function with respect to t (or τ), and $\overline{Z}_s \equiv \partial\overline{Z}/\partial s$, $D_t \equiv \partial D/\partial t$, and the such.

From (22) and (23), the optimal development location $d(t)$ and the corresponding lot size $s(t)$ at each time t (>0) must satisfy the following conditions:

$$Y(t) - D[d(t),t] - Z[s(t),u(t)] = s(t)R^A(t), \text{ and} \quad (25)$$

$$\overline{Y}(t) - \overline{D}[d(t),t] - \overline{Z}[s(t),t] = -s(t)\overline{Z}_s[s(t),t]. \quad (26)$$

To examine how an optimal combination $[d(t),s(t)]$ changes with time t, we take the derivative of each side of (25) and (26) with respect to t; then, solving them for $\dot{d}(t)$ and $\dot{s}(t)$, we have

$$\dot{d}(t) = \frac{(\dot{Y}-D_t-Z_u\dot{u}-s\dot{R}^A)s\overline{Z}_{ss} - \gamma s(Z_s+R^A)^2}{s\overline{Z}_{ss}D_d + (Z_s+R^A)\overline{D}_d}, \text{ and} \quad (27)$$

$$\dot{s}(t) = \frac{(\dot{Y}-D_t-Z_u\dot{u}-sR^A)\overline{D}_d + \gamma s(Z_s+R^A)D_d}{s\overline{Z}_{ss}D_d + (Z_s+R^A)\overline{D}_d}, \quad (28)$$

where the right hand sides of (27) and (28) are evaluated at $[d(t),s(t),t]$.

From (24) we see that the numerator of the right hand side of (27) must be positive. Also, by using (24) we can easily show that if the denominator of the right hand side of (28) is non-positive, then the numerator of the right hand side of (28) is positive. Therefore the following five different cases are possibilities for the signs of $\dot{d}(t)$ and $\dot{s}(t)$: (a) $\dot{d}(t) > 0$, $\dot{s}(t) > 0$, (b) $\dot{d}(t) > 0$, $\dot{s}(t) = 0$, (c) $\dot{d}(t) > 0$, $\dot{s}(t) < 0$, (d) $\dot{d}(t) = \infty$, $\dot{s}(t) = \infty$, (e) $\dot{d}(t) < 0$, $\dot{s}(t) < 0$.

Each of these five cases theoretically is possible, depending on the parameter values. However, we can show that the sign of d(t) is usually positive under the set of reasonable assumptions [see Fujita (1978)]. Generally we have only three possibilities [i.e., (a), (b) and (c)]. In the rest of this section these three cases are demonstrated by taking an explicit example of each. These examples show that, though the model considered in this section is quite simple, the model can generate a variety of different development patterns. Since the solution is very difficult to obtain in an explicit form in the case of *closed cities* [i.e., a situation where population N(t) is exogenously given at each t ($0 \leq t \leq \infty$) and equilibrium utility level u(t) is an endogenous variable at each t], here we take the case of an open city described below.

Imagine a city surrounded by the (big) national economy. The utility level of people in the national economy is given as a function of time, $u(\tau)$, $0 \leq \tau < \infty$. Migration is free between the city and the rest of the nation. Thus, the utility level of the households in the city is always adjusted to that of the national economy through migration, and housing rents are determined such that the utility level of the households in the city is the same as the utility level of the national economy. Developers determine the optimal construction plan of houses under accompanying equilibrium housing rents.

We adopt here the following simplifying assumptions:

$$D(d,\tau) = \lambda d \text{ for all } d \text{ and } \tau, \text{ where } \lambda > 0; \tag{29}$$

$$R^A(\tau) = 0 \text{ for all } \tau; \tag{30}$$

$$B(s) = B \text{ for all } s, \text{ where } B > 0; \tag{31}$$

$$U(z,s) = z^a s^b, \text{ where } a,b > 0; \text{ and,} \tag{32}$$

there is no residential house before time 0, and hence $N(0) = 0$. (33)

From (32), $z = u^{1/a} s^{-b/a}$, and thus if we set

$$U \equiv u^{1/a}, \quad \frac{b}{a} \equiv k, \tag{34}$$

then

$$z = Us^{-k}. \tag{35}$$

Next, suppose that the national utility level is given by $u(\tau)$ for $0 \leqq \tau < \infty$. Then, from (34),

$$U(\tau) = u(\tau)^{1/a}, \quad 0 \leqq \tau < \infty. \tag{36}$$

Hereafter, we call $U(\tau)$ the *utility level* at time .

Under these assumptions, by using (25) and (26), we obtain the following:

$$s^k(t) = \left[\frac{(k+1)U(t) - U(t)}{\overline{Y}(t) - Y(t)}\right]^{1/k}, \text{ and} \tag{37}$$

$$d(t) = \left(\frac{(k+1)\overline{U}(t)Y(t) - U(t)\overline{Y}(t)}{(k+1)\overline{U}(t) - U(t)} - \gamma B\right)/\lambda, \tag{38}$$

where $\overline{Y}(t)$ is defined as in (24), and

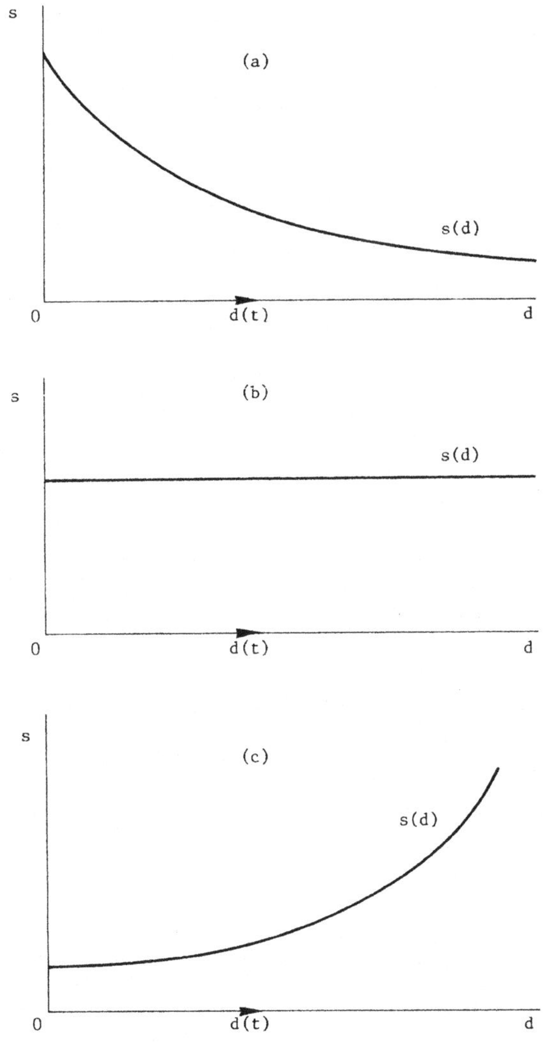

Fig. 10. Development Patterns

$$\bar{U}(t) = \int_t^{\infty} \exp(-\gamma(\tau-t))U(\tau)d\tau . \tag{39}$$

In the following three examples we commonly assume that

$$U(\tau) = \begin{cases} U_0 \exp(\alpha\tau) \text{ for } 0 \leqq \tau < T , \\ U_0 \exp(\alpha T) \text{ for } \tau > T, \end{cases} \tag{40}$$

$$Y(\tau) = \begin{cases} Y_0 \exp(\beta\tau) \text{ for } 0 \leqq \tau < T , \text{ and} \\ Y_0 \exp(\beta T) \text{ for } \tau \geqq T , \end{cases} \tag{41}$$

where $U_0 > 0$, $Y_0 > 0$, $0 \leqq \alpha < \gamma$, $0 < \beta < \gamma$. Also we will examine the character of the residential development pattern under various conditions on T, α and β.

Example 1. Assume that

$$T = \infty , \alpha < \beta . \tag{42}$$

Then from (37) to (41) we have

$$s(t) = \left(\frac{\gamma-\beta}{\gamma-\alpha} \cdot \frac{\alpha+\beta\gamma}{\beta} \cdot \frac{U_0}{Y_0} \right)^{1/k} \exp((\alpha-\beta)t/k) , \tag{43}$$

$$d(t) = [Y_0 \exp(\beta t)(1 - \frac{\gamma-\beta}{\gamma-\beta} \frac{\beta}{\alpha+k\gamma}) - \gamma B]/\lambda , \tag{44}$$

and thus

$$s(d) = \frac{\alpha+k\gamma}{\beta} \left(\frac{\lambda d+\gamma B}{Y_0(\gamma-\alpha)/(\gamma-\beta)}\right)^{(\alpha-\beta)/k} . \tag{45}$$

Therefore we conclude that

$$\dot{d}(t) > 0, \dot{s}(t) < 0 \text{ and } s_d(d) < 0 .$$

Accordingly, the residential development pattern under condition (42) can be depicted as in Figure 10(a). That is, when the growth rate of the utility level is slower than that of income, then the spatial pattern of the residential development is such that the house size s(d) decreases with distance from the city center, and hence population density 1/s(d) increases with distance.

Example 2. Suppose that

$$T = \infty \quad \text{and} \quad \alpha = \beta \quad . \tag{46}$$

Since (43) to (45) are valid also in this case, we have

$$\dot{d}(t) > 0 \ , \ \dot{s}(t) = 0 \quad \text{and} \quad s_d(d) = 0 \quad .$$

Hence, we can depict the residential development pattern under condition (46) as in Figure 10(b). That is, when the growth rate of the utility level is the same as that of income, then both house size s(d) and population density are constant with distance.

Example 3. Assume that

$$T < \infty \quad \text{and} \quad \alpha = \beta \ . \tag{47}$$

Then from (37) to (40) we have

$$s(t) = \frac{U_0}{Y_0}\left(1 + \frac{k\gamma/\alpha}{1 - \exp[-(\gamma-\alpha)(T-t)]}\right) \ , \text{ and}$$

$$d(t) = \left\{\frac{k\gamma Y_0 \exp(\alpha\tau)}{\alpha+k\gamma-\alpha \ \exp[-(\gamma-\alpha)(T-t)]} - B\right\}/\lambda \ .$$

So,

$$\dot{d}(t) > 0, \ s(t) > 0 \quad \text{and} \quad s_d(d) > 0 \ .$$

Therefore, when utility and income increase at the same α (=β) for a finite period of time, the development pattern of the residential area can be depicted as in Figure 10(c). In this case house size s(d) increases with distance from the city center, and hence population density 1/s(d) decreases with distance.

EXTENSIONS AND ALTERNATIVE MODELS

In the previous sections we proposed a dynamic model of urban land use, and explained the solutions for three special problems. We might notice that though the model is relatively simple and is based on a set of extreme assumptions (i.e., perfect foresight, no uncertainty, no depreciation of building stocks), the model can generate a variety of interesting spatial phenomena. For example, leap-frog development and the resulting mixture of different buildings, urban renewal, non-monotonic spatial variations of house size and house rent. Nevertheless, our model is only one out of many conceivable models; hence, it is important to examine the alternative models of dynamic urban land use. In the following discussions, we first briefly set forth the possible extensions of our study using the framework of the basic model (presented in the previous sections), and then propose alternative models.

Extensions Within the Framework of the Basic Model

Interesting extensions of our study within the framework of the *basic model* include the following:

(i) Introduction of vintages of building stocks. One of the most serious shortcomings of our model is that vintages of building stocks are not considered [i.e., buildings stay (or are kept) in the same condition through time unless they are destroyed]. Conceptually, there is no difficulty in incorporating vintages of building stocks into our model; but, the resulting increase in analytical complexity is great since in this case each building must be identified by three parameters (location, structure and age).

(ii) Consideration of land use for transportation. In our model, though transport costs play an important role, development of a transport network is not explicitly considered. That is, transport structure and land use for transportation are exogenously determined. Since development of a transport network and land use development for urban activities are very closely related to each other, the complete model must enable us to determine the two simultaneously.

(iii) Introduction of externalities. Another important extension of our model is the introduction of externalities (i.e., transport congestion, neighborhood externalities and local public goods). The consideration of neighborhood externalities is particularly important in determining the zoning restriction for achieving efficient spatial development.

(iv) Generalization to non-monocentric city models. Our model is based on the assumption of a monocentric city, which enables us to avoid explicit consideration of direct interactions between urban activities. The limitations of the assumption of a monocentric city are clear. One possible way to generalize our model to include non-monocentric dynamic urban land use would be to combine our model with the model developed in Fujita and Ogawa (1980).

(v) Combination with an urban growth model. In our model all the determinants of urban growth were exogenously given, and we restricted our attention to the spatial structure of the city. To study the interactions between urban land use and urban growth we must combine our model with urban growth models.

Alternative Models

Even if we achieved all of the above extensions, our model still would represent only a very special type of dynamic urban land use models. To understand why this is so it would be helpful to refer to Figure 11 [after Anas (1976a)], which depicts a conceptual framework for modeling dynamic urban land use. As noted before, one of the key elements in modeling the dynamics of urban land use is the specification of the expectation behavior of market participants (or the planner) about the future (endogenous and/or exogenous) variables. It is assumed in our model that

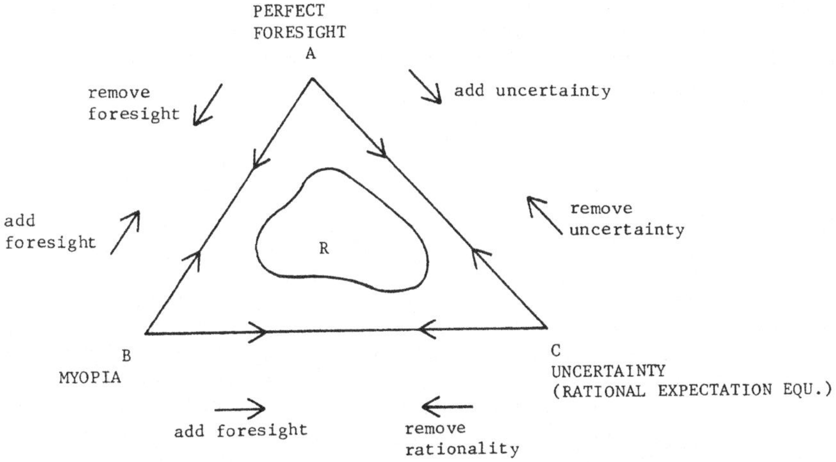

Fig. 11. Conceptual framework for modeling urban spatial growth [modified version of Anas (1976a)].

all developers have perfectly accurate foresight about future market variables (e.g., land price, housing rent). This perspective places our model at vertex A of Figure 11. Though our model located at A might be unrealistic, it represents a good general starting point for building a theory of dynamic urban land use for the reasons mentioned in the Introduction. An alternative starting point for a theoretical study of urban spatial dynamics is vertex B, where all developers are assumed to be completely myopic and to base their decisions on currently known land prices, building rents and other variables. Anas (1978) presents a model located at B. Most of the current studies of urban spatial dynamics are confined to the segment AB, and thus no element of uncertainty is considered. However, the future is essentially uncertain and expectations should take the form of stochastic processes. When uncertainty is considered, models must specify not only the expectation behavior of market participants (or the planner) but also the particular manner in which these participants weigh alternative uncertain prospects. One of the good general starting points for constructing the theory of dynamic urban land use with the uncertainty dimension is vertex C, where all developers are assumed to have an identical stochastic expectation (about all the future variables) under which plans made by all developers are mutually consistent at every time in the future (i.e., the market is cleared under every possible event at every time). That is, we study the *rational expectation equilibrium path* of urban spatial growth. Fujita (1980) already has represented such a study. As a future strategy for developing a realistic model of dynamic urban land use we may start at one of the vertices A, B or C, and proceed towards region R by appropriately specifying processes of expectation formation and learning by market participants (or the planner) as well as the market adjusting mechanism under changing environments.

REFERENCES

Akita, T. and Fujita, 1979. Spatial Pattern of Urban Growth and Renewals, Working Papers in Regional Science and Transportation 20, University of Pennsylvania.

Alonso, W., 1964. Location and Land Use, Harvard University Press, Cambridge, Mass.

Anas, A., 1976a. Comments on M. Fujita's "Optimum Spatial Patterns of Urban Growth," mimeographed, Northwestern University.

Anas, A., 1976b. Short-run Dynamics in the Spatial Housing Market, in: Mathematical Land Use Theory, G.J. Papageorgiou (ed.), Lexington Books.

Anas, A., 1978. Dynamics of Urban Residential Growth, Journal of Urban Economics 5(1).

Anas, A., 1979. The "Long-Run Equilibrium" Fallacy in Static Urban Land Use Models: A Critical Essay, Geographia Polonica, 42.

Ando, A. and Fujita, M. 1979. Dynamics of Residential Development with Multiple Income Classes, Working Papers in Regional Science and Transportation, 19, University of Pennsylvania.

Arnott, R., 1980. A Simple Urban Growth Model with Durable Housing, Regional Science and Urban Economics 10(1).

Brueckner, J., 1980. A Vintage Model of Housing Production, Journal of Urban Economics, forthcoming.

Fujita, M., 1976a. Towards a Dynamic Theory of Urban Land Use, Papers of Regional Science Association, 37.

Fujita, M., 1976b. Spatial Patterns of Urban Growth: Optimum and Market, Journal of Urban Economics, 3.

Fujita, M., 1978. Spatial Patterns of Residential Development, Working Papers in Regional Science and Transportation, 6, University of Pennsylvania.

Fujita, M., 1979. Spatial Patterns of Urban Growth and Contraction: Porblem A, Geographia Polonica, 42.

Fujita, M., 1980. A Multiperiod Model of Urban Land Market Under Uncertainty, Working Papers in Regional Science and Transportation, 27, University of Pennsylvania.

Fujita, M. and Ogawa, H., 1980. Multiple Equilibria and Structural Transaction of Nonmonocentric Urban Configuration, presented at the First World Regional Science Congress, Harvard University, June 14-23.

Herbert, J. and Stevens, B., 1960. A Model for Distribution of Residential Activity in Urban Areas, Journal of Regional Science, 2(2).

Hestenes, M., 1966. Calculus of Variations and Optimal Control Theory, John Wiley, New York.

Hochman, O. and Pines, D., 1980. Costs of Adjustment and Demolition Costs in Residential Construction and Their effects on Urban Growth, Journal of Urban Economics, 7(1).

Muth, R., 1969. Cities and Housing, University of Chicago, Chicago, Ill.

Pines, D., 1976. Dynamic Aspects of Land use Pattern in a Growing City, in: Mathematical Land Use Theory, G.J. Papageorgiou (ed.), Lexington Books.

Pontryagin, L., Boltyanski, V., Gamkrelidge, R.L. and Mischenko, E.F., 1962. The Mathematical Theory of Optimal Processes Nauka (English translation from Wiley-Interscience).

Radner, R., 1972. Existence of Plans, Prices and Price Expectations in a Sequence of Markets, Econometrica, 42(2).

Ripper, M. and Varaiya, P., 1974. An Optimizing Model of Urban Development, Environment and Planning A, 6.

EPILOGUE

Geography has quite recently moved from a dominantly descriptive discipline to one that encompasses complex theoretical models, not only of spatial covariation, but also of substantive social, economic, and physical processes, emphasizing of course, the locational implications of such systems. As the discipline evolves in these directions, it is only natural that questions of dynamics arise and are studied within the geographical tradition. Technical and theoretical reasons often make these initial attempts at incorporating dynamics overly simple--simple either in terms of how time or space is handled, or the number of variables included in the analysis. There appears to be a rather sharp trade-off between spatial, temporal, and system complexity. For example, ordinary difference equation theory has been successfully applied to studies of change through time [Miller (1968)]. Attempts to translate its solutions for two-dimensional situations have yielded rather artificial results. Dacey (1976) identified numerous solutions of this sort. These are artificial in that directionality of influence has to be constant. Relaxing this constraint results in a partial difference equation in terms of one dimension that is insoluble in both dimensions [see Griffith (1978)]. Adding a third dimension of time merely exacerbates this problem. Thus, many geographical models are currently rather naive in terms of their spatial detail or structure. One can only hope that these deficiencies will diminish as geographers and those in related disciplines become more adept at coping with multivariate spatial dynamics. In this regard, it is likely that more powerful spatial simulation models will be developed to predict, analyze, explain, and control social, economic and ecological systems in the future.

Recurrent themes of contemporary spatial dynamics include the inseparability of space and time. The first attempt to incorporate time into geographical models paralleled the manner in which space was first introduced into numerical analyses, namely a subscript, neglecting feedback and feedforward effects. Findings presented here consistently indicate there is more to spatial dynamics than adding a locational subscript i to a temporal subscript t.

Other parallels from the history of geography pervade the dynamic spatial models literature. For instance, the success of potential theory and catastrophe theory in the physical sciences has led social scientists to adopt them for use in their studies. It is hoped that geographers have learned their lesson from the pitfalls of the gravity model analogy, and heed Beckmann's (1981) warnings.

Two emerging approaches that are not adequately represented in this volume are Marxist geography and the so-called "time geography." Both of these frameworks are explicitly concerned with change in geographical systems. Moreover, they are both concerned with theoretical issues. The reasons for non-inclusion relate to the current lack of formalism of these approaches. It is, however, likely that over the next decade time and Marxist geography will develop mathematical models such that they would be represented at any future conference on this topic.

It is hoped that those papers included here provide useful stepping stones to future developments in a rapidly emerging field of analysis.

REFERENCES

Beckmann, M., 1971. Continuous Spatial Models of Income Diffusion and Commodity Trade, in: Dynamic Spatial Models, D. Griffith and R. MacKinnon (eds.), Plenum, New York, 8-19.

Dacey, M., 1976. Solutions to 136 Simple Recurrences that Generate Spatial Trends, unpublished manuscript, Department of Geography, Northwestern University.

Griffith, D., 1978. The Impact of Configuration and Spatial Autocorrelation on the Specification and Interpretation of Geographical Models, unpublished doctoral dissertation, Department of Geography, University of Toronto.

Miller, K., 1968. Linear Difference Equations, W.A. Benjamin, New York.

INSTITUTE PARTICIPANTS

Country	Participant	Affiliation
Belgium	P. Allen	Campus Plaine U.L.B.
	J. Deneubourg	Campus Plaine U.L.B.
	H. Beguin	Universite Catholique De Louvain
	J. Thisse	S.P.U.R. Universite
Canada	G. Barber	University of Victoria
	A. Carson	Saskatchewan Wheat Pool
	L. Curry	University of Toronto
	C. Marchand	Universite de Montreal
	G. Papageorgiou	McMaster University
	O. Tangri	University of Manitoba
	R. White	Memorial University of Newfoundland
England	R. Bennett	University of Cambridge
	R. Flowerdew	University of Lancaster
	R. Haining	University of Sheffield
	R. Sandford	
	J. Thornes	University of London
	S. Trussler	University of Cambridge
France	A. Faure	Cite Universitaire de Rangueil
	B. Marchand	Universite de Paris - VIII
	L. Sanders	
Germany	L. Bach	University of Dortmund
	M. Beckmann	Institute for Statistik und Unternehmensforschung
	H. Bischoff	Universitat Bremen
	J. Nipper	Geographisches Institut
Greece	K. Koutsopoulos	National Technical University
Israel	M. Sonis	Ruppin Institute
Malaysia	K. Tan	Universite Sains Malaysia
Netherlands	J. Ancot	Nederlands Economisch Institut
	J. Paelinck	Nederlands Economisch Institut
Norway	S. Flaam	CHR Michelsens Institutt
Sweden	T. Puu	UMEA University

Country	Participant	Affiliation
Turkey	I. Tekeli	Middle East Technical University
United States	J. Aron	Princeton University
	E. Casetti	The Ohio State University
	W. Clark	University of California, Los Angeles
	J. Desbarats	Northwestern University
	G. Elmes	West Virginia University
	M. Fujita	University of Pennsylvania, Philadelphia
	D. Griffith	State University of New York at Buffalo
	V. Hetrick	University of Florida
	J. Huff	University of Illinois at Urbana-Champaign
	R. Lee	University of Iowa
	R. MacKinnon	State University of New York at Buffalo
	J. Odland	Indiana University
	B. Ralston	The University of Tennessee
	V. Robinson	Hunter College
	P. Rogerson	U.S. Census Bureau
	R. Sokal	State University of New York at Stony Brook
	T. Smith	University of California, Santa Barbara
	D. Wartenberg	State University of New York at Stony Brook

The manufacturer's authorised representative in the EU is Springer Nature Customer Service Centre GmbH, Europaplatz 3, 69115 Heidelberg, Germany. If you have any concerns regarding our products, please contact ProductSafety@springernature.com

Printed and bound by CPI Group (UK) Ltd, Croydon, CR0 4YY

30/09/2026

02225505-0016